DRESDEN

Frederick Taylor

DRESDEN

TUESDAY, FEBRUARY 13, 1945

HarperCollins*Publishers*

HarperCollins books may be purchased for educational, business, or sales promotional use. For information, please write: Special Markets Department, HarperCollins Publishers Inc., 10 East 53rd Street, New York, NY 10022.

FIRST EDITION

Designed by Nancy B. Field

Maps by Götz Bergander, adapted by Digby Atkinson

Title page photo: Polaris Images

Printed on acid-free paper

Library of Congress Cataloging-in-Publication Data

Taylor, Frederick
   Dresden, Tuesday, February 13, 1945 / Frederick Taylor.—1st ed.
     p. cm.
   Includes bibliographical references and index.
   ISBN 0-06-000676-5 (alk. paper)
   1. Dresden (Germany)—History—Bombardment, 1945. 2. World War, 1939–1945—Aerial operations, British. I. Title.

D757.9.D7T39 2004
940.54'2132142—dc22   2003057139

04 05 06 07 08 ❖/RRD 10 9 8 7 6 5 4 3 2 1

To Alice

How lonely lies the city that was full of people. All her gates are desolate. The holy stones lie scattered at the head of every street. From on high He sent fire, into my bones He made it descend. Is this the city that was called the perfection of beauty, the joy of the earth?

She took no thought of her doom; therefore her fall is terrible, she has no comforter. For this our heart has become sick, for these things our eyes have grown dim.

Why do you forget us forever, why do you so long forsake us? Restore us to your self, O Lord, that we may be restored. Renew our days of old, O Lord, behold my affliction, O Lord, and behold my distress!

— LAMENTATIONS OF JEREMIAH, 1, 1, 14, 13; 2, 15; 1, 9; 5, 17, 20-21; 1, 9

Wie liegt die Stadt so wüst, die voll Volks war. Alle ihre Tore stehen öde. Wie liegen die Steine des Heiligtums vorn auf allen Gassen zerstreut. Er hat ein Feuer aus der Höhe in meine Gebeine gesandt und es lassen walten. Ist das die Stadt, von der man sagt, sie sei die allerschönste, der sich das ganze Land freuet?

Sie hätte nicht gedacht, dass es ihr zuletzt so gehen würde; sie ist ja greulich heruntergestossen und hat dazu niemand, der sie tröstet. Darum ist unser Herz betrübt, und unsre Augen finster geworden.

Warum willst Du unser so gar vergessen und uns lebenslang so gar verlassen? Bringe uns, Herr, wieder zu dir, dass wir wieder heimkommen. Erneue unsre Tage wie vor alters. Herr, siehe an mein Elend, ach Herr, siehe an mein Elend!

— *KLAGELIED JEREM.* 1, 1, 14, 13; 2, 15; 1, 9; 5, 17, 20-21; 1, 9

Verses from Luther's translation of the Bible, arranged in: Funeral Motet for mixed choir *a cappella:* "Wie Liegt die Stadt so Wüst." Introduction to the *Dresden Requiem* by Rudolf Mauersberger (1889–1971).

CONTENTS

PART TWO TOTAL WAR

PART THREE AFTER THE FALL

Illustrations follow page 250.

PREFACE

"**WHEN THE FACTS** become the legend, print the legend!" So says Dutton Peabody, the cynical newspaperman in the classic Western movie *The Man Who Shot Liberty Valance*.

As a student in the 1960s, I knew only the legend of Dresden, because it was just about all that was ever printed. Like so many others of my age, I had learned of the city's destruction principally through a work of fiction: Kurt Vonnegut's acidly surreal masterpiece, *Slaughterhouse-Five*. A brilliant novel, written partly from the perspective of his own grim personal experiences as a prisoner of war who shared the city's fate—but a figment of the imagination nonetheless.

For thirty years Vonnegut's bestselling work—and books by David Irving and Alexander McKee—sufficed to describe the catastrophic air raid on Dresden in February 1945, which for most readers in the English-speaking world and elsewhere came to represent not just the savage apogee of the conventional bombing war but something far worse: a senseless crime. The message these works conveyed for us, the next generation in Britain and the United States, was one of almost entirely unmitigated shame. Dresden was the unforgivable thing our fathers did in the name of freedom and humanity, taking to the air to destroy a beautiful and, above all, innocent European city. This was the great blot on the Allies' war record, the one that could not be explained away.

Perhaps there were always parts of the legend that didn't ring entirely true. The vast casualty figures cited—rising into the hundreds of thousands—so much more horrifying than the consequences of any other conventional air raid, and greater, some claimed, than the numbers killed at Hiroshima or Nagasaki. The notion that Dresden, a city of almost three-quarters of a million hardworking human beings in one of the oldest-established industrial regions of Europe, concerned itself only with harmless cultural pursuits and the making of luxury goods and china, even in the middle of the Nazi regime's self-proclaimed "Total War." The almost complete lack, wherever one looked, of any background information on the city, its political passions, economic problems and social anxieties: its ugly and intolerant aspects, which must be considered along with its beautiful, cultured side.

Part of the problem was always that, less than three months after its destruction, Dresden exchanged one set of totalitarian masters for another, when the Communists replaced the Nazis. What records remained after 1945 of the city's former life were less than fully open to scholars and investigators, and most of its surviving people kept the silence of official conformity. Versions of what happened between ten P.M. on the night of February 13, 1945, and noon on February 14, 1945—many originating from the fertile brain of Hitler's propaganda minister, Josef Goebbels—became set in cold-war stone, and further reexamination of the circumstances was not encouraged by a communist government eager to blacken the names of the Western Allies. The liberating moment came in 1989, with the collapse of the Soviet Union and the end of communism in East Germany. At last the people of Dresden could write, discuss, and access their collective memory without hindrance or fear of official persecution, and so could outside scholars and investigators.

The most objective work previously available concerning the destruction of Dresden has all been in German. Götz Bergander, Dresden-born and a teenage witness of the bombing, later a radio journalist and writer based in Berlin, wrote his book *Dresden im Luftkrieg* (Dresden in the Air War) in the 1970s and after 1989 revised it extensively in the light of the new information becoming available. Scandalously—considering the heedlessness with which apocalyptic legends of Dresden's fall continue to be printed in the English-speaking world—Herr Bergander's scrupulous, rich, and fascinating account of the attacks on his home city has never been translated into English. Likewise in the case of another Dresden historian, Matthias Neutzner, whose books *Lebenszeichen* (Signs of Life) and

Martha Heinrich Acht (Martha Heinrich Eight—Dresden's code name on the German air defense grid) manage the almost impossible task of setting Dresden's destruction in wartime perspective while at the same time heightening to an all but unbearable level of intensity the tragic human loss it involved.

It was after I read these books, and came into contact with their authors, that my own journey began. The journey was, of course, a physical one: to Dresden and Berlin and London and Washington to consult records and documents; from an RAF veteran's cottage in Norfolk to a former slave laborer's house on the edge of the Bavarian forest; from interviews with Dresden's survivors in hotel rooms to emotional conversations in neat apartments built on the very rubble of the districts where eye-witnesses had grown up. It was also, however, a mental journey, confronting evidence that did not fit my old idea of what Dresden had been, and forcing myself to see the wartime years, not through the eyes of the pacifistically inclined baby-boomer I had been and remain, but as it might have been regarded by those who lived and fought, suffered and struggled at the time, when the future was unknown and thousands of innocents were still dying every day.

The picture that emerged for me was not by any means one of an "innocent" city but of a normally functioning city (both in the universal sense and in the context of Nazi Germany), made extraordinary by its beauty. This is not to go to the other far extreme and say that Dresden "deserved" to be destroyed, but that it was by the standards of the time a legitimate military target. The question is whether enemy cities, necessarily containing large numbers of civilians and fine buildings but also many vital sites of manufacturing, communications and services of great importance to that nation's war effort, should be bombed despite the probability of high casualties among noncombatants. This issue remains one that can and should unleash passionate moral and legal arguments—even in the age of the so-called "smart bomb."

Dresden: Tuesday, February 13, 1945 will not settle any such arguments, but my belief is that it will reveal a more complex and ambivalent moral framework than has hitherto been generally recognized. The final moral judgment about the city's fate in February 1945 remains, as it must, the readers'.

Perhaps if there is a moral conclusion it can only be found in the

German phrase that I heard again and again from the lips of Dresdeners, spoken with a passion born of terrible experience: *Nie wieder Krieg*. Never again war. With the terrible weapons of mass destruction at its disposal, humanity can no longer afford intolerance and war, and that is the ultimate lesson of the bombing of Dresden. May it eventually be heard loud and clear, even though sixty years have passed.

ACKNOWLEDGMENTS

WITHOUT THE HELP of many individuals and institutions in Britain, Germany, and the United States, researching and writing this book would have been at best difficult, at worst impossible.

In Germany I especially should like to express my thanks for the advice of four of Dresden's most distinguished historians—Götz Bergander and Matthias Neutzner, Günter Jäckel and Karl-Ludwig Hoch—three of whom, as eyewitnesses of the bombing of Dresden, also kindly granted me frank and moving interviews about their experiences. Dr. Helmut Schnatz of Koblenz, chronicler of the Rhineland city's wartime history and author of the remarkable book *Tiefflieger über Dresden: Legende oder Wirklichkeit?*, also gave me invaluable information and advice and, along with Frau Schnatz, generous hospitality. Like the other historians mentioned, he also allowed me to read and draw upon material from his own private archive. In Dresden, the staff of the City Archive (Stadtarchiv), Main Saxon State Archive (Sächsisches Hauptstaatsarchiv) and City Museum (Stadtmuseum) were kindness and patience incarnate. In the case of the City Museum, I should especially like to thank Herr Friedrich Reichert and Herr Holger Starke, who gave me the benefit both of their curatorial knowledge and, as working historians, their labor in the field of Dresden's history, prewar, wartime, and postwar, pointing me in the direction of information and sources I might otherwise never have managed to locate.

To all those survivors of the Second World War in Germany, Britain, the United States, and Israel who granted me interviews—and in many cases the use of private photographs and documents—my heartfelt gratitude.

In Britain, I am very grateful to the secretary of the Bomber Command Association, Douglas Radcliff, for his early advice and help in contacting RAF veterans of the Dresden raid. Equally important was the military historian Robin Neillands, whose contacts among those same veterans he willingly granted me permission to pursue. I trust that he will consider I have done right by them. Dr. Noble Frankland also gave much useful and enlightening advice at an early stage in my research. Sir Martin Gilbert put me in touch with survivors of the German slave labor system and also read parts of the unrevised manuscript, making many invaluable suggestions. I must also thank the efficient and helpful staffs of the Imperial War Museum and the Public Record Office.

In the United States, the staff of the National Archives and Records Administration in College Park, Maryland, were superbly helpful, enabling me to cover a great area of ground during a relatively brief stay. And to Dr. Gary Shellman of the University of Wisconsin (Milwaukee) and Sally Shellman for their hospitality and for organizing a richly informative trip to the Midwest, especially my lecture at the Oshkosh Air Show, which brought a number of interesting experiences to light.

The skillful liaison work of my agents, Jane Turnbull in London and Emma Parry in New York, enabled me to concentrate totally on the book. All right-minded authors value their agents, but Jane and Emma are jewels beyond price. My editors, Bill Swainson at Bloomsbury Publishing in London and Dan Conaway of HarperCollins Publishers, New York, were stern taskmasters and intrepid companions on the journey of writing this book. Without their keen observations and diplomatic suggestions, this book would have been considerably longer and made appreciably less sense. The hard work of their assistants, Sarah Marcus in London and Jill Schwartzman in New York, enabled this manuscript to reach its final destination at the printer with astonishing ease. Also thanks to copy editor Eleanor Mikucki, and to production editor Sue Llewellyn and designer Nancy Field at HarperCollins, for going the extra mile.

Had it not been for a lunchtime stroll though the Cornish countryside with the poet and critic Derrek Hines, this book might never have been conceived in the first place.

And last but not least, eternal gratitude to my wife, Alice, who read every word and put up with those long absences—whether in my study or in foreign parts—for almost three years.

DRESDEN

13th / 14th February 1945 Dresden Primary Target

RAF Bomber Command 5 + 1 3 6 8 Bomber Groups

Map of the damage inflicted by the two night attacks in the city area
Inside the unbroken line: 75% - 100% destroyed
Inside the broken line: 25% - 75% destroyed
Circles: Stray bomb drops
Quarter-circle: (Wedge) Aiming sector of the first attack
In Johannstadt and Striesen, a few housing blocks remained standing in the area
of total destruction

Hospital

Assigned approach routes.

5 Group 1st attack   1 3 6 8 Groups 2nd attack

Neustadt

Elbe

Weisser
Hirsch

Blasewitz

Loschwitz

Johannstadt

Striesen

Grosser Garten

chertnitz

Strehlen

Seidnitz

Dobritz

1000 yds

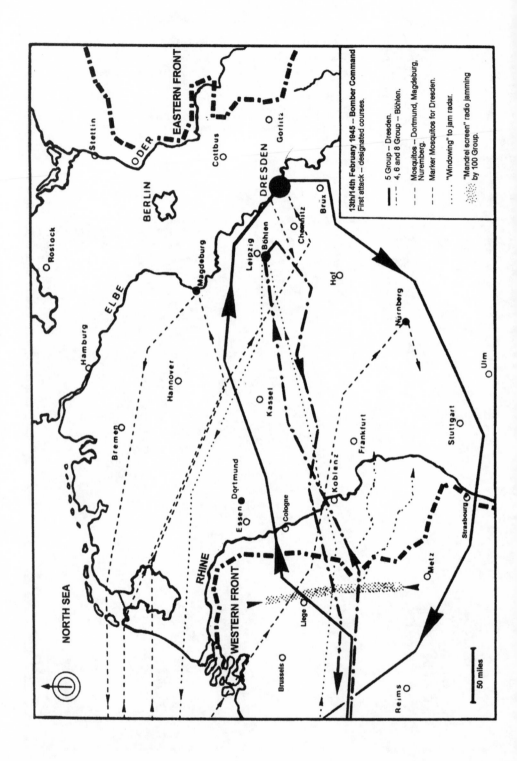

NORTH SEA

EASTERN FRONT

ODER

Stettin

Cottbus

Görlitz

BERLIN

DRESDEN

Rostock

Magdeburg

Brüx

Böhlen

Chemnitz

ELBE

Leipzig

Hof

Hamburg

Hannover

Nürnberg

Bremen

Kassel

Ulm

Frankfurt

Stuttgart

Essen Dortmund

Koblenz

Cologne

Strasbourg

RHINE

WESTERN FRONT

Metz

Liège

Brussels

Reims

50 miles

13th/14th February 1945 — Bomber Command
First attack — designated courses.

————— 5 Group — Dresden.
—·—·— 4, 6 and 8 Group — Böhlen.
— — — Mosquitos — Dortmund, Magdeburg, Nuremberg.
— — — Marker Mosquitos for Dresden.
··········· "Windowing" to jam radar.
░░░░░ "Mandrel screen" radio jamming by 100 Group.

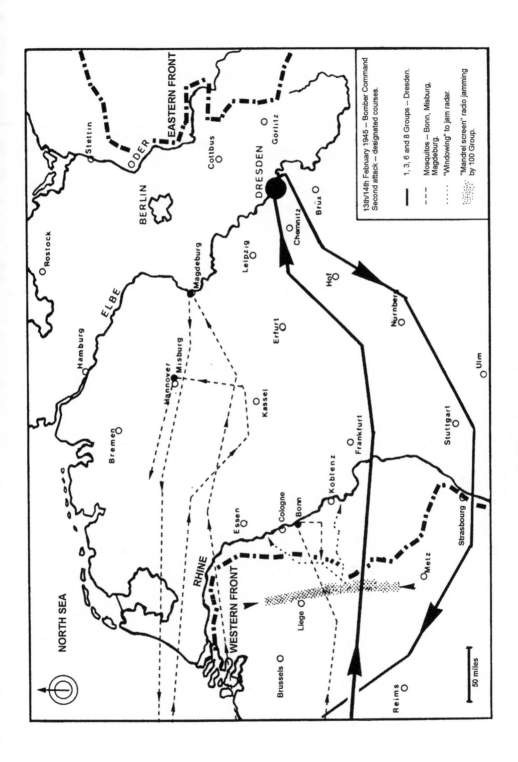

NORTH SEA

EASTERN FRONT

Stettin

ODER

Rostock

BERLIN

Cottbus

Görlitz

DRESDEN

Brüx

Chemnitz

ELBE

Magdeburg

Leipzig

Hamburg

Hof

Hannover

Misburg

Nürnberg

Erfurt

Bremen

Kassel

Ulm

Essen

Frankfurt

Stuttgart

Cologne

Koblenz

Bonn

RHINE

WESTERN FRONT

Strasbourg

Liège

Metz

Brussels

Reims

50 miles

13th/14th February 1945 — Bomber Command
Second attack — designated courses.

1, 3, 6 and 8 Groups — Dresden.

Mosquitos — Bonn, Misburg, Magdeburg.

"Windowing" to jam radar.

"Mandrel screen" radio jamming by 100 Group.

PROLOGUE
Saxons

SILVER GRAY CLOUDS blurring into slate gray sea on the misty horizon. Villages crouched between marshlands and shifting sands, hostage to the pitiless easterly winds and to the restless, fierce invaders from the northwestern part of Germany, known as Saxonia. It was their ships that those cold winds brought lurching across the North Sea and scudding onto its sandy beaches.

More than a millennium and a half ago, when the Romans still ruled the island of Britain, they called its vulnerable, low-lying eastern coast the Saxon Shore, and appointed a count to the onerous task of organizing its defenses. This was around A.D. 350.

In the end, the wooden forts and the ditches and the local forces were too few, and the invaders from the sea too many. Even before the Romans left, the Saxons had gained a foothold. Within a century they had expelled or absorbed the Celto-Roman natives and given their own flat names to the great, flat counties of Eastern England. Norfolk. Suffolk. The kingdom of Lindsey, later to be known as Lincolnshire. The transplanted Saxons, along with the other German tribes who had also settled in the former Roman province, with a mixing-in of Dane and Norman from later continental incursions, became the peoples known all over the world as the English, or the Anglo-Saxons.

As for those continental Saxons, the ones who had not sailed west to Britain all those centuries ago, they instead moved farther into the heart of Europe. They infiltrated the Slav lands, expelling or absorbing the original natives, just as their relatives had done in Britain. These East Saxons too finally converted to Christianity, founded towns, fought wars, traded and farmed and worked minerals, produced great art, and hoarded great riches. Partly against their will, they became part of a united Germany toward the end of the nine-teenth century.

By the fifth decade of the twentieth century, the English have vast wealth, an empire spread throughout the globe, great cities, a navy, army, and air force. Latterly the air force has become a new symbol of the nation's power. Much of it is based in Lincolnshire and Norfolk and Suffolk, the subdivisions of the Saxon Shore. "Bomber country," as it has been dubbed. Where fifteen hundred years ago these flat, exposed counties presented a reception platform, open for invaders to swarm ashore, by the 1940s they are crisscrossed with hastily built concrete runways. They are now the launch decks of an island aircraft carrier facing east toward the continent.

It is the year 1945. The Anglo-Saxons are engaged in war to the death with their distant family across the North Sea, from whom they had divided so long ago and far away. As a consequence, aircraft laden with bombs are taking off from the wintry fields of Lincolnshire on a belated reverse voyage this February afternoon.

The fliers' mission is to wreak a terrible revenge on those German relatives' most proud and beautiful treasures, as well as on the most precious thing they have—their lives.

TWILIGHT. FEBRUARY 13, 1945, Shrove Tuesday. The first waves of air-craft are taking off. The ponderous flocks further darkening the winter sky are made up of Avro Lancaster heavy bombers, attached to 5 Group of Bomber Command. They are bound for a rendezvous point over Berkshire. Their armor—perfunctory to start with—has been fur-ther depleted to save weight, and the planes have been equipped with extra fuel tanks because of the exceptional seventeen-hundred-mile round-trip distance to the target. Each Lancaster, big with a seven-ton load of bombs and incendiary devices, wields twice the destructive capacity of the famous American Flying Fortresses and Liberators. By 6 P.M., in the gathering darkness, a total of 244 bombers are circling together in the air, the town of Reading blacked out thousands of feet beneath them, ready to set course.

The aircrew have been routinely briefed that afternoon, and their target described as follows:

The seventh largest city in Germany and not much smaller than Manchester, is also by far the largest unbombed built-up area the enemy has got. In the midst of winter with refugees pouring westwards and troops to be rested, roofs are at a premium, not only to give shelter to workers, refugees and troops alike, but also to house the administrative services evacuated from other areas . . .

The target sounds workaday, scarcely worthy of special notice in the busy schedule of the massive, sophisticated machine of destruction that Bomber Command has become. This is deceptive, and perhaps the briefing's tone is deliberately disingenuous. This particular city has been renowned over centuries for its architectural beauty and *douceur de vivre*, and in that respect the war has until now changed surprisingly little. Its name is Dresden.

Although the spearheads of the Russian armies have temporarily halted about seventy miles to the east, and a stream of refugees from the eastern front has recently begun to tax the city's housing resources, the situation remains surprisingly calm. The theaters and opera house—where works by Webern and Wagner and Richard Strauss saw their first performances—are temporarily closed under orders from Berlin, but Dresden's famous cafés are still open for business. Tonight the Circus Sarrasani is staging a show at its famous domed "tent" just north of the river, drawing hordes of sensation-eager spectators.

How can Dresden know that for some time it has been marked out for destruction? Weeks of bad weather, making accurate bombing difficult if not impossible, have saved its people until now. Today over the target city, conditions have cleared. Unlike other parts of Germany farther to the west, the area has experienced a pleasant day leading into a cold, dry night with only light cloud. What the Germans call *Vorfrühling*. Pre-spring. This simple, cruel meteorological fact has finally sealed the city's fate.

The Lancasters have reached their cruising speed of around 220 miles per hour, flying in layers at between seventeen thousand and nineteen thousand feet to avoid collisions. They maintain a southeasterly course at first, breasting the French coast over the Pas de Calais, continuing over northern France until they reach a point roughly halfway between Reims and Liège. There the formation banks north-east, heading for the border city of Aachen—now in Allied hands—before setting course due east, over the front line into enemy-held territory. Soon the bombers pass, as planned, just to the south of the Ruhr industrial area, avoiding its massive concentrations of well-practiced antiaircraft batteries. Here is where they also pass beyond the protective shield of "Mandrel," the jamming screen provided by the RAF's 100 Group to fog the enemy radar defenses. "Window" devices are dropped in the thousands to further confuse the enemy. En masse, these small strips of metal appear on German radar screens as a wandering bomber fleet while the real aircraft do their work elsewhere. These measures will be especially effective tonight, for the area west of the target remains blanketed with thick cloud, making visual tracking of aircraft movements impossible.

At 9:51 P.M. in Dresden the air raid sirens sound, as they have so often during the past five years, and until now almost always a false alarm. The city's people, and especially its children, have spent the day celebrating a somewhat toned-down, wartime version of Fasching, the German carnival. Many of them, and again especially the children, are still in their party costumes. There are, perhaps, even more than the usual laughter, the usual jokes, as families sigh and head for their cellars. Stragglers, hearing the alarm, scuttle home through the winding, cobbled streets of the old town, or pick up their pace as they make their way past the grand buildings of the Residenz.

There are remarkably few major public air raid shelters for a city of this size. One of the largest, beneath the main railway station, built for two thousand people, is currently housing six thousand refugees from the eastern front. The Gauleiter—the local Nazi Party leader but also the province's governor and defense commissioner—has consistently failed to divert the necessary resources to remedy this situation, although (as his subjects are well aware) he has commandeered SS engineers to build sturdy shelters beneath both his office and in the garden of his personal residence. In the latter case he has put a reinforced concrete shield several meters deep between himself and the bombs he has always insisted will never fall.

Meanwhile, after four hours in the air, the bombers are reaching the end of their outward flight. The Luftwaffe has not contested the airspace (the only enemy aircraft shot down that night will be an unlucky German courier plane which crosses the path of the British air armada en route between Leipzig and Berlin). The visibility remains poor, even as the Lancasters begin their final approach to the target. Only now, as they track the southeastward curve of the river Elbe, does the cloud cover begin to disperse. The aircrew, shivering in subzero temperatures despite their thick flying gear, can look down and glimpse landmarks, roads and railways, occasional lights three miles or more below them. They wait, watch for enemy night fighters, and fly on over darkened woods and fields, over the icy ribbons of the country roads that link the neat, slumbering villages of Middle Germany.

So far—and the aircrew know to be thankful for this—it is an altogether uneventful trip. Even now, few in the target city have any inkling of what is to follow. There had, after all, been no "prealarm." In the industrial center of Leipzig, fifty miles to the west and already subjected to heavy bombing earlier in the war, specific warnings have already been issued over the radio. But in the target city the authorities have chosen not to place their fellow citizens on any kind of special alert. At this point the controllers at the Luftwaffe's tracking stations know that one of the major eastern population centers is being targeted. However, they share Germany's and much of the world's conviction that Dresden will never be subjected to serious bombing.

One of the myths, then and now, is that this city has been completely spared until tonight. Over the previous few months there had been a scattering of daylight raids by American formations on the suburban industrial areas and on the marshaling yards just outside the city center. Adjacent residential blocks had actually been hit a month previously, at the cost of more than three hundred civilian fatalities. But most citizens put these incidents down to mischance or poor navigation, and still consider the city inviolable. There are many rumors about why Dresden has not been, and will not be, subjected to the massive destruction meted out to other towns in the Reich.

Ten minutes after the first air raid alarm, an advance guard of fast, light RAF De Havilland Mosquito Pathfinders from 627 Squadron swoop unchallenged over the darkened buildings. Their job is to identify and mark the target. They begin to drop the marker flares— known to German civilians as "Christmas trees." These will enable the huge force of following bombers to find their targets. Their focus is

the stadium of the city's soccer club, just to the west of the old city. It
is suddenly clear from this that the bombers are aiming not just for the
city's industrial suburbs and adjacent marshaling yards but its treasured historic heart. Only when they hear the Mosquito's engines overhead do the local civil defense authorities, on alert in their bunker
beneath police headquarters, realize that their city is actually going to
be bombed. The frantic voice of an announcer comes on the local
cable radio, telling citizens to get off the streets on pain of arrest—to
take the best cover they can.

There are no antiaircraft guns to be readied. No searchlights
probe the skies. Just a few weeks before, the sparse flak defenses of the
city—much of it light guns and captured Soviet pieces not thought
highly of by their crews—were dismantled and shipped away, some
westward to the heavily bombed Ruhr industrial districts and others
to the hard-pressed eastern front. The city is completely unprotected.

One man's diary for February 13 also describes "perfect spring
weather" in Dresden during the daylight hours. But there is nothing
else cheerful about his report of the day's events. This citizen,
Professor Victor Klemperer, changed careers in early middle age from
journalist and critic to become a distinguished academic. A decorated
veteran of World War I and a firm German patriot, in the past ten years
he has lost his job, his house, and his savings. He is not permitted to
own or drive a car or a bicycle, or to use public transport. He cannot
keep pets. There are certain streets he cannot walk along, or can cross
only via specific junctions. This is because Klemperer is Jewish. He
has been saved from "deportation" until now, not because his family
has been established in Germany for two hundred years, but because
he is married to an "Aryan."* And today, he reports in the pages of his
journal, he has been touring the homes of the few other Jews still
remaining in the city (about two hundred out of a prewar total of six
thousand), to tell them that they will be deported to an undisclosed
"labor task" in three days' time, on February 16. Every one of them
knows what this means.

---

* The Nazi jargon phrase for allegedly racially pure Germans. From now on in the text, quotation marks will no longer be added. This in no way implies acceptance or validation of the
term.

That evening, arriving back at the house he and his wife share with other Jewish survivors since their own home was confiscated, he eats a modest dinner. Klemperer sits down to coffee just as the air raid sirens sound. Then they hear aircraft overhead. One of his companions says with bitter prescience: "If only they would smash everything."

The Lancasters are over their target; their bomb doors have opened. The raid is under way. The first wave of destruction lasts between fifteen and twenty minutes. The second, two hours later and featuring even more aircraft, lasts slightly longer. The time lag is a deliberate, cold-blooded ploy on the part of Bomber Command's planners, who have become expert at such pieces of business. By this time many who survived the first raid will be back above ground, and there will be firefighters, medical teams, and military units on the streets—including auxiliaries who have raced along frozen roads from as far away as Berlin. Now a fresh hail of high explosive and incendiary bombs will suck the individual existing fires into one, and the firestorm will begin to build. In the morning Flying Fortresses of the American Eighth Air Force will finish the work of destruction. Dresden is doomed.

The next morning aircrew from the 796 Lancaster bombers that flew to Dresden have almost all landed safely back at their bases in England. The young fliers enjoy hearty breakfasts and wait to be debriefed. Used to horrendous losses over Germany during the previous three years, they have reason to celebrate a mission that was, for them, a bloodless affair.

Not so for Dresden. Overnight, those same aircrew dropped more than twenty-six hundred tons of high explosives and incendiary devices on the target city, utterly destroying thirteen square miles of its historic center, including incalculable quantities of treasure and works of art, and dozens of the finest buildings in Europe. At least twenty-five thousand inhabitants are dead, and possibly many more: blown apart, incinerated, or suffocated by the effects of the firestorm. Bodies will be piled up in one of the main squares. They will be placed on huge slatted shutters, salvaged from the display windows of one of the city's department stores, then burned in the thousands to stop the spread of disease.

Victor Klemperer and his wife, two out of a relatively small group of human beings for whom the horrors of mass destruction represent not a cataclysm but a miracle of deliverance, have taken advantage of the chaos. The professor has torn the telltale yellow star from his coat, and they are on their way to safety and freedom. But that is another part of the story, and certainly a different side of the moral equation . . .

The day is Ash Wednesday, February 14, 1945. It is eighty-four days before the end of the Second World War in Europe. Almost a lifetime later, the name of Dresden continues to echo uneasily in our collective memory, and controversy about the city's destruction has not ceased to rage.

PART ONE

FLORENCE ON THE ELBE

1

Blood and Treasure

"THE ENGLISH WERE TREASURED. I think it was only after the raid that there was a hatred of the English in Dresden, not before."

Pastor Karl-Ludwig Hoch, Lutheran man of God, architectural historian, and community leader, is in his early seventies now. A profoundly spiritual man, he is saved from otherworldliness by a wry, almost cynical sense of humor. His patrician features are folded in a sad smile as he describes his fellow citizens' lost love affair with England.

"People just knew that the British and the Americans loved Dresden so much . . . St. John's was the English church on the Wiener Platz, and the American church was All Saints."

In the garden of the Hoch family's suburban waterside villa is a stone monument, from which it is possible to look downriver and view the skyline of Dresden two or three miles distant. It was built by some long-ago Francophile to commemorate the afternoon when Napoleon, on headlong retreat from Moscow and considering where to make a stand, was led to that same height, at that same spot, so that he too could examine Dresden from a distance. The year was 1813. Saxony was one of the few allies Napoleon had left. The French emperor was thinking of having a battle on its territory. In the event, he liked the idea so much, he had several. The Saxons, as the pastor often points out, have never been especially clever in their choice of friends.

In 1945 Pastor Hoch's family were spared the total destruction visited upon the inner city. Isolated stray bombs scarred their leafy neighborhood, but the Hochs and their lodgers and neighbors just took refuge in the shelter in the garden until the raid was over. Then—

when the roar of aircraft engines had faded—they emerged, to be presented with a grandstand view of their native city, two miles or so distant, being devoured by flame. A woman who lived up the hill, a fervent Nazi, spotted them out on their balcony and called out, "So, Frau Hoch! Was Goebbels right or not? Are the English criminals or not?"

Josef Goebbels. In many ways, the legend of the destruction of Dresden was the dark, agile Nazi propaganda minister's last and grimmest creation. For Goebbels the city's near-annihilation was both a genuinely felt horror and a cynical opportunity.

Most Germans had realized at the time of the fall of Stalingrad that talk of victory was hollow. By the winter of 1944–45, even Nazi fanatics realized that to all practical intents the war was lost. Ever resourceful, Goebbels now made a characteristically bold and cunning decision: Instead of putting a positive gloss on the German position, he would hammer home the horrors in store if the Third Reich was defeated. The Bolshevik hordes pressing from the east, raping and looting as they advanced into the neat, untouched towns of East Prussia and Silesia; the treacherous, hypocritical Anglo-Americans with their pitiless bomber fleets and their cosmopolitan (read Jewish) contempt for Germany's unique cultural heritage. These were the threats to German—and European—civilization.

The only answer was to nobly resist these enemies, totally and to the end—and wait for the miracle that might come any day from the new wonder weapons that Germany's scientists and engineers would soon bring to devastating application, or from the growing cracks in the impossible, artificial alliance between communism and capitalism. Meanwhile, the worse the crimes that could be laid at the door of the Reich's enemies, the more powerful the spell this twilight masterpiece of Goebbels's black art would cast. Failing the élan of everlasting victory, Germany must summon up the courage of temporary despair.

Therefore no attempts were made to minimize the atrocities being committed by the advancing Russians. On the contrary, unsparing accounts of the horrors that German forces had discovered during brief reoccupations of East Prussian towns during the ebb and flow of battle were broadcast and rebroadcast on the radio. Refugees still in shock were interviewed, and horrifying atrocity articles appeared in the thin newssheets that had now replaced the Reich's once-voluminous press. The newsreels showed devastation and ruin—and the brave determination of those still eager to resist the enemy. It was a grim route to final victory, *Endsieg*, but (so the propaganda implied) that route remained open despite all the setbacks.

So, in the early days of 1945, Dresden waited; but for most of the city's people, the arrival they feared was not that of Allied air forces, but of the Soviet Red Army. A hundred and more miles to the east, the capital of the neighboring province of Silesia, Breslau, had been all but encircled by the Russians. From the air base at Klotzsche just north of Dresden the Luftwaffe was running an airborne supply shuttle to the beleaguered Silesian metropolis. The eastern defenses of the Reich were threatening to crack, and after Breslau the next major German city in their path was Dresden.

Camera in hand, on February 13, 1945, Karl-Ludwig Hoch met his brother, and together they took a number 11 tram to Postplatz, in the heart of the Altstadt, the old town. Their plan was to snap photographs of the proud city of Dresden to remember it by. This was because their mother had said that, as an aristocratic family, they might soon have to flee the Communist advance, and so might never see Dresden again. The weather was wintry-mild under slight cloud. The brothers wandered through familiar streets and alleys, passing landmarks they had seen most days of their lives. They returned to their suburban home late that same afternoon, as the twilight crept over the valley of the Elbe, not knowing that they had just seen Dresden for the last time in its historic form, or that many, if not most, of the fellow citizens they had mingled with in the familiar streets would be killed during the night. The Hoch brothers had, to all intents and purposes, seen the dead, walking.

But just for now, during these final hours, Dresden remained itself. Frightened, thronged with refugees despite efforts to speed them on to the west, and aware of the possibility of coming suffering, but still itself. The "Florence on the Elbe" that for two hundred and more years had attracted artists and aesthetes and lovers of culture, not just from Germany but from all over Europe, Britain, and the Americas. And which, everyone agreed, would undoubtedly survive to dazzle future generations even if Hitler's Reich was liquidated during the months to come.

There were some who comforted themselves with the supposition that the Allies were preserving Dresden for the future, that they had earmarked it as their administrative capital after the war. The fine buildings, the Baroque palaces and Art Nouveau apartment houses, the pleasant landscapes and the roomy villas. All these things, so peculiar to Dresden with its rich past and continuing tradition of enlightened architectural planning, spoke in favor of this theory. And there were other rumors whispered around the city during those days. Churchill had a favorite aunt who lived in Dresden. He had spared the city for her sake. And the English, the English had a special love for Dresden. But then that was where we started . . .

Better to start again, perhaps. Not with the doomed architectural jewel, but with the simple settlement in the floodplain of a river valley that the city had been long ago. The village known unromantically to the Slavs who originally lived there as Drezdzány: "forest dwellers in the swamp."

MARSHY, BUT WELL-SHELTERED by hills to the north and south of the river Elbe, Drezdzány represented the first easily navigable river crossing as the great central European waterway snaked down through gorges from its origins in the great Bohemian forest and flattened out for the long run northwestward to the sea. Even today the determining geography is absolutely clear: within a few miles' travel, progressing downstream, we move from rocky sandstone overhangs with spectacular views of water far below to wide water meadows and balmy, fertile lowland.

And fertility was what the Saxon migrants were seeking as they moved east into the Slav lands. Defeated and subjugated by the Christian king of the Franks, later to be crowned as the Emperor Charlemagne, Saxon tribes who had been pagan just a few decades before built strongholds, and within those strongholds, churches. Christianity quickly became their badge in the struggle against the still godless Slavs. Their restless energy made the Saxons rich and successful.

In the eleventh century a Saxon count of the Marches, or margrave, took up residence on a great hill dominating the Elbe and fortified himself thoroughly there, as a base for colonization. He called the place Meissen and himself count of the same—a corruption of the Slavic kingdom of Misni, which had previously existed here. This was the Wild East, the Germans' ever-expanding frontier.

A final Slavic revolt in 1147 provoked the calling of a Crusade under the auspices of the emperor himself. The massed armies of Germanic Christendom suppressed the natives with a fury as bloody as it was righteous. Thereafter the colonists encountered little opposition, at least in the area around the Elbe and Oder. The land was productive. It proved to be rich also in minerals. The Germans developed mining techniques that were more advanced than anywhere else on mainland Europe, laying the basis for future wealth. As in Africa, Asia, and the Americas in later centuries, some place names were introduced by the invaders, some adopted from native forms. As in Africa, Asia, and the Americas, trade and conquest went hand in rapacious hand.

The first mention of Drezdzány as a Germany colony-town comes in records of the counts of Meissen dating from 1216. The German settlement, on the south bank of the river, became the "old town" (Altstadt), and the Slavic areas across the Elbe—though older—the "new town" (Neustadt), presumably because it was still awaiting the pleasure of Germanization. In 1270 the place now becoming known as Dresden entered History with a big "H." Count Henry the Illustrious of Meissen moved his seat twelve miles upriver to Dresden. The site was pleasant, mild in winter and not too hot in summer, and was already becoming well known for its vines, orchards, and market gardens. Not as spectacular as the rocky crag of Meissen but, shall we say, more civilized.

Dresden didn't stay a Marcher lord's capital for long. After the death of Henry the Illustrious, following the vagaries of princely DNA, it passed in quick succession to the rulers of nearby Bohemia and Brandenburg, before passing back to the Meissen jurisdiction just before the Black Death swept through Europe. It would now remain continuously in the possession of members of Meissen's ruling family—the Wettin dynasty—for six hundred years.

The spread of German influence to the east and the south ran into limits as the fifteenth century took its course. The Czechs of Bohemia, smart Slavs with their own sturdy social hierarchy, not forgetting industrial and mineral resources that rivaled those of their neighbors, had already started to chafe at the continuing influx of German immigrants, especially in Prague and the mineral-rich Sudeten mountains. Bohemia's kings might encourage the settlement of hardworking outsiders in their towns and villages—rulers have a tendency to think in terms of tax and skills bases—but the old-established natives were not so favorably impressed. Mutual massacres followed, in which Germans were terrorized and forcibly expelled from their strongholds in the towns and, in return, troublesome Czechs tossed down mineshafts—the usual bloodstained small change of ethnic conflict. Decades of war between Germans and Czechs, Catholics and Hussites, finally ended in uneasy peace. Dresden lies less than thirty miles north of the Bohemian borderlands where the Elbe rises and where this agonizing ethnic drama was being played out. In 1429 a Czech army reached the gates of Dresden and laid waste to the suburbs.

Nevertheless, around the turn of the sixteenth century, Dresden acquired the status that it would never again lose: capital of Saxony. The Wettin dynasty split the Saxon lands between two competing sons. The richer part, including Dresden, was given to Albert, thenceforward prince elector (*Kurfürst*)—acknowledged master of eastern Saxony and member of the committee of princes that chose each Holy Roman ruler—not just a kingmaker but an emperor maker. There was a fire, which destroyed substantial parts of the town, not for the first or last time. As a result, architects were instructed to build in less readily combustible stone—in most cases the local sandstone—a feature that became characteristic of Dresden's architecture.

Through the sixteenth century Dresden developed into an artistic capital as well as a political one. Martin Luther threw out his Protestant challenge to the pope, and the elector of Saxony became his chief protector. Soon Saxony was predominantly Lutheran and the leading Protestant power in Germany. With religious divisions making Europe in general, and Germany in particular, a more dangerous place, the elector built an elegant *Schloss* on the Elbe and surrounded it with substantial fortifications. Despite all the splendor and pride, during all that time the city retained the faint, beleaguered aura of a frontier encampment. Soon it was time for a new dispute over who ruled Bohemia and by what religious law. The old, still smoldering enmity lit the fuse that started the Thirty Years' War.

Europe was ravaged by marauding armies on a scale not seen since the fall of the Roman Empire. When peace came in 1648, Bohemia lost its remnants of independence, and both Catholicism and Germandom were strengthened there. Nevertheless, the Czech lands were left with a mix of Slav majority and German minority that would always be combustible. On the positive side, the reputation of Dresden as a military stronghold had led to a rapid influx of anxious country folk seeking to escape the rampaging armies. Most of Europe might be exhausted and depopulated, but the Saxon capital had grown into a community of twenty-one thousand souls. By the standards of the time, it was a very substantial city.

Yet Dresden is, as we approach the end of the seventeenth century, still a provincial center. A *Residenzstadt* or residence town where an important but nevertheless, in European terms, middle-ranking regional magnate holds court. Literally. Almost every advantage the people of Dresden enjoy—jobs, industries, trade, arts, and amusements—is conditional on the elector of Saxony continuing to choose this place to live in. It is a factor that will shape, perhaps unconsciously, many attitudes in the city even when it has outgrown that status as royal residence and found itself transformed into a teeming modern metropolis.

However, at this stage, royal decisions definitely remain final. So the first step toward this distant destination is taken by a new prince elector, a bull-necked young adventurer—Wettins have rarely matched the beauty of their capital—named Frederick Augustus. When he unexpectedly succeeds his sickly brother to the Saxon throne in 1694 at the age of twenty-three, he has already shown a precocious interest in mistresses, architecture, and the art of war. Frederick Augustus has money and gold and jewels, he has an inbuilt confidence and seemingly limitless ambition. Saxony has prospered since the end of the Thirty Years' War. Frederick Augustus, sent on the grand tour by his princely parents as a teenager, had fallen in love with the glories of Renaissance Italy. Above all, though, he had been deeply impressed by the splendor of Versailles and the towering, absolutist figure cut by the great King Louis XIV of France.

Frederick Augustus is determined to carve out an important position for himself not just in Germany, but in Europe as a whole. He will shamelessly exploit his Saxon dominions to achieve this goal. But, truth be told, he needs a larger power base than Saxony alone can provide.

And it so happens that, not far to the east, there is a kingdom for sale.

2

The Twin Kingdom

THE PRIZE THAT ATTRACTED Frederick Augustus's notice was Poland—or, to give it its proper title, the Commonwealth of Poland and Lithuania.

Just over a decade earlier, in 1683, its gallant king, John III Sobieski, had led a heavily outnumbered Catholic-Protestant army against the Turkish siege of Vienna. He put the sultan's troops to flight and, in effect, saved the capital of the Holy Roman Empire from becoming an outstation of the Islamic caliphate. One of his comrades-in-arms was Frederick Augustus's father, Prince Elector George of Saxony.

Now, in 1696, the old Polish king lay dying at his family's castle of Wilanow. Emperor Leopold of Austria had repaid Sobieski's help with arrogant disdain; the notoriously fractious nobility of Poland had greeted their king's attempts to revive a weakened Polish state with its usual combination of jealous suspicion and kamikaze arrogance. John had a son, James, but James would never succeed him as king. For Poland was not a hereditary but an elective monarchy, and the nobles who chose their own master wanted a foreigner for the throne. A rich one.

Within months of John III's death, the Polish parliament—called the Sejm—seemed ready to choose the French prince of Conti. However, young Frederick Augustus of Saxony was looking for a kingdom, and he had cultivated powerful friends, including Emperor Leopold of Austria and Czar Peter the Great of Russia. Between them, they had already agreed that the Sejm's decision was not, after all, final. They had large armies waiting on Poland's borders to hint that the nobility should think again. More than that, Frederick Augustus had amassed huge sums of money by mortgaging his country, imposing new taxes, and selling quantities of precious metals and stones. His representative, the Count von Flemming, busied himself distributing it to influential individuals in Poland itself. A lot of country gentlemen all over Poland suddenly found themselves, it is reasonable to suspect, able to contemplate a new stable block on the estate or a new mistress in town.

All the same, when the election was run, Frederick Augustus and the prince of Conti achieved equal shares of the chaotically organized vote. Each side duly declared itself the winner. Frederick Augustus, twenty-seven and in no doubt, despite all this election nonsense, as to where power really comes from, marched his Saxon troops over the border and on to Warsaw. Settled? Not quite.

There was a final problem to be solved before Frederick Augustus could give his soon-to-be-adopted country the push it needed to recognize the overwhelming, not to say intimidating, justice of his claim. He had to change his religion. This meant a minor political earthquake in central Europe. Since Luther's time the prince elector of Saxony had been recognized as the predominant ruler in Protestant northern Germany, the Reformation's shield and defender. Saxony's sacrifices in the Thirty Years' War were still just within living memory. But under the Polish constitution, only a Catholic could be elected king, and a king is what Frederick Augustus was determined to be.

There were mutterings of rebellion in Saxony at the prince elector's proposed conversion. Frederick Augustus's wife, already forced to put up with his womanizing, stubbornly refused to abandon her Protestant heritage. They separated. She withdrew to a remote castle, where she died thirty years later. All this made no difference. Frederick Augustus stuck to his new faith—though he was politically shrewd enough to reassure his Saxon subjects that no one would be forced to submit to Rome, and set up a council of Protestant worthies to guarantee this. The flames of revolt subsided.

And so a Catholic Frederick Augustus became. He was crowned King Augustus II of Poland amid pomp and celebration on September 15, 1697. The Cathedral of Cracow, traditional city of Polish coronations, was surrounded by Saxon troops.

• • •

"THE STRONG" may be what his subjects called him—a reference to his physical robustness and amatory prowess—but the reign of Augustus II would not be considered a success in political terms. The general historical verdict was that the final death throes of Poland began with the accession of its first Saxon ruler, and by the time the second was done with it, almost seventy years later, the country was coughing its last.

While this might not turn out to be Poland's greatest hour, or Saxony's easiest (Augustus uses it as a military cash cow—and that is before he starts on the mistresses, the hunting lodges, and the collections of really nice things), it is certainly good news for the craftspeople, merchants, artists, and influence peddlers of Dresden, the privileged Saxon royal city or *Residenzstadt*. Because, if he is going to be a European monarch, Augustus needs a capital that gives the right impression, he needs it in short order, and he is determined to spend whatever it takes to make that happen.

As a contemporary wrote: "Augustus the Strong can boast of having found Dresden a small city made of wood, but to have left it a large, glorious city built of stone." These few exhilarating, some would say crazy, opening decades of the eighteenth century are when pretty, postmedieval Dresden becomes grand, iconic "Dresden": a visitor destination, center for the arts and crafts, and perforce (given the monarch's delicate religio-political situation) a showcase for Protestant-Catholic understanding.

The capital that Augustus the Strong built. The Florence of the North. City as work of art.

Forty years later, on the king's death hundreds of miles away in Poland, Dresden glowed with sophisticated sandstone palaces and churches, surrounded by Baroque apartment buildings and squares, mostly made of that same native stone. It is curious—to rush forward more than two hundred years to Dresden's nemesis—that a British RAF pilot could be said to have observed, as he swooped over the darkened city center to mark it for bombing, that he had glimpses of the river Elbe lined with "old half-timbered houses." There were in fact none in the heart of the city, only in the outer suburbs where hardly any bombs were dropped. This was not quaint, wooden Hildesheim or Würzburg.

Augustus the Strong had made sure of that. The British pilot was imagining, based on his own preconceptions about how old German cities ought to look. Dresden was and is different.

It was the combination of royal interference and municipal pride that was to make Dresden unique. Dresden was a case where, well into the twentieth century, the useful and the beautiful—function and fashion—nourished each other. Saxony's rulers wanted a fine showcase, but they also wanted a functioning capital that supplied them with all the practical needs of war and manufacture that even the most aesthetically obsessed monarch requires.

BUT HOW TO PAY for it all? Augustus the Strong's wars ate up more money than even a rich manufacturing and mining state like Saxony could provide. Like a lot of rulers between the late Middle Ages and the first dawning of the Age of Reason, the king elector believed—or maybe half believed—in alchemy. Were it really possible to transmute base substance to gold, clearly the budget could be balanced no matter the circumstances. So when Augustus was introduced to someone who could, it was claimed, turn base metal into gold, he was understandably tempted to look into it.

It should have been a clue that the young man in question, an apothecary's assistant and goldsmith by the name of Johann Friedrich Böttger, was at the time a fugitive. He was in full flight from the king of Prussia, to whom he had already promised the secret of alchemy, and thereby eternal solvency, but who had proved impatient. Augustus II, though a natural optimist, was no fool. He gave Böttger protection, but at the cost of his freedom. From his cell, first in Dresden but soon at the Albrechtsburg, the ancient Wettin fortress perched above Meissen, the fast-talking would-be miracle worker was instructed to press on with the experiments he had already been conducting in Berlin. The king elector sat back and waited. And waited.

Three years passed while Böttger boiled and burned and battered combinations of lead and mercury and other traditional alchemical materials. He was always on the brink of a breakthrough, but the crucial transmutation somehow never came. No gold. There was an unfortunate escape attempt by Böttger in 1703. Augustus hung the threat of execution over his recalcitrant protégé. Still a result remained just out of reach.

Infuriated, the king elector put Böttger under the supervision of a trusted servant of the Saxon state, Count von Tschirnhaus. Tschirnhaus, a prolific mathematician and scientist, was eager to develop new industries that would make the king elector's realm more prosperous and self-sufficient. For years he had been investigating ways of producing a hard, vitreous porcelain to match the famous white China pottery that had long been imported into Europe (at enormous expense) from its country of origin. A softer porcelain had been manufactured in Florence since the sixteenth century and imitated elsewhere, including Germany. Although it superficially resembled the famous "china" kind, it could be cut with a file and still, unfortunately, absorbed dirt—while true porcelain did neither.

Count Tschirnhaus and Friedrich Böttger began experimenting with mixtures of native Saxon earth and minerals. Clearly, the count was pushing the research away from the cul-de-sac of classic alchemy and more in his own preferred direction. The harnessing of the, shall we say, overimaginative craftsman with the stolid pillar of science produced a breakthrough. Together they discovered the secret of making real hard paste porcelain. By 1707 Böttger's workshop, now in Dresden once more, was producing pottery made from red-brown stoneware. It was hard and glasslike, and possessed all the characteristics of classic china except its white coloring. Sadly, with the work at this crucial stage, and Augustus scraping the bottom of his already depleted treasure chest, Tschirnhaus contracted dysentery, curse of the early modern city, and died.

What happened next is not entirely clear. One popular story claims that Böttger had a "eureka" inspiration when his manservant used a hitherto unknown type of native Saxon china clay as an experimental wig powder. Another, more cynical, explanation has it that, after months of desperate inactivity following Tschirnhaus's death, Böttger finally managed to lay hands on his late colleague's scientific notes. Anyway, not too long afterward, Böttger came up with a kaolin-based white porcelain identical in every respect to the Chinese variety. It did the trick.

In 1710 a factory was opened under conditions of extreme secrecy within the confines of the Meissen castle, where it was to remain for more than a century. Five years later, the king elector even deigned to release Böttger from fortress arrest. It was a little late. Böttger died in 1719, still not yet forty years old. The incessant exposure to trapped fumes, the close contact with poisonous chemicals—and perhaps the stress of serving a capricious and remorseless master—had consigned him to an early grave.

Meanwhile, what became known as Meissen porcelain (and abroad as Dresden china) had been born. It was dubbed the "white gold" of the eighteenth century, much imitated elsewhere but rarely equaled for its perfection of material and its delicacy of style. After Böttger's death, efficient managers of money, men, and talent greatly expanded the factory's production. Dresden figurines were everywhere. As a royal monopoly, the State Porcelain Manufacture became hugely profitable for the Saxon crown. So perhaps the king elector's wayward alchemist fulfilled his brazen promise after all.

WHEN AUGUSTUS II came to the throne of Saxony in 1694, whole areas of Dresden had recently been laid waste in a great fire. True to his energetic, controlling nature, a few years later the king set up a building administration responsible to the city's governor—the same Count Flemming who had organized his election to the throne of Poland. In other capitals, such as London and Paris, where great fires gave rise to attempts at rational city plans that were never fulfilled, medieval chaos was simply replaced by Renaissance chaos. In the case of Dresden, regulations were strict and—it has to be said—wise. Wooden buildings disappeared from the center of the city. Churches, palaces, and private houses alike were subjected to planning laws and limitations unique in eighteenth-century Europe.

In the heart of the city, south of the river, lay the Zwinger (outer ward of the castle). In 1710 a pleasure palace began to be constructed there in stone. Matthäus Daniel Pöppelmann, the brilliant architect who served Augustus during the first part of his reign, and his court sculptor Permoser combined their talents to create a fabulous festive sculpture garden and outdoor stage contained within an extravagant, wonderfully lighthearted set of ornate buildings. These could be entered from the city side via a bridge and a magnificent covered gateway, the Kronentor, which was topped with a golden sculpted crown supported by four golden eagles, symbols of Polish royalty.

The Zwinger (as it continued to be called) housed the royal art gallery, a collection of amazing scientific instruments, a theater, an opera house, and an orangery. In the extravagantly domed assembly rooms, balls and festivities were housed in the winter. It became one of the most famous buildings in Europe, mentioned in the same breath as Versailles in Paris and Schönbrunn in Vienna.

When the king elector was resident in Dresden, the constant round of court festivities centered on this place. Within twenty years the other major buildings of Dresden had been created. The new royal palace, still named the castle, or *Schloss*. The Dutch Palace just across the river in the area known as the Neustadt (new town), which as fashions rapidly changed, became the Japanese Palace. The other great Baroque town palaces, built for the king's advisers, mistresses, and bastard and legitimate children alike. The Catholic Court Church was also begun, with its crypt where the bones of the Wettin rulers of Saxony would rest until 1945, though it was not finished until the reign of Augustus's successor. A bridge, akin to the Bridge of Sighs in Venice but crossing a street rather than a canal, connected it with the royal palace, so that the royal family and its Catholic courtiers need not expose themselves to the vulgar Protestant gaze on their way to Mass. In wary deference to the feelings of the Protestant majority, the doors were locked during the consecration of the church in 1751, and no bells were rung in the Court Church's tower until 1806.

But the greatest, most defining religious building of Dresden's Baroque era was not built by either of the king electors. There had been disquiet at the monarch's conversion to Catholicism in 1697. In 1722 began the construction, on the site of a demolished Gothic church, of the Frauenkirche (Church of Our Lady), which would become Dresden's Protestant cathedral, proof that Luther's faith was still dominant in Saxony. The building was paid for not by the king but out of city funds. Its designer, George Bähr, was official city master builder. Under his supervision arose an extraordinary design, unlike anything built elsewhere in Europe before or since. Plain on the outside, brilliantly ornate within, the church may have been cruciform in shape in the traditional way, but the actual area of worship was circular, with eight high, astonishingly slender columns supporting not just balconies where worshippers were seated but a great cupola of Saxon sandstone that would, when finally finished twenty years later, soar to a height of well over three hundred feet.

The Frauenkirche was consecrated in 1734, the year after Augustus the Strong's death. The acoustics were superb. In 1736 Johann Sebastian Bach traveled from his home in Leipzig and gave the first public performance on the church's organ. It had been built by Gottfried Silbermann—the Stradivarius of organ building—who even-handedly also created a marvelous—some say unsurpassable—instrument for the nearby Court Church. The Frauenkirche, this great, bright domed masterpiece, like St. Paul's in London dominated the new skyline of the city for more than two hundred years, a symbol both of Protestant self-assertion and municipal wealth.

Dresden became a fixture on the European grand tour. The paintings of the Venetian Bernardo Bellotto showed the glorious architectural vista along the Elbe, the smart squares and thriving markets. By the 1750s, with a Wettin still on the thrones of both Saxony and Poland in the form of Augustus III, the Catholic succession secured, and Dresden solidly established as one of Europe's finest capitals and most prominent centers of culture, it seemed that a golden future beckoned for the erstwhile "forest dwellers in the swamp."

Nothing, however, could have been further from the truth. And the danger came not from the Turks, or the Cossack hordes, or any of the other "enemies of civilization." The man who was about to order the large-scale destruction of the Wettins' newly built jewel on the Elbe was a fellow German, renowned as a patron of the arts and friend of French philosophers—the flute-playing king of Prussia, Frederick II, later known to those who admired him, from a safe distance, as "the Great."

• • •

It was the most terrible day of my life. By three that afternoon, the Church of the Cross, the administrative building, and my apartment were all in flames. I hastened over to the governor's offices . . . and gazed at this horrific prospect of destruction. I stayed there for a while, and at around 5 o'clock my honest servant came with the news that my house had burned down. The attic had been smashed open by bombs and everything in it destroyed by fire . . .

The date was July 19, 1760. The horrified letter was written by a senior civil servant and author, Gottlieb Wilhelm Rabener, to his friend, cabinet secretary Ferber, in the king's other capital, Warsaw. There Augustus III, son of Augustus the Strong, and his entourage had for some time been enjoying immunity from the immediate effects of the conflict that would be known to history as the Seven Years' War. And a fearsome conflict it was. Those who think of the eighteenth century as a time of elegant dynastic maneuverings punctuated by set-piece battles between men in wigs and highly colored uniforms, leaving the average citizen little affected, should look at the destruction of Dresden. Or, perhaps it should be said, one of the destructions of Dresden.

Frederick II of Prussia was extremely clever, highly if selectively sensitive, and a dyed-in-the-wool child of the European enlightenment. The French philosopher Voltaire was his regular houseguest, as were other famous thinkers and scientists. The Prussian king was also an emotionally warped, ruthless plunderer and wrecker of kingdoms who let nothing stand in the way of his military needs or his territorial ambitions. Frederick had first cast an eye over his wealthy, glittering neighbor during the first of his wars. He cynically compared Saxony with "a sack of flour. You can thump it as often as you like, and something will always come out."

Taking advantage of disputes over the Hapsburg succession, Frederick seized the prosperous neighboring province of Silesia from the Austrians. At first Augustus III, hoping for easy gains, supported the Prussians. Unfortunately, peace gave the Prussians Silesia but the Saxons nothing.

New diplomatic cards were dealt. When war broke out again in 1745, Augustus III and his Saxon forces turned up on the Austrian side. It was winter. Fighting in subzero temperatures on a frozen plain just a few miles from the Saxon capital, the Prussians annihilated Augustus III's army at the Battle of Kesselsdorf, and in short order occupied Dresden. Saxony's Austrian allies had thoroughly plundered the city themselves before leaving his capital to the mercy of the enemy. The Prussians looted what was left, then levied a huge reparations bill, causing galloping inflation and widespread hunger among civilians—an unpleasant taste of the "philosopher-king's" methods. But the next time, it would be bitterer still.

The Seven Years' War broke out in August 1756. Meanwhile the strength of the Saxon army had been reduced by almost half, to mend a huge hole in the state's finances, though the king elector's expenditure on pleasure, building, and ceremonies had not been correspondingly cut. He and his ministers had planned a policy of strict neutrality—as if this was within their choice. On August 28 the Saxon chief minister Count von Brühl received a note from Frederick the Great demanding free passage through Saxon territory. The next day, without waiting for a reply, Prussian troops crossed the border. The powers that be were reportedly astonished by Frederick's characteristic use of surprise.

Within two weeks the Prussians were back in Dresden, treating the capital as a Prussian city and Saxony itself as a conquered province. After first abandoning Dresden, then his army, Augustus III had shut himself inside the fortress at Königstein on the right bank of the Elbe—to which many of the royal house's most precious treasures and paintings had already been removed. In October he made a deal with the Prussians that gave Frederick the use of the Saxon army and in return allowed the Saxon king elector and his ministers safe passage through to his other realm, Poland. There he was obliged to stay for the duration. And he did.

Saxony's mineral wealth, trading riches, and agricultural plenty were ruthlessly siphoned into Frederick the Great's war effort. Young Dresdeners were press-ganged into the Prussian army.

The artists, craftsmen, performers, and singers who had flocked to Dresden to fulfill the Wettins' seemingly inexhaustible need for pleasure during the boom years now disappeared as quickly as they had come. At one point the Prussian commandant of the city ordered hundreds of houses razed to the ground to provide a free field of fire in case of siege. The homeless were forced to beg shelter and support from their already hard-pressed fellow citizens. Disease and ruin stalked the city. In 1757 in Dresden, 4,454 burials were registered, and only 1,647 baptisms.

So the fortunes of war ebbed and flowed. In 1759 the Prussians left in a hurry. An Austrian army descended on Dresden. More requisitions. More taxes. More unruly soldiery billeted in already overcrowded dwellings. And the next year, the return of those persistent, unkillable Prussians.

The siege, in the summer of 1760, saw fourteen thousand Austrian troops holding out against a large Prussian army. On July 19 began the massive, relentless bombardment described by Rabener in his letter. The Prussians fired not just thirty-pound cannon balls but, as Rabener described, oil-filled incendiary bombs, in a grim premonition of Dresden's fate in the twentieth century. Prussian grenadiers tossed hand-held bombs into upper rooms. Whole streets were set on fire.

Half the city's built-up area was destroyed. Many of the city's finest palaces—the Turkish Palace, where Augustus III had celebrated his wedding, the mansion of the king elector's clever, corrupt chief minister, Count Brühl—and a host of fine burghers' houses, were reduced to rubble and ash. The Frauenkirche survived—the Prussians' projectiles bounced off its cupola "like peas off a tortoise's back"—but the treasured, and more ancient, Church of the Cross (Kreuzkirche) succumbed as its great wooden pinnacle caught fire and toppled over, pouring flame onto the surrounding buildings and spreading the conflagration still further.

The Austrian defenders, grimly dug in under fire and subjected to the usual brutal eighteenth-century discipline, held out. Not so the Dresdeners. Panic-stricken refugees poured northward across the Elbe. Once again, starving citizens were cast on the charity of their luckier or more prosperous neighbors for food, clothing, and shelter. From Warsaw, King Augustus III sent yet another helpful message urging all Dresden's citizens to take courage, continue their business, and not to desert the city in its time of need.

Clearing the ruins was a slow, painful process. An average of fifty cartloads of rubble per day passed through the city gates after they were reopened in 1761, and things continued that way for years. Crippling inflation, food shortages, endemic street crime, and debasement of the coinage disfigured the once-proud capital. The Seven Years' War ended in 1763. A hundred thousand subjects of Augustus III had died during the years of war and occupation. The population of Dresden itself had been drastically reduced, from sixty-two thousand to thirty-six thousand.

Augustus III returned to his shattered residence in the summer of 1763. He was deeply shocked by the destruction he saw. The city that had been among the wonders of Europe was no more. Within months the king elector and Count Brühl were both dead. The throne of Poland passed out of Saxon possession to Catherine the Great's former lover Stanislaw Poniatowski. The Wettin heir, Friedrich Christian, died a few months later of smallpox at the age of forty-one, to be succeeded by his thirteen-year-old son, another Frederick Augustus. He was to reign for more than sixty years.

Five years later a keen-eyed young visitor from Frankfurt climbed the spire of the Church of the Three Kings in the Dresden Neustadt. He looked out over what was left of the city, and later reported what he saw:

> Dresden no longer exists in its entirety. Its best and most beautiful places have been reduced to ashes. Its greatest palaces and streets, where art and pomp competed for ascendancy, are heaps of stones . . . its wealthiest citizens have become poor, for what little was spared by the flames, robbery has now taken from them . . . Whoever saw this Residence before, in its full flower and glory, and looks at it now, would have to possess no heart were he not to be moved to the extreme by its present wretched circumstances, and stirred to tears of pity.

Johann Wolfgang von Goethe, Germany's greatest writer and humanist, was not yet twenty years old when he wrote this letter. Not for nothing did he make famous the story of the brilliant Faust, who to gain the world made a pact with the devil. Young Goethe was there to witness Dresden's fall from the heights of wealth and glory to the depths of poverty and despair. Its rulers had reached for glory, only to be left with a handful of dust.

Who said the past is but a distant mirror?

3
Florence on the Elbe

ASTONISHINGLY, within a few short decades Dresden was more famous and more favored that ever. Starting from the mid-eighteenth century, the old late-medieval fortifications that had protected Dresden in the religious wars of the previous era began to be dismantled. A whole swathe along the Elbe was donated by Augustus III to his chief minister, Count Brühl, who leveled the walls and turned the area into a private riverside terrace and pleasure garden. Fifty years later, when the Russians occupied Dresden, the czar's governor of the city, Prince Repnin-Volokovski, insisted on opening up the terrace gardens along the river for public use. Soon cafés and restaurants had opened there, with sculpted staircases connecting it to the Altstadt. The garden walk, known to this day as the Brühl Terrace (Brühlsche Terrasse) became one of the great promenades of Europe, like the Bois de Bologne or Hyde Park or Unter den Linden, where dandies and doxies, families and flâneurs alike, could take the air and their pleasure.

So, at the beginning of the nineteenth century, Goethe's friend, the writer Johann Gottfried Herder, was able to remark: "In respect of its cultural treasures, Dresden has become a German Florence." The description stuck. Elbflorenz—Florence on the Elbe—became the accepted version, found everywhere from tourist guides to political speeches

It was not only the buildings that earned Dresden its flattering title. The middle of the nineteenth century saw such cultural giants as Carl Maria von Weber, Schumann, Wagner, Caspar David Friedrich, and Ibsen settle on the Elbe, many with salaried jobs paid for by the crown. Dresden came back from the depths of 1760.

If anything, the loss of the Polish kingship turned the attention of the electors (kings in Saxony since 1806) back to their ancestral turf, with positive results. There was an aberration during Napoleonic times, when the Wettin addiction to bad alliances infected the new generation. For a few years, in pale echo of his father and grandfather, Frederick Augustus of Saxony (ruled 1768–1827) also became grand duke of Warsaw, a kind of mini-Poland. This title he held by grace of Napoleon Bonaparte, whose loyal ally he remained to the end.

As punishment, when France surrendered in 1814, Frederick Augustus lost not only his short-lived Polish grand duchy but also almost half of his ancestral kingdom to the new Prussian superstate that emerged from the French revolutionary wars. He was lucky not to lose it all.

The first half of the nineteenth century in Dresden, as elsewhere in Germany, was a time of gradual but decisive economic and social change, against which the established rulers defended their prerogatives with a grim, doomed determination. Nevertheless, after 1830 a two-chamber parliament (voted for by a select electorate that excluded the dangerous masses) came into being. Dresden's small Jewish community was granted civil rights. A few years later began the building of the famous Dresden synagogue, one of several masterpieces in the city designed by the great Gottfried Semper, architect and passionate democrat. The first paddle steamer, the *Queen Maria,* took to the Elbe in 1837. Others followed, the foundation of a fleet that still plies the river today. In 1839 the railway line between Leipzig and Dresden was finished, the first between two major German cities.

In many respects, however, Saxony remained backward. It was not until 1861, long after Prussia and other more dynamic states had done so, that the Saxon government allowed freedom of profession. This finally abolished the medieval caste restrictions and guilds that had for centuries secured a living for craftsmen but held back the innovation, enterprise, and personal mobility essential to real economic development.

The chief aim of Saxon external policy remained, as before, to limit Prussian encroachment. One more disastrous alliance followed before such matters were, in effect, taken out of Wettin hands. The brief and decisive 1866 war between Prussia and Austria for control of Germany found Saxony siding with Austria, along with most of the other states, large and small. Saxony ended up on the losing side. Again. But soon there followed a transformation. The war of 1870–71 between Prussia and France, into which the other German armies were also dragged, led not only to a decisive victory, but to a wave of national patriotic fervor that finally swept away the historic divisions between the states.

On January 18, 1871, in the Hall of Mirrors at conquered Versailles, the assembled representatives of the German states acclaimed the grizzled, seventy-three-year-old King Wilhelm of Prussia as their emperor. The new Germany contained four kingdoms, six grand duchies, five duchies, seven principalities, and three free cities. They kept their thrones or senates and many of their ancient rights, but overall power moved decisively to Berlin. The larger states, including Saxony, were allowed to keep their own armies, uniforms, and traditions, but in case of war were duty bound to answer the Reich (read: Prussian) commander in chief's call. The emperor and his chancellor dictated foreign policy and shaped much of the new Reich's economic, military, and political direction.

Suddenly, where for centuries there had been weakness and discord and jealous particularism, there was a united nation of more than sixty million placed in the heart of Europe. Rapidly industrializing, its sovereignty stretching from Alsace-Lorraine in the west to Posen (Poznan) in the east, the new German Empire could also boast an army that had recently proved itself invincible against France and Austria, two of the three other continental superpowers of the time.

True, the Reich had been united through blood and iron rather than persuasion and porcelain—the following three-quarters of a century would play out the consequences of that with almost sadistic attention to detail. All the same, this was a moment of real rejoicing. Past hostilities and failures were forgotten (or reinterpreted) and the novel sense of power, of possibility, enjoyed without reservation.

The contest between Saxony and its old rival, Prussia, had finally been decided in Prussia's favor. Since the Battle of Kesselsdorf more than a century previously, this had been an increasingly unequal struggle. For the moment, though, the effects of "defeat" for Saxony, and especially Dresden, seemed entirely positive. The lingering responsibility of being the capital of a state with warlike ambitions finally fell away, and as a result the city became a much more easygoing place, known all over Europe and in the Americas for its beauty, its civilized amenities, and a general style of life that would toward the end of the twentieth century be described as "laid back."

THE FOUR DECADES between German unification and the outbreak of the First World War witnessed a new blooming of Augustus the Strong's historic city that paralleled, even exceeded, the development of other parts of the Reich. A fourfold increase in population was mostly accommodated in carefully proportioned apartment blocks and fast-growing but spacious and rustic suburbs. With railways now crisscrossing Europe like vital arteries and veins, there was a boom in mass tourist traffic that put Dresden even more firmly at the heart of European artistic and cultural life. And the time witnessed a rapid expansion of the modern consumer industries, which Dresden not only adopted but also, in a disproportionate number of cases, actually originated.

The brassiere was invented in Dresden by a Fräulein Christine Hardt in 1889. (Even more piquantly, the first and last ruling Gauleiter of Nazi Saxony was a failed lingerie manufacturer.) The city could also claim to have been the first place in Europe to manufacture the cigarette (initially by hand, later by machine), the coffee filter, the tea bag, squeezable toothpaste ("Chlorodont")—and the latex condom. Oh, and it became a key center of the typewriter and the camera industries. Seidel and Naumann's classic Erika portable typewriter became world famous. Carl Zeiss may have ground his special lenses and mirrors in Jena, but when it came to producing cameras for the public, it was the nimble fingers and sharp eyes of thousands of Dresden workers he relied on. Many other companies would build camera factories in Dresden, not just Zeiss, making it the city's most important single industry.

The common element was affordable luxury, artifacts conceived to provide pleasure to the vastly increased number of relatively ordinary people who had money, leisure, and taste—or at least aspiration. The dominance of royal court requirements had shaped the workforce. Patient, delicate making of objects was the forte of the Dresden journeyman or woman. *Feinarbeit*, as the Germans say. Precision work. Dresden's skillful workers had served the whims of kings and nobles for centuries; now they catered to a prosperous middle class that wanted nice things and comfort with a passion that, in its way, matched that of Augustus the Strong. But by the end of the nineteenth century the sovereign was no longer the consumer. The consumer was sovereign.

So came the turn of the twentieth century. Dresden seemed symbolic of the best of old and new.

Largely owing to the early imposition of strict planning—Dresden was the first city in the world to accept the notion of zoned development—it had grown to a metropolis of four hundred thousand (half a million by 1920) while still retaining its reputation as a garden spot among European cities. New factories were not allowed in the oldest part of the city, though some workshops and yards quietly survived, and business went on. Some of the newer buildings—the new royal ministries on the north bank of the Elbe, the neo-classical museum and archive building known as the Albertinum after the king who ordered its construction in the 1880s—were criticized as disproportionately massive in the Prussian style and made from hard alien stones, but generally Dresden was not ruined or made ugly in the process of accommodating its increased population and supplying it with employment.

Although the heart of the city had been densely built upon, the royal parks and the municipal green spaces had survived. It was decided back in the 1840s that the new railways would not be allowed either to drive through the historic city center or to ring Dresden and deform its suburbs as they had elsewhere. Of course there were poor people. Of course there was plenty of housing that a few decades later would be considered unhealthy and primitive—respiratory and lung problems were a quiet scourge in historic Dresden—but there were no exclusively working-class ghettos, no dehumanizing slums to match those in Glasgow or New York—or, for that matter, booming, sprawling, ugly Berlin, where even by the most favored measure the population density was twice that in Dresden.

Particularly in the heart of Dresden, the prosperous and the struggling citizen coexisted side by side to a remarkable degree, as they had since the Middle Ages, but in far more salubrious conditions. This was a good place to live. By the standards of the early twentieth century elsewhere in Europe and North America, very good indeed.

So Dresden became a popular tourist destination, leading to the building of a host of comfortable hotels and boardinghouses, as well as places of entertainment and restaurants. Moreover, many well-off Europeans and Americans came for long visits, or even settled permanently. With its enviable architecture and lively (but not too avant-garde) cultural traditions, pleasant climate, magnificent surrounding countryside, and relatively low cost of living, the Saxon capital attracted thousands of such foreigners as long-term residents. There were—as Pastor Hoch pointed out—British, American, and Russian churches. There were also international finishing schools for young ladies. This was connected to the fact that in winter, much of the central European haute bourgeoisie and aristocracy visited Dresden for society balls—which doubled as discreet marriage markets.

But again, just as Dresden seemed to have found its way onto a sunlit plateau of best-of-all-worlds pleasure, the dark horsemen who last galloped this way in the 1760s were once more turning their steeds' heads toward the Florence on the Elbe.

4

The Last King of Saxony

BEFORE 1914 Dresden had a host of attractive features, especially for those who enjoyed comfortable incomes, but it was no light-opera idyll. "Many households had to exploit every imaginable source of income," according to a recent history of the city, "including child labor, second jobs, and subletting rooms. Especially bad for thousands of citizens were the unacceptable, often unhygienic and crowded living conditions." Here the historian is describing the picturesque, much-praised Dresden Altstadt.

There was widespread unemployment and labor unrest, most famously the 1903 cigarette workers' strike. Between 1903 and 1909, a total of 185 separate labor disputes affected 1,400 factories, along with twenty-two lockouts in another 250 factories. By 1907 there were sixty thousand trade union members in Dresden. The Marxist Social Democratic Party took the majority of Saxony's Reichstag seats and the state became known as the "Red Kingdom."

By the same token, Dresden might provide a haven for the arts and culture, but not all of what nourished this richness was attractive or even palatable. The young Richard Wagner had been living in Dresden for some years when he was appointed conductor of the Royal Opera there in 1843. *Rienzi* and *The Flying Dutchman* were both premiered there. *Tannhäuser* and *Lohengrin* were written there. He started to map out what was later to become the *Song of the Nibelungen* there. At that stage of his life a convinced democrat, Wagner made the mistake of throwing in his lot with the revolutionaries during the events of 1848–49. Even though his seditious activities had been confined to watch duty on the tower of the old Kreuzkirche, after the old regime was restored with the aid of Prussian bayonets, he became a fugitive from Dresden with a warrant out for his arrest. An amnesty was not officially granted until 1861, when Wagner was already becoming the most famous composer of his age. And of course, also a notorious, poisonous anti-Semite.

Along with the rise of the left, Dresden also witnessed a less conspicuous but no less portentous rise of the far right. Dresden's long-serving (1895–1915) high burgomaster, Gustav Otto Beutler, was himself a decided sympathizer with "national" causes. Such organizations as the (anti-Polish) Society for the Eastern Marches, the Navy League (Flottenverein), and the vigorously, rabidly racist and social Darwinist Pan-German League (Alldeutscher Verband) were exceptionally strong in the city. The Pan-German League especially included "astonishing numbers of dignitaries from the middle and upper classes of the population, among them several Reichstag and State Parliament (Landtag) deputies and numerous city officials and councilors."

In the minds of many Dresdeners, their city remained what it had been hundreds of years earlier: a fortress on the Slavic frontier. In a continuing echo of ancient conflicts, successful boycotts of foreign— especially Czech—products were organized. Rules forbidding the city authorities from employing itinerant Polish and Czech workers were introduced, and the Association of German Students in Dresden proudly declared itself "Jew free" in 1900.

A huge new group of salaried clerks and retail assistants, who felt themselves superior to the proletariat but as wage earners were socially excluded from the old, self-employed and professional middle class, came into being in Dresden, as in other cities of the Reich. This class compensated for its uncertain economic and social status by embracing the new politics of power and national aggrandizement with extra enthusiasm. Between 1898 and 1907 a conspicuously large number of congresses and conferences of far-right organizations took place in Dresden, often subsidized by the city authorities and with High Burgomaster Beutler acting as greeter and keynote speaker.

Before 1914 Dresden's superficial social mode was artistic and relaxed, but much of its politics was authoritarian, with ancient intolerances seething hidden beneath the city's perfect, well-cared-for skin. All this would become even more apparent when the old rulers went and the hard times came.

• • •

BY 1914 the Saxon royal family had been on the throne for 825 years. In 1889, on the Wettins' eight hundredth anniversary, there had been grandiose, colorful celebrations in Dresden and all over Saxony. Now they had four years and a few months left. The present king of Saxony would be its last.

In July 1914, Frederick Augustus III and his family were on a climbing vacation in the Austrian Alps when a telegram arrived, warning of a European crisis. By the beginning of August 1914, Germany and its ally Austria were at war with Britain, France, and Russia.

Dresden suffered, like the rest of Germany, from shortages and hunger, caused by the ruthless British naval blockade—which killed, it is said, many more German civilians than the Allied bombing campaign in the Second World War. With its great palaces and lavish public buildings, the capital became a city of hospitals and convalescent homes. And at least 120,000 of Saxony's young men, out of a total population of around five million, died in the trenches for kaiser, king, and fatherland. Of Dresden's half a million inhabitants, around fourteen thousand were killed—proportionately above the state's average—with many more permanently disabled or psychologically damaged.

If we are to believe his son's account, King Frederick Augustus was among those in Germany who floated the idea of a compromise peace with Britain and France late in the war. Perhaps he realized that this slaughter was destroying not just the German army but also the country's entire social and political system, with monarchs such as himself at its apex. Unfortunately, by 1917 Germany was little more than a military dictatorship. Generals Ludendorff and Hindenburg—both figures who were to play key roles in the rise of the ex-corporal Adolf Hitler—exercised decisive power. Except for Kaiser Wilhelm II, Germany's individual monarchs counted for almost as little as the politicians in the Reichstag, with their concerned speeches and peace resolutions.

Toward the end of 1918, with the German army in inexorable retreat and food shortages reaching crisis level, there were leftist-led protests and mutinies in Leipzig and Dresden, in common with other large German cities. The diehard right, also radicalized by the war, responded accordingly. As late as November 2, 1918, there was a grand rally of the self-styled People's Committee for National Defense (Volksausschuss für Nationale Verteidigung) in Dresden, which united all the patriotic, racist, and conservative groups. Bloodcurdling calls were made for "total war."

The calls went unheard. In the second week of November revolution broke out in Dresden. The post office and telegraph offices, the police headquarters, and the main government buildings were occupied by armed left-wing insurgents, many of them soldiers and led by rebellious airmen from the Grossenhain airfield. They were in touch with revolutionary soldiers and sailors elsewhere in Germany via the new communications phenomenon, radio. The red flag was hoisted on the royal *Schloss*. A hastily assembled Council of Workers and Soldiers crammed into the landmark big top of the famous Circus Sarrasani on the banks of the Elbe. The delegates declared the king deposed.

The conservative-nationalist elements, which had talked boldly just a week or so earlier of repression and national salvation, seemed to melt away. The king fled by boat to the royal pleasure palace at Moritzburg, an hour or so from Dresden, then on to the castle of a relative farther east.

On November 13, 1918, Frederick Augustus III Wettin, the last king of Saxony, officially abdicated, ending 829 years of rule by a single dynasty. A few days later he crossed the border into Prussian Silesia, where he settled at Sibyllenort, one of his family's ancestral estates. There he lived quietly as a private citizen until his death in 1932, respected even by many of those who had overthrown him. His former subjects still recount a well-known—though probably apocryphal—story that Frederick Augustus remarked wryly in Saxon dialect as he left his capital for the last time: "*Also, Kinder, dann machd eiern Drägg alleene!*" So, children, make your muck alone!

The "children," excited as they might be to find themselves in charge at last, were not to experience the time after 1918 as a happy one. The mix of revolutionary idealists and nervously hopeful democrats who had declared the "Social Republic of Saxony" had only short-term aims in common. In the longer term, they were ill-suited for cooperation.

As for the far right in Dresden, it was not dead but lying low, waiting for the tide to turn.

• • •

AFTER THE 1918 REVOLUTION there was a joke in circulation that went like this: If you told German workers to seize the railway station, they would, of course do so—but first they would obediently get in line to buy platform tickets. Considering the enormity of events in Germany and Saxony in the last months of 1918 and the beginning of 1919—national defeat, the overthrow of dynasties, rioters on the streets, radical political change—the process was relatively orderly and bloodless compared with other times and places in history.

Not that there were no casualties. In May 1919, after the moderate socialists had won the elections, a war veterans' demonstration against rising prices and lack of state help got out of hand in Dresden. Neuring, the Social Democrat minister of war in Saxony, was dragged from his office by rioters and tossed off the historic Augustus Bridge into the river Elbe. When the unfortunate politician tried to swim to the bank, an unidentified marksman from among the veterans gathered on the bridge shot and killed him as he floundered in the water. Within days martial law was declared, and with the aid of the army the remaining "workers' councils" in Dresden and Leipzig were dissolved.

Saxony was now supposedly a model parliamentary regime, founded on the moderate social democracy that for many years dominated politics there. All the same, conflict between extreme left and extreme right was as bitter as elsewhere in Germany, perhaps more so. The Saxon middle classes, humiliated by the Versailles Treaty, with their savings wiped out by hyperinflation, resentful of the new workers' rights enshrined in law after 1918, fretting for their lost privileges, and worried about their children's future, were alienated from the new status quo. Most longed for the old days of order and authority, of kings and generals and professors who knew how to rule and workers who knew their place. This was nowhere more true than in Dresden, where so many conservative-minded lawyers, civil servants, and education professionals were concentrated. To overcome the hostility of this key group, democracy needed stability. That was precisely what it did not get.

There was a short period of relative prosperity in the mid-1920s. Tourism in Dresden revived a little. No longer the wealthy independent travelers of the prewar period, or the long-stay foreign *rentiers*, but new trippers, who came in by train and made quick group tours of the city and its surrounding countryside in the new motor buses. It was a healthy contribution to the city's economy, but not the continuous subsidy that Dresden had been used to before 1914, when there had been a permanent, wealthy expatriate community numbering many thousands.

The artistic and architectural heritage of Dresden was still a great draw. In the 1920s Dresden even flourished briefly as a haven for the avant-garde. Before 1914 the Brücke (Bridge) group of artists, based in Dresden, had blazed a trail for the Expressionist movement. After the war many young artists, who had been through the hell of the trenches, were not satisfied with anything less than what they saw as absolute artistic truth. Searing paintings from this period include Otto Dix's *The War Cripple* and *Prager Strasse* (which featured war wounded begging on Dresden's most exclusive shopping street), and *The Unemployed Man* by Otto Griebel. The Austrian Oskar Kokoschka was professor at the Dresden Academy from 1919 to 1924, though he eventually left, finding Dresden "suffocating."

Toward the end of the 1920s many stalwarts of the avant-garde left for Berlin or Paris. Dresden remained provincial at heart, they decided. People flocked there to see old, beautiful things and an aesthetically comforting vision of history, not to endure too much reality.

Dresden lived to a great extent in the past, which was where its visitors and citizens alike were most comfortable. It was no accident that two of the earliest acts of Dresden's Nazis after taking power were these: the dismissal of Otto Dix from his teaching post at the academy, and the demolition of the modernistic, glass-and-steel Kugelhaus (the Globe House), built for the Health and Hygiene Exhibition in 1928 and loathed by the city's numerous artistic conservatives.

AFTER THE FIRST WORLD WAR, Dresden's manufacturing industries were saved by the fact that they were, by and large, of the more modern sort: cameras and optical instruments, typewriters, sewing machines, cigarettes, and toward the end of the 1920s, radios. In other parts of Saxony, traditional industries predominated, and they proved brutally vulnerable to foreign competition: Textiles, toys, bicycles, musical instruments (which also suffered from the advent of radio and cinema) all declined. During the temporary revival in world trade, the Saxon capital benefited more than the countryside.

But then came the slump. Dresden slid into the same economic quicksand as the rest of the world. The number of wholly dependent welfare recipients in the city increased from fewer than twenty-eight thousand in 1927 to almost seventy-four thousand in 1932. In Saxony, unemployment rates were the highest in Germany. The Social Democrats remained the largest party but continued their slow decline. In 1930 elections brought increases in both the Communist and the Nazi vote. Parliament was paralyzed.

Gertraud Freundel's father was one among tens of thousands of bankrupt and unemployed men who threw in their lot with the extremes of left or right.

> Hitler was a bringer of hope.... My father had a shop, he was a sup-
> plier of bicycles and gramophones and all that. Things went so
> badly that he had to close it. Every day he used to take me to the
> kindergarten and there was another man there who was also unem-
> ployed. My father was an optimistic sort of man, but the other one
> said, look, at thirty-three we'll never get another job in our entire
> lives.

Herr Freundel joined the Nazi Party—and got work. His was not an untypical case.

With the left and center unable to unite against the right, a cabinet of "unpolitical" conservative civil servants ran Saxony for the next three years. The Communists could have combined with the Social Democrats to form an anti-Nazi bulwark but refused, claiming that the Social Democrats were, in effect, the real fascists (one of the main Communist taunts was to call them "social fascists"). The Nazis, on the other hand, were simply a symptom of the decline of the wretched petty bourgeoisie. Therefore the important thing for loyal Communists was to oppose the Social Democrats—not the Nazis, who were doomed in any case, along with the class they represented.

This bizarre and fatal stalemate continued until Hitler came to power. Saxony remained the most depressed area in the Reich. Even as slight improvements in the national economy started to become apparent elsewhere, in the autumn of 1932 Saxony was still flat-lining.

In the Reichstag elections of July 1932 the Nazi Party won 37.7 percent of the overall vote in Dresden, becoming the largest party by a big margin. Of the thirty-nine cities in Germany with populations over a hundred thousand, Dresden was in sixteenth place, but of the seven metropolitan cities with between four hundred thousand and six hundred thousand inhabitants (the others being Cologne, Leipzig, Munich, Breslau, Essen, and Frankfurt), the Nazi vote in Dresden came second only behind Breslau, a city on the Polish frontier and long a refuge of the far right.

Dresden was a Nazi stronghold even before Hitler took power. Certainly it harbored a large proportion of Hitlerite activists. Less than four years after its founding, the local Nazi Party newspaper, *Der Freiheitskampf* (*The Struggle for Freedom*), had achieved the largest circulation of any daily newspaper in the city, overtaking the liberal *Dresdner Neueste Nachrichten*. Gauleiter Mutschmann's mouthpiece, *Der Freiheitskampf* was packed with violent political polemic. It routinely featured rabid headlines such as "Red Plague Is Ruthlessly Stamped Out" (the Reichstag fire) or (on election day) "Defeat the Jews' Machinations!" Even so, by the beginning of 1933 it had more than a hundred thousand regular subscribers in the city and its immediate surroundings.

In the city council elections of November 13, 1932, the Nazis gained a slight edge in percentage of the popular vote over the Social Democrats, who had hitherto formed by far the largest single group. In terms of seats, the result was a dead heat (twenty-two each) out of seventy-five between the two big parties, with the Communists also booking a large increase (to thirteen), but the middle-class parties that had formerly supported democracy voted for a Nazi chairman.

On January 25, 1933, Communists in the industrial inner suburb of Dresden-Friedrichstadt called a meeting at a pub cozily named Keglerheim (Home of the Bowler). Bands of Nazi storm troopers had gathered threateningly outside, with a line of police standing between them and the building. The Reds assumed that the police were there to protect their perfectly legal meeting. But when fights predictably started, the police fired not at the Nazis out in the street but into the crowded assembly room of the inn. There hundreds of Communists, including women and children, were still gathered, thinking themselves safe. Nine were killed, dozens injured. It was an indication of how far the Nazis had infiltrated the sources of power even before their formal call to government.

By the turn of the year 1933, one thing had become clear: the real struggle for power in Saxony's cities, including Dresden, was between the Communists and the Nazis.

And in that contest, the Nazis were already emerging as the winners.

5

The Saxon Mussolini

ON MAY 23, 1933, in the sandstone Landtag (State Parliament) Building by the river Elbe, the Saxon people's freely elected representatives voted themselves out of existence. All but a handful of the deputies gathered there gave assent to an Enabling Law for Saxony. This mirrored one already passed by the Berlin Reichstag, giving absolute power to a central government dominated by Adolf Hitler's National Socialist Party. Landtag deputies from the Social Democratic Party cast the only dissenting votes. Once the mighty overthrower of emperors and harbinger of the German welfare state, the SPD was now reduced to a brave but pathetic rump of six members.

The tragic farce of the May 23 sitting had long been a foregone conclusion. In March the last freely elected government of Saxony had been dissolved by decree and replaced by a Nazi "Reich commissar." In April the distribution of seats in the Landtag had been rearranged on the lines of the recent Reichstag results to give the National Socialists and their allies an absolute majority. The Communists, who even in the Nazi-controlled national elections of March 1933 had gained 17.4 percent of the vote, had quickly been banned and their leaders arrested.

Of the Social Democrats "permitted" to remain in the Saxon parliament, ten were already under arrest, two undergoing medical treatment for injuries inflicted by Nazi thugs, and another five had fled over the border into Czechoslovakia. Otto Neubrig, on behalf of the remaining handful, gave a courageously defiant speech, opposing the dictatorial powers claimed by the Nazis and calling on the remaining representatives of the middle-class parties to join him in voting against them. In vain. The constitution was abolished and the principle of parliamentary government in Saxony abandoned. Within days the Social Democratic Party had been banned. The nonsocialist parties dissolved themselves. That was how quickly democracy could be dismantled.

The man who now became the chief—soon the only—power in the land was unprepossessing in both appearance and style. Martin Mutschmann, appointed by Hitler as *Reichsstatthalter* (Reich governor) of Saxony on May 5, 1933, was also the long-serving Gauleiter (provincial leader) of the National Socialist Party. A brutal-looking man with staring eyes, he was not popular even among fellow Nazis. It was never suggested even by those who kept him in his job that he was an especially competent administrator. But he was a fanatic—which appealed to Hitler and his entourage. Mutschmann was also a man who knew how to accrue and handle power, to such a single-minded extent that within two years of the Nazi "revolution," he ruled Saxony in an absolute way unheard of in other areas of Nazi Germany—which is why the fate of Dresden was inextricably linked with the character of its Gauleiter, and why Mutschmann's personality loomed uncomfortably large in Dresden's story before and after February 1945.

Mutschmann was born on March 9, 1879, and educated in the west Saxon textile center of Plauen, making him fifty-four in 1933—ten years older than Hitler, and the best part of a generation senior to most of the other leading Nazis such as Goebbels, Himmler, and Göring. He left school at the age of fourteen, undertaking a commercial apprenticeship that saw him enter the textiles business as a foreman/manager. He took various management positions in lace and underwear companies before starting up his own factory at the age of twenty-eight. This factory stayed in business, through various vicissitudes, including its owner's war service between 1914 and 1916, until 1930, when it finally closed down owing to the economic crisis (or, as Mutschmann's supporters claimed, the dastardly machinations of Jewish competitors). By then, as a leading member of the National Socialist Party—joined 1922, assigned the high-ranking membership number 35 in the reorganization of 1925—Gauleiter and Nazi member of the Reichstag, Mutschmann could make a living solely from politics.

In the years before 1933 the Saxon Nazi Party under Mutschmann had not been an especially united organization. Repeatedly,

Mutschmann's brutal personal style—and the lack of transparency in the party's financial dealings under his leadership—caused major upsets in the party. There was a drip-drip of important defectors who loudly proclaimed their disapproval of his arrogance and vanity. In July 1932 Arno Franke, the editor in chief of the main Nazi newspaper in Saxony, *Der Freiheitskampf*, left the Nazi Party and published a searing critique of its "quite incredible mismanagement," which he blamed on the party bosses' untrammelled power: "In no other party do so many dishonest elements make it to the top as in the NSDAP . . ."

This came from a man who knew Mutschmann better than almost anyone else. As Goebbels, also an old acquaintance, would later confide to his diary: "Mutschmann will allow no gods other than himself. In this way he loses a lot of prestige." Nevertheless, his being liked or not seemed to make little difference to Mutschmann's inexorable rise.

Mutschmann's main rival was the radical SA (Brownshirt) leader Manfred von Killinger. On March 8, 1933, with a non-Nazi government still technically in power in Dresden, von Killinger was appointed Reich commissar for the Saxon police, essentially placing him as the Nazi government's controller within the Ministry of the Interior. The effect was immediate and terrible. All over Saxony, from the back streets of the cities to the smallest village, local Nazi Party and SA groups took swift, sometimes bloody revenge on their democratic and communist opponents.

Dresden too was turned into a playground for SA thugs. In the square opposite the school of music, the Nazis made a huge bonfire of the books they disapproved of. That same week the internationally famous German conductor and director of the Dresden Opera, Fritz Busch (1890–1951), had been prevented from conducting *Rigoletto* by hordes of rioting SA men. His crime was to object to the dismissal of fourteen Jewish musicians. The reasons given included "private intercourse with Jews" and the hiring of Jewish and foreign musicians. Busch immigrated to America and became director of the Metropolitan Opera in New York. He refused to work in Germany ever again. Not just Otto Dix, but other modern or left-wing artists were dismissed from their teaching posts. Others hastily made plans to leave. The much-respected liberal high burgomaster of Dresden, Dr. Wilhelm Külz, was "sent on leave." Later in the summer he was quietly dismissed.

> Again it's astounding how easily everything collapses . . . complete
> revolution and party dictatorship. And all the opposing forces as if
> disappeared from the face of the earth . . . No one dares say anything
> any more. Everyone is afraid.

So wrote Victor Klemperer, a Jewish professor at the Dresden Technical University, in his diary entry for March 10, 1933. With control of the police, and a thousands-strong paramilitary organization under his command in the shape of the Saxon SA, von Killinger seemed to be the man in charge. Mutschmann, however, had the ear of Hitler, and soon, as governor, a brief to "cleanse" the state's government of unreliable elements. This he exploited to the full. The conflict had already begun that would not be resolved until the so-called Night of the Long Knives. The national head of the SA, Ernst Roehm, the second most powerful man in Germany after Hitler, was keen to carry through a "National Socialist revolution" with the aid of his Brownshirts, who would shadow government agencies and act as "guarantors" of Nazi purity. By mid-1934 the government in Berlin was squaring up for a showdown with the SA and Roehm. Whatever the contradictions in Mutschmann's behavior, he was a fanatic, unquestionably loyal to the Führer, and he seemed to be delivering the goods. Von Killinger, as an SA leader, was doomed.

When the moment of reckoning came, Von Killinger was, in a way, lucky. On June 30, 1934, on the pretext of suppressing an attempted coup, Hitler personally organized the slaughter of Roehm and his acolytes in the SA. Several of Dresden's SA leaders were also taken out onto the Heller, an area of heath land on the northern outskirts of the city, and shot during the night of June 30–July 1 by a squad of SS killers brought in from Chemnitz specifically for the purpose.

Von Killinger escaped the fate of so many of his SA comrades. He was arrested, spent four days in a cell at Hohnstein, but was then released. The former Reich commissar never regained his exalted position, despite desperate petitions to Hitler. He continued to serve the Nazi Party, however, acting as a confidential emissary to General Franco in the Spanish Civil War, then being appointed German consul-general in San Francisco—a posting that reeked of espionage. In 1939 he was transferred to "special duties in the Foreign Ministry," first becoming German representative in Bratislava, capital of the puppet state of

Slovakia, and then in Romania, one of Germany's most important allies in the war against the Soviet Union.

By the end of 1934 Mutschmann was not just Gauleiter and Reich governor but also prime minister of Saxony. At the outbreak of war he also acquired for himself the post of Reich defense commissar for Saxony (Defense District IV), putting him in charge of matters such as air raid measures and general defense coordination. Mutschmann became and remained for more than ten years a kind of bonsai-sized dictator of Saxony.

The populace secretly referred to him as "the Saxon Mussolini" (he had a similar strutting manner to the Italian dictator, and the same bullet-shaped head), or as "King Mu."

THE NIGHT OF THE LONG KNIVES on June 30, 1934, in which the regime had dealt with the radicals in its own ranks, also represented an opportunity to settle accounts with its non-Nazi opponents.

Prince Ernst Heinrich of Saxony, a conservative democrat and third son of the last king of Saxony, had always loathed the Nazis. He had joined those who tried at the last moment to influence the circle around President Hindenburg not to hand over power to Hitler. "Only the most stupid calves choose their own butcher," as he maintained, but even though his royal status guaranteed the ear of senior advisers, he failed. Once Hitler came to power, Ernst Heinrich was soon to learn that even royal blood did not guarantee his safety.

June 30, 1934, was a sunny, warm summer's day. The prince's car emerged from a royal hunting lodge on the outskirts of the town of Moritzburg, which had remained in his family's hands. Suddenly a car containing two SS officers cut across in front of him. It was followed by a troop carrier manned by thirty SS soldiers in steel helmets and carrying rifles. The senior SS officer was the equivalent of a major general, in honor of the prince's status, but his request was not notably polite: "By order of the Gestapo, you are under arrest," said the *Gruppenführer*.

"Why?"

"That is none of your concern. You will find out later."

At police headquarters in Dresden the prince was led before the president of police, an old acquaintance, and heard of the supposed "revolt" by the SA. Ernst Heinrich assumed there had been a mistake, since he had no relationship whatever with the Brownshirts, but he was wrong. The prince stayed in custody, though he was allowed to send to Moritzburg for his things and make a phone call to a sympathetic minister in the Saxon government. Then the Gestapo arrived and things turned bleak.

Ernst Heinrich spent the next hours in a tiny cell. During that time he heard a muffled shot—a fellow prisoner had been murdered out of hand. Toward dawn he was roused and dragged into the yard with other bleary-eyed prisoners. They were loaded onto an open truck, warned they would be shot if they made any untoward movement. The prince described their drive through the still-dark streets of Dresden:

> No one dared even to sneeze. A magnificent day was starting to dawn in the sky. We roared through the city where 16 years before my father had still ruled as King. Now his son was being dragged through the former royal seat as a political prisoner.

Hohnstein was a forbidding sixteenth-century fortress set high in the rocky heights of the "Saxon Switzerland," upstream from Dresden. It had long served as a prison for captured enemy soldiers and troublesome subjects of the Saxon monarch. In the early days after the Nazi seizure of power, the castle had been turned into a prison-cum-torture chamber by the SA, and then taken over by the Gestapo as an official concentration camp. The prince and his fellow prisoners arrived there at first light, to be greeted by an armed SS squad.

The squad were ordered to release the safety catches on their rifles and take a bead on the new arrivals as they lined up in the entrance courtyard. Then the SS commander called out: "The prisoners will turn around!" Immediately one of the prisoners, an arrested Dresden SA leader, shouted out: "Shoot us from the front, we won't turn around!" It was a warning to his fellow captives. As a Nazi bruiser, he knew the system: You made the prisoners turn around so that wounds in their backs would show that they were killed "while trying to escape." There was a minute-long standoff. Then the SS officer turned away and bellowed for the prisoners to be escorted into the fortress.

Over the next three days, the SA leaders were tortured by their SS former "comrades." Ernst Heinrich and another two civilians, both civil servants, were merely subjected to threats. Then, without explanation, the prince was suddenly released. As he left the fortress, some SS sentries turned their backs on him, while the rest presented arms like royal honor guards of old. A perfect expression of the contradiction between the two irreconcilable elements with the Nazi movement—the yen to build a terrible new world, held back by nostalgia for the old.

Within the day Ernst Heinrich was back at Moritzburg with his wife and children. He later discovered that out of lingering class solidarity, an aristocratic Gestapo commander had removed the prince's name from the list of the condemned. The fate of the disgraced SA men is unknown. Judging from other, less privileged accounts of what happened in Hohnstein during those days of horror, they were unlikely to have been spared much. Otto Griebel, Communist and painter, described in his autobiography the shock of meeting someone who had survived Hohnstein:

> Silently, he showed us his bare back, which was covered with deep scars and welts, and finally spoke: "At every interrogation, those beasts abused me with steel rods and bull-whips. Some people, who couldn't take it anymore, hanged themselves in their cells. When a Jewish prisoner lay unconscious, they beat his exposed private parts with a rolled-up wet towel until he came to. No, those people are not human, not anymore . . ."

The Social Democrat politician Hermann Liebmann, once minister of the interior in Saxony and for years chairman of the party's faction in the state parliament, was arrested in Dresden in April 1933. Mutschmann had a special fate in mind for a man who had spent a considerable part of his political career attempting to stem the Nazi tide. On May 20, 1933, after a meeting of the Nazi district and local leaders in nearby Bad Schandau, Mutschmann drove with an entourage of these worthies to Hohnstein, where Liebmann was imprisoned. Liebmann was led in front of the gauleiter and his cronies and handed a manuscript containing anti-Nazi speeches he had made in the Saxon parliament. He was forced to read aloud from the text amid uproarious laughter. He was then beaten up so badly that he died of his injuries, apparently on the same day.

Another prominent Dresden social-democrat politician and writer, Dr. Max Sachs, was murdered in Sachsenburg, the purpose-built concentration camp established by Mutschmann southwest of the city. A fellow prisoner described the horror of Sachs's final hours:

> When I had to fetch water from the washroom . . . I saw how the SS had Comrade Sachs on the floor, naked, and were working on him once more with scrubbing brushes and water. His body was green and blue from head to toe, livid with red welts. A short time later it was announced that Sachs had died of a heart attack.

Both these men were singled out as Jews. Those Dresden Social Democrat leaders who were not arrested in the first days of the Nazi regime managed to flee over the border into Czechoslovakia. The surviving activists distributed smuggled anti-Nazi material into Dresden, often using the resulting income to support the families of comrades held by the Gestapo. In October 1933 they opened a tobacconist's shop, which they planned to use as a camouflaged meeting place for the resistance.

The Social Democrats were betrayed from within their own ranks. Weeks later the Gestapo rounded up the entire three hundred–strong underground structure of the party in Dresden. Many were sentenced to long terms of prison and hard labor. The Nazis had moved ruthlessly against the Communists even before taking full power, arresting many of the party's leaders before they could disappear abroad or underground. Later the Gestapo succeeded in infiltrating most of the Communist cells that remained. Again, there were show trials and heavy sentences "to encourage the others."

The secret police in Dresden possessed only two hundred salaried employees but was able to rely on an army of informers and its indispensable helpers in the regular police. By the middle of 1935 it had all but extinguished organized opposition in the city. In April 1937, adding the former political section of the old police force to its strength, the Gestapo moved into the former Hotel Continental in the Bismarckstrasse, which extended along the southern rim of the main station, the Hauptbahnhof. The new headquarters' notorious warren of interview rooms and detention cellars were to become part of the nightmares of Dresden's Jewish citizens, as well as any other political and racial undesirables.

Otto Griebel bears witness in his autobiography to the growing sense of helplessness in the face of the new regime's ruthless efficiency and—it has to be said—widespread popularity. For Griebel, as an avant-garde artist, the setbacks were also not purely political.

As early as September 1933 the Nazi-appointed Commissar for the arts in Saxony, Walter Gasch (himself a painter) and the head of the Dresden Academy, Richard Müller, staged a show of modern art at the city hall, under the title "Reflections of Decay in Art." The show contained works in the possession of the Dresden City Museum. There were paintings by Otto Dix (which included his horrific panoramas of the trench warfare he had witnessed as a soldier in the First World War), Schwitters, Georg Grosz, Kokoschka, and many other well-known artists of the time, including Griebel himself. The Dresden show attracted Goebbels's approving attention. It was later expanded, moved to Munich, and then sent out on a national tour as the infamous "Exhibition of Degenerate Art" (*Ausstellung der Entarteten Kunst*).

Otto Griebel's life was meanwhile punctuated by financial problems and worries for his family. The regime controlled public purchases and support of the arts, vetted the appointment of teaching staff in cultural institutions, and had infiltrated the clubs and associations that were such a lifeline to individual artists. Artists they disapproved of found it hard to make a living. A fair proportion of Dresden's artists had, in any case, always looked to the right. But now many of Griebel's friends had either withdrawn from politics or, with a frequency that surprised and disgusted him, hurried to curry favor with the Nazi enemy.

In common with all who try to survive under totalitarian regimes, Griebel learned to compartmentalize his life. He remained under occasional surveillance by the Gestapo. The woman who ran the fruit stall opposite his family's apartment building would tell him of men who sidled up to her and asked if the artist Griebel had a lot of visitors, and if so what kind of visitors. All the same, he kept a circle of trusted artist friends—they called themselves "the Seven Just Men"— and with them he could relax. They would walk in the country, set up their easels and paint together in the open, and only then talk freely about art and politics. This routine made survival tolerable, at least until world war came and sucked the Just Men into its maw along with tens of millions of others—just and unjust alike—throughout Europe.

By the mid-1930s in Dresden, active opposition to Hitler had been all but crushed. In July 1935 the largest single group of political offenders imprisoned in Saxony was made up not of organized resistance workers but of private individuals arrested for "utterances hostile to the State"—anti-Nazi political jokes, or the spreading of (accurate) rumors that the party leadership around Mutschmann in Dresden was corrupt and lived in luxury at the people's expense.

Gauleiter Mutschmann's Dresden had by now become one of the great regional capitals of Nazi Germany. Politically it belonged among the regime's strongholds. It remained so until the terrible consequences of that regime were brought terribly close to home, this time not just for a few hundred unfortunate dissidents but for hundreds of thousands of "normal" citizens.

6

A Pearl with a New Setting

ON MAY 30, 1934, Hitler declared to a huge and enthusiastic crowd: "Dresden is a pearl, and National Socialism will give it a new setting."

The Führer was in Dresden to open the "Reich Theater Festival Week." It was his first visit since the Nazi seizure of power. The last time, in July 1932, he had spoken to an audience of a hundred thousand cheering supporters during his campaign for the Reich presidency. The theater week two years later, with the Nazis now in power, was a triumph for Mutschmann, proud to host not just the Führer but also Propaganda Minister Goebbels and a selection of other major chieftains of the new Third Reich. The city was bedecked with swastika flags and banners. In the Führer's honor, the city's major architectural monuments were floodlit at night for the first time.

The atmosphere of the time was, for many, euphoric. Every day, seven-year-old Günter Jäckel and his schoolmates had to recite a special prayer before starting classes:

And bless the deed that has liberated our homeland
For our glorious Führer we thank You
Bless him and the Fatherland forever and ever
Amen.

The brash new regime, with its harsh yet colorful theater, appealed to a child's sense of order, of joyous repetition. As Jäckel recalls:

An ordered world, then, this Dresden after 1933! A city on a constant high of festivals, of celebration, of roll calls and parades. Banners fluttered as the SA formed ranks, we were supposed to salute them with our arm outstretched—we did this gladly and often . . .

On the long-awaited day when the Führer showed himself in Dresden, Jäckel was present with his mother. They had walked from their basement apartment south of the main station all the way to the city center. Once there, they weaved among the vast crowds controlled by SA and SS marshals, searching for a place that would give a view of the Führer. They found one near the city theater, the Schauspielhaus (for was this not a theater festival?) and waited. Alas, the excited little boy had not listened to his mother's advice about going to the bathroom before leaving for the demanding excursion. For a while he was stoical enough. Then the pressure became too much. After some frantic darting around, checking of ornamental bushes by the side of the avenue—rejected as not providing sufficient cover—they pushed their way out into the Theaterstrasse. Since everyone was in the crowd awaiting Hitler's motorcade, the usually busy street stood empty.

Just as little Günter was finding relief against a shady wall of the theater building, a band struck up, the shouts of the crowd shaped themselves into hoarse cries of "Heil Hitler," and the Führer roared by in his convertible, to stop moments later at the luxury Hotel Bellevue, opposite the opera house. "So, I never got to set eyes on the beloved Führer. But we would all be able to trace his trail for many, many years to come."

One thing was clear. Within a fairly short time of Hitler's accession to power, most of Dresden was definitely on his side.

There were good reasons for this, and they were not just political ones. Unemployment declined swiftly after 1933. The Nazis had discovered deficit spending, and the masses felt at least some of the benefits. A modern airport was constructed at Klotzsche, just north of the city. Both the tram network and the rail and bus links were improved, allowing easy access to and from the surrounding country. By 1936 the new *Autobahn* network had reached Dresden, sucking in unemployed labor like a peacetime war. Works included a massive road bridge over the Elbe that was one of the wonders of the age.

Tourism recovered. The economic upswing also boosted demand for the leisure-oriented consumer goods in which Dresden specialized. The first 35mm single reflex camera (the Kine Exakta) began production in 1936 at Ihagee's factory in the Schandauer Strasse. The design would later be produced all over the world, so widely that few would be aware of its Dresden origins. All this, by the mid-1930s, brought cheer to the burghers of Dresden and moved them to acclaim the Führer as he inspected their theaters and made sweet speeches about their city and its future. They were not to know that the price of this was more than just the loss of political freedom. Hitler's government was running up debts that could be repaid only by the fruits of conquest, and that almost certainly meant war.

Almost from the first, the new government had started to favor industries relevant to rearmament. This was a planned, conscious direction throughout the Reich, and Dresden was no exception. There were no smokestack industries here; unlike in the Ruhr district, home of Krupp, or the coal and steel hells of Silesia. No guns, ships, tanks, or trucks were made in Dresden. Dresden's specialty was what the Germans call *Feinarbeit*—precision work. There was Zeiss-Ikon (cameras and lenses), Radio-Mende (radios, fuses, communications equipment), Sachsenwerk (electrical products), Seidel and Naumann and Clemens (sewing machines and typewriters), along with famous cigarette manufacturers whose brands were known throughout Germany and central Europe.

Germany before World War II had become the largest and most advanced producer of machine tools in Europe, its output far exceeding that of Britain. This meant that switching machines from harmless consumer goods to war equipment—and thereby changing the nature of entire factories and firms—was relatively easy. Cigarette machines could (and would) be adapted to produce bullets. Radio assembly lines could be adapted to produce communications equipment and electrically operated fuses for the Wehrmacht. And nowhere were the skills of the lensmaker more useful than in guaranteeing the deadly accuracy of a bombsight. These Dresden factories, along with their machines and workers, could with relative ease be turned around and equipped to produce, in many cases, the "cutting-edge" apparatus of modern war, 1940s style. On sea, on land, and in the air.

The location in Dresden of such resources, with the workers whose skills were required to exploit them most efficiently, was not the city's only advantage. Geography was crucial. Saxony lay in the eastern part of Germany, and Dresden was located in the eastern extremity of Saxony—even farther beyond the then-practical range of British or French bombers.

Finally, Dresden, as a cultural and architectural monument of great renown, might be spared major destruction. After all, the British still loved Dresden, did they not?

THE LARGE ALBERTSTADT military and industrial complex was named after King Albert of Saxony, who had ordered its construction in the 1870s. It sprawled on either side of the Königsbrücker Strasse, just to the north of the Neustadt.

With its barracks, training grounds, ordnance factories, and warehouses, Albertstadt had fulfilled an important role in the First World War. Under the terms of the Versailles Treaty that ended the war, most of the factories and warehouse buildings were compulsorily privatized and let on commercial leases to local businesses. Facilities deemed ineligible for civilian conversion (including all but a few of the rail sidings and loading ramps) were to be destroyed. The victorious Allies' International Military Control Commission was set up to monitor this process here and throughout Germany's other military-industrial centers. It made regular inspections of the Albertstadt and repeatedly requested action on this issue.

Despite years of wrangling, almost none of the buildings or facilities was demolished. Three ammunition-filling facilities, for instance, were a constant bone of contention. After the final visit by an IMCC inspection team led by the British general C. Walch, the Allies grudgingly accepted that the planned letting of the buildings to a sugar company represented a change to civilian use. On October 25, 1925, the director of the complex's administering company wrote with a hint of triumph:

> Although the Commission did not take a final position on this question, we nevertheless believe that the Albertstadt industrial area will now be spared further visits from the Entente Commission . . .

So the military-industrial area escaped with reversible changes of use but without significant loss. Its reclamation and expansion became a priority again as soon as the Nazis took power. Dresden already housed the largest garrison in Germany in 1933, with five thousand troops stationed there, amounting to 5 percent of the entire postwar army. The total strength of the Dresden command would quadruple to twenty thousand before war broke out again. By then the second largest permanent concentration of troops in the entire Reich outside Berlin, it included a Luftwaffe training battalion and a company of SS engineers. These forces, for years held in reserve like a powerful chess piece within easy reach of both Czechoslovakia and Poland, played a major role in the incorporation of the nearby Sudetenland into Germany in October 1938, and then the forcible occupation of the remaining Czech lands as the Protectorate of Bohemia and Moravia just a few months later.

As the decade advanced, the ancillary industrial area was slowly remilitarized. The administering company remained technically independent until 1941, when it was formally taken over by the army. However, long before that the main private companies leasing manufacturing and warehousing space there had been integrated into the rearmament program, and their freedom of activity accordingly limited.

A "PEARL" Dresden may have been, but the city had never been simply a collection of pretty buildings. Nor were all those buildings sacrosanct. Strict planning laws had kept large-scale industrial development to a minimum in the old center, the Altstadt, but even before 1933 historic areas had been redeveloped or demolished to accommodate the needs of a fast-growing city. To build the grandiose New Town Hall in the early 1900s, an extensive area of eighteenth-century dwellings had been cleared, and the expansion of the Alsberg department store complex in the late 1920s had also meant the demolition of a number of historic buildings in the Altstadt.

It is ironic that these new commercial buildings, though built in a consciously "fitting" style, were steel-framed in the manner developed in America around the turn of the century, and therefore survived the bombing of Dresden in far better condition than the surrounding area. In the early Nazi years, before all civilian construction was postponed due to the war, several high-density, decaying districts were redeveloped, with entire blocks of picturesque but unsanitary houses being pulled down to provide more air and light and allow for street widening. Among the historic areas subjected to total demolition was the Frohngasse, which fronted a tumbledown network of dark yards and alleys notorious through the centuries as the haunts of prostitutes.

Other observers—not just the city's sex workers—were less than delighted at Hitler's promise to reset their "pearl." Several "garden suburbs" were built on the outskirts of the historic city, designed in the steep-roofed, traditional "Germanic" style favored by the Nazis. Like the new buildings, their tenants were politically sound—large families, many of whose breadwinners were employees of the National Socialist Party and affiliated organizations.

Among the genuinely alarming elements of the "improvements" Hitler had promised was the building of a gigantic new Nazi Party conference hall and headquarters, the Gauforum. This was to be constructed in the overbearing style developed by Albert Speer—inflicting a disparity of scale with the surrounding city that Dresden's adroit planners had spent centuries determined to avoid. The planned building would have meant demolishing hundreds of dwellings in the eastern part of the city center, and hacking off a great chunk of the Bürgerwiese, the carefully sculpted municipal park created in the middle of the nineteenth century by the landscaping genius Paul Joseph Lenné. What remained of this precious area of greenery would have been demoted to little more than a decorative access area for the Nazi Party's enormous congress building. Only the coming of war prevented this act of vandalism.

There was also much thought of how Dresden, like Germany's other historic cities, would cope with the increased motor traffic that the Reich's new prosperity and dynamism would inevitably bring. In the Altstadt alone, it was planned that twenty-six hundred dwelling units would have to disappear to "open up" Dresden to cars and trucks. In 1937 Dresden, along with Hamburg, Augsburg, Bayreuth, Breslau, Graz, and Würzburg, was declared one of the so-called Führer cities singled out for direct intervention by Hitler and Speer. From now on, the fuddy-duddy conservationists who had hitherto prevented widespread changes in the face of the city were rendered powerless.

On September 1, 1939, Gauleiter Mutschmann officially established the Durchführungsstelle für die Neugestaltung der Stadt Dresden (Implementation Office for the Reshaping of the City of

Dresden). The man named to head it was Professor Hermann Martin Hammitzsch. By a happy coincidence, Hammitzsch was not only an architect, but a few year earlier had also married Adolf Hitler's elder sister and former housekeeper, Angela.

In an altogether grimmer constellation of circumstance, Hammitzsch's appointment occurred on the day that Germany invaded Poland and the Second World War began.

7

First the Synagogue Burns, Then the City

JEWS HAD NEVER SETTLED in Dresden in the numbers typical of other central and eastern German cities. In 1933 there were 30,000 Jewish citizens in Breslau, 12,000 in Leipzig, and 160,000 in Berlin, but only just over 6,000 in Dresden.

The first Jews were said to have settled in Dresden around 1300. The exact date is unknown. What is certain, however, is that Jews were burned alive on the Altmarkt (old market) after being blamed for the Black Death that in 1348–49 swept through Europe. This was no isolated incident—in Prague, three thousand were massacred. Throughout the rest of the Middle Ages and into the Renaissance period, Jews were by turn permitted, then expelled or persecuted, then tolerated again at the whim of successive rulers. Most were moneylenders, welcome when the powerful needed cash, less so when the cash had to be repaid. In 1705 a Dresden merchant complained that Jews were "every day to be seen walking all the streets and thoroughfares, openly and shamelessly plying their trade."

It was only under Augustus II and III that Jews became permanently tolerated (though not yet given equal rights). A Jewish banker, Behrend Lehmann, was instrumental in collecting the vast sums that Augustus the Strong needed to purchase the Polish throne. A decade later, in 1708, recompense for Lehmann came in the form of a royal proclamation:

> We are desirous in estimation of the services performed to us over many years to allow him the special mercy and freedom, that he may

settle with wife, children, and necessary servants at our Residence
here and purchase a house and garden and become in essence resident in this place . . . under our protection . . . and further shall pay
an annual protection money of eight Reich talers in currency, this to
be rendered to our treasury . . .

The Dresden Jews were labeled *Hofjuden* (court Jews), in the
mildly contemptuous phrase of the time. They accepted the limitations of this role. Later in the eighteenth century, it was the energetic,
mercantile Jews in Leipzig—vital to the functioning of the annual fairs
that made the city rich—who pressed for full legal emancipation.

By 1763 there were around eight hundred Jewish residents,
mostly trading in some way or other, most still associated with supplies for the court and the army. Nevertheless, until the early nineteenth century a sign forbade Jews and dogs access to the pleasure gardens of the Brühl Terrace. Only when a general's wife's inability to
take the air without her pet led to a lifting of the animal ban were the
authorities shamed into relenting in the case of Jews as well. The sign
was, however, restored in 1935.

Things gradually improved, aided by the French Revolution and
the spread of the doctrine of human rights and liberty. Nevertheless,
though in Prussia and the West German states Jews were emancipated
in 1806, their Saxon coreligionists had to wait another thirty years or
more. A measure of equality in commercial, economic, and religious
matters was somewhat reluctantly granted by the Saxon legislature in
the laws of 1837–38. Full civil rights had to wait until the 1848 revolution. The next year, 1849, Bernhard Hirschel—in 1825, the first
Jewish boy to be admitted to a *Gymnasium* (high school)—was also
elected as the first Jewish city councillor.

One definite advantage, however, was that Jewish communal worship became permitted (until 1837, religious devotions were confined
to private houses). The Jews of Dresden could now realize their dream
of building a synagogue to match those that had existed in Prussian
and West German cities for decades. Even while the law of emancipation was being discussed, a committee had been set up with this in
mind. It began raising money among the community. In 1838, with
the law now passed, representatives approached the young gentile
architect Gottfried Semper. At thirty-five, Semper was already profes-

sor at the Dresden Academy of Arts, and running a busy practice whose current projects included two museums and alterations for plans to the Royal Theater. He was also well known (perhaps even notorious) for his liberal views. Exactly who first contacted him is no longer clear, for most of the committee's papers have been lost, but it is known that the fee offered (500 talers) was fairly modest. All the same, Semper did not hesitate. He saw the symbolic importance of the project immediately, and wanted to be involved.

A master builder and a master carpenter were hired. A site was purchased behind the Brühl Terrace, not far from the Elbe. Soon the prolific young architect had draft plans ready: an Oriental-style building complex, topped by a dome that had the look of a Byzantine or Romanesque cathedral. From the middle of the cuboid main building rose a polygonic structure topped by a pyramid roof. The design allowed plenty of room for the congregation with, by tradition, in the center a clearly defined area where the *Almenor*—the table on which the Torah scroll was set—the candelabra, and the eternal light would be proudly placed. The height of this central area (with its carefully located windows) would allow extra light into the synagogue that would otherwise be limited by the intrusion of the two women's galleries overlooking the main area of worship. Two towers, each with a Star of David on its pinnacle, would be built either side of the entrance lobby.

The imposing structure began to rise, soon clearly visible from the river Elbe. There were those who regarded it with respect, those who harbored envy toward a community that could fund and build such an extraordinary house of worship in so short a time, and those who felt a little of both. What outsiders did not know was that the financing was quite quickly in real trouble. Rabbi Fränkel, who had pushed so hard to make this project happen, was soon criticized for having overestimated the wealth of his congregation. A third of the 119 Jewish heads of households in Dresden were in receipt of poor relief, and only a further third earned enough to pay taxes. Some of those who had pledged funds were now pleading financial problems, or withholding contributions because of sudden reservations about aspects of the design.

Nevertheless, the money was found. On May 8, 1840, the solemn consecration of the Dresden synagogue took place. It was now the largest Jewish house of worship in Germany, providing room for three hundred men and two hundred women to attend divine service. All the king's ministers attended. The Jewish community in Dresden at last had a highly visible focus to its spiritual and cultural life.

Semper's synagogue, as part of the skyline along the Elbe, was also constant, living evidence of the achievement of equality for the Jewish religion and people in the city. It would survive for ninety-eight years and almost exactly six months, and its destruction would anticipate the fate not just of Dresden's Jewish community but also of the city itself.

FROM THE HUNDRED or so households of 1840, the Jewish population grew steadily, though not quite in proportion to Dresden's overall expansion. While there were 1,279 inhabitants of Jewish origin in 1876 out of a city total of 187,500, thirty years later the figures were 3,510 and 517,000 respectively—representing in percentage terms a modest decline. Compared with foreigners resident in Dresden—at twenty-eight thousand representing over 5 percent—the Jews were a fairly small group, outnumbered by the four thousand or more British and American residents. Between 1905 and 1925, with an influx of so-called *Ostjuden* (eastern Jews) from Poland and Russia, there was almost a doubling of the Jewish community to just over six thousand. The proportion increased, but only to 1 percent.

The arrival of the *Ostjuden* brought both an increase in numbers and the introduction of more orthodox methods of worship. The first-generation immigrants were also often more conspicuous, with their traditional black gabardine and their heavy foreign accents. In Dresden, as elsewhere in Germany, the established Jewish population was inclined to be a little embarrassed at any association with the "foreigners," especially as the latter were seen as a constant reminder of bad old stereotypes.

Despite the low number of Jews in Saxony and Dresden, levels of anti-Semitism were always relatively high. This is complex, because in many ways the actual active discrimination against Jews—that is, expressed in boycotts and direct protests of various kinds—was limited. The Dresden branch of the radical right Pan-German League, for instance, was preoccupied much more with anti-Slav than anti-Jewish propaganda (a result of Dresden's proximity to the border with Bohemia and Moravia), even though anti-Semitic clauses remained imbedded in their program and anti-Semitic rhetoric routinely featured in their agitation. Before the First World War there was no aggressive discrimination against Jews, or campaigns against Jewish businesses, whether banks or property brokers or hotels (although there may have been private discrimination). What definitely grew was the ideological rootedness of anti-Semitism, giving a low-level noise of hatred even when the conscious attention was involved elsewhere.

Respectability was lent to anti-Semitism by the centrality of the work of the nationalist historian Heinrich von Treitschke. Born in Dresden to a prominent Saxon military family and for many years professor of history in Berlin, Treitschke was famous for his vivid rhetorical skills and his influential five-volume *History of Germany in the Nineteenth Century*. He shaped, arguably corrupted, the mindset of a whole generation of students, sending them out into the world not just worshipping power (that is, the Prussian monarchy) but infected with a lethal, slow-release dose of anti-Semitism. Treitschke's essay "A Word about Our Jewry," published in 1879, provided an academic underpinning to the notion of the Jews as racially alien to "Germandom," no matter their actual religious allegiance. "The Jews are our misfortune," Treitschke proclaimed, in a phrase that echoed down the decades. It would be used to lethal effect by both Josef Goebbels and the notorious Gauleiter of Nuremberg, Julius Streicher, who emblazoned the brilliant, dangerous professor's words on the masthead of his pornographic anti-Semitic sheet, *Der Stürmer*.

The influence of von Treitschke and his disciples proved insidious. Racial (as opposed to religious) anti-Semitism gradually became more acceptable. By and large, before the turn of the twentieth century, a Jew who converted to Christianity was accepted as no longer Jewish. This began to change before the First World War. Even in quite "respectable" circles, "blood" came to condition everything. This change was part of the polarization in German society, which fed internal extremism and external aggression, and arguably played its part in the genesis of the 1914–18 European conflict.

The growth in prejudice varied to some extent according to region, and Saxony did not feature on the philo-Semitic side. In Dresden and Leipzig, because of widespread anti-Semitic feeling, privileged Jews tended to marry within their own group much more than in other parts of Germany. Historical conditions created a climate relatively favorable to Jewish participation in public life in, for example, Hamburg and Altona, Frankfurt-on-Main and Baden, but virtually disbarring Jews in the kingdom of Saxony or the city of Bremen.

Between 1914 and 1918 the nationalist right's dreams of Germany as a world power vaporized in the realities of the trenches. Meanwhile, German Jews appeared to figure prominently in the antiwar movements and later in the left-wing parties of the new republic. The right identified them as the "enemy within," whose subversive activities had led to the otherwise inexplicable defeat of the Reich. This view found widespread support among the educated classes. It was easy to ignore the large numbers of Germans of Jewish descent who had fought and died in the First World War, or who shared the right's patriotic, socially conservative views in every other respect except their anti-Semitism.

The Nazi accession to power led to torchlight processions, demonstrations, and improvised anti-Semitic outrages in Dresden too. Gauleiter Mutschmann could not resist the call of his soul mate (and geographic near neighbor), Gauleiter Julius Streicher, for a one-day boycott of Jewish shops and businesses. This took place on April 1, 1933. Even before the day, many non-Jewish-owned businesses in the city had posted signs in their display windows declaring messages such as: "Recognized German-Christian Enterprise."

The day of the boycott dawned. Columns of Brownshirts marched from the Schützenplatz to the Altmarkt, where an SA orchestra was playing. From here groups swarmed out and blockaded Jewish-owned shops, doctors' surgeries, and lawyers' practices. SA troops also stood at the entrances to the Justice Building, denying access to Jews. In some cases Jewish lawyers were beaten and dragged from the building. The Nazi thugs brandished signs such as: "He who gives his money to Jews/Makes the German economy lose!" or "Anyone who buys from Jews is supporting the Jewish boycott of German goods abroad!"

Soon a special division of the Dresden Gestapo, Section IIB3 ("Freemasonry, Émigrés, Jewry") was set up to oversee action against the Dresden Jews. For some years this was led by a police professional. In 1941 SS-Untersturmführer Henry Schmidt took over, becoming head of what became known as the Judenreferat, or Jewish Department. Schmidt would remain in charge of the surveillance, deportation, and persecution of the Jews in Dresden until the bitter end. He led a small staff. In interrogating the unfortunates summoned to his offices in the Bismarckstrasse, he was assisted by the boss of the SD (SS intelligence service) in Dresden, Hans Max Clemens, and another Gestapo official, Arno Weser. The few Jews who survived the war remember that Weser, Clemens, and Schmidt were nicknamed "the spitter, the hitter, and the shouter" according to the specialty each espoused during interrogation sessions. One Jewish wife of an Aryan businessman recalled standing in the corner of the interrogation room and being spit at for two hours between the questions and accusations.

Dresdeners of all classes thought they lived in one of the most beautiful, cultured, and well-administered cities in twentieth-century Europe. These things remained in most respects true for those Aryan Dresdeners who gave the regime no trouble. Not so for their Jewish fellow citizens, no matter their political beliefs. What the latter began to realize—some sooner than later—was that, against this seemingly unchanging backdrop of time-honored beauty combined with judicious modernity, their circumstances were about to return to those of the Middle Ages.

HENNY WOLF had been eight when the Nazis came to power. Blond and blue-eyed, the pretty daughter of a prosperous movie theater owner and his wife, she had led a carefree existence within a loving family until then. After 1933 everything changed. The reason was simple: Her father was Aryan, but her mother was a Russian-born, religiously observant Jew. Henny counted as a "mixed-race" or *Mischling* child.

For a teenager, as Henny had become by 1938, certain aspects of life—such as choice of school and the attitude of friends—loomed especially large. Henny was still at a normal state school and living a relatively normal life, but already she was suffering from the first pinpricks of official persecution. One was the ban on Jews (and *Mischlinge*) using park benches in the Grosser Garten, where she and her parents had always loved to walk and feed the squirrels. She had also noticed the increasing numbers of anti-Jewish signs, some specific in their meaning (such as in the park and along the Brühl

Terrace), some purely insulting. The Nazi newspaper in Dresden, *Der Freiheitskampf*, reported with satisfaction that in the inner suburb of Johannstadt the local Nazis had put signs on all the advertising pillars in the area bearing messages such as "The Jews Are Our Misfortune" or "He Who Buys from Jews Is a Traitor to the *Volk*!" Henny Brenner's family's house lay just east of Johannstadt; the abusive posters and signs lined their regular route into the city.

Again in Johannstadt, at the beginning of 1938 notices started to appear on the doors of apartment blocks and houses saying, "Jews live in this building." Soon a competing notice becames widespread: "In this building live no Jews." At the beginning of February, as part of the carnival (Fasching), a special procession toured the entire city under the motif: "The Children of Israel Move Out." On March 4 alone, a hundred anti-Semitic meetings were held in the Dresden District. In two different large venues, Mutschmann and Julius Streicher addressed crowds. Other speakers who took part in the campaign—brought in from all over the Reich—were rewarded with a gift of wall plaques of pure Meissen porcelain inscribed with the Hitler quotation: "In warding off the Jews, I fight for the Word of the Lord."

Soon the flood of discrimination became a torrent. In the spring several other suburban communities followed the spas at Weisser Hirsch and Bad Schandau in refusing admission to Jews. The Dahlener Heide (an area of heathland popular with Dresdeners) was closed to Jews. Jews were banned from regional bus services, from occupying land belonging to the city, and in June Jews were not allowed to move to Dresden without police permission. In July 1938 it was decreed that all Jewish businesses in Dresden must be marked with a uniform sign declaring in black letters on a yellow background: "Jewish Business."

The climax to the hate campaign approached. During the night of October 27–28 1938, some 724 men, women, and children, born in Poland and/or without German citizenship—90 percent of the Jewish-Polish population of Dresden—were arrested in a lightning operation personally supervised by the Dresden chief of police. They were assembled at the mainline station in Dresden Neustadt and at around 1:15 in the afternoon of October 28 were sent by special train, under police guard, to the border town of Beuthen in Silesia. From there, despite the fact that many had lived in Germany all their lives, they were literally forced over the border into Poland. This was part of a planned, nationwide orgy of expulsions. Dresden's figures easily beat those of other Saxon cities such as Leipzig (50 percent of the Polish-Jewish population) and Chemnitz (78 percent).

The mass expulsion of Polish-born Jews from Germany led directly to the cataclysm that would reveal the true direction in which Hitler and his underlings were heading. Many Jews were refused entry by the Polish authorities and remained in no-man's-land for weeks before their hosts relented.

Among the Polish-born Jews deported from the Reich were the Grynszpan family, parents and two sisters. Soon a postcard detailing the Grynszpans' pitiful plight in no-man's-land got through to their son Herschel, a seventeen-year-old illegal immigrant, living alone in Paris. The young man acquired a gun. On November 7, 1938, he arrived at the German embassy and talked his way inside. He had wanted to see the ambassador, but was directed to a junior official, Legation Secretary Ernst vom Rath. Alone in an office with the diplomat, Herschel Grynszpan pulled out his gun and shot him.

The irony was that vom Rath was no Nazi. Tragically, this assassination, carried out by a desperate, lonely youth, pitched Germany into a final devastating eruption of violence. It was for this outrage that the past year of escalating anti-Semitic agitation in towns and cities all over the Reich had surely been preparing the way.

Kristallnacht—the Night of Broken Glass—began in Dresden, as elsewhere, early on the evening of November 9, 1938, following vom Rath's death in Paris. It was initiated by the Nazi Party's local organs, themselves centrally directed from Munich, where Hitler and the other prominent Nazis were gathered to celebrate the fifteenth anniversary of the attempted Bürgerbräuhaus coup of 1923. Goebbels embarked on a hate-filled speech before the leadership returned to their hotel and orders were given for a nationwide pogrom. In Dresden, a "spontaneous" open-air protest meeting against the murder of vom Rath was staged in the Rathausplatz—a few hundred yards from where Semper's Dresden synagogue, now nearing the centenary of its consecration, stood proudly by the river Elbe. The crowd listened to inflammatory speeches by Nazi orators.

The rampage began under cover of darkness. Soon came the sound of shattering shop windows in the near-deserted streets of

Dresden. Jewish-owned businesses in the Prager Strasse, the city's main shopping thoroughfare, were worst hit. The mob swept through, leaving destruction its wake. Finally the rioters turned to the visible sign of Jewish presence in Dresden, the synagogue.

Gottfried Semper's grand, half-Romanesque, half-Oriental house of worship by the river Elbe had become a well-known, even well-loved feature of the city's skyline—a fascinating architectural addition, giving a touch of exoticism to the place where the old city walls had once opened to the east. The synagogue acted as a meeting place for what until 1933 had been a flourishing Jewish community—even after Hitler's rise to power, plans were carried forward for a long-projected extension to the main building, allowing room for an extra 150 worshippers. Despite the Nazis' growing encroachment on the community's freedoms, this new section was completed and opened in 1935, a touching example of courageous, if misplaced, optimism.

Perhaps no one could quite believe that what happened all over Germany on the night of November 9, 1938, could also take place in somewhere like Dresden, a *Kulturstadt* (city of culture), where intolerance had always been kept within polite bounds, where, as every Dresdener will proudly declare, "art has precedence" (*die Kunst hat Vorrang*). But that night art was irrelevant. A group of SA men broke into the synagogue even before the main mob arrived, scattering gasoline and setting alight the interior, with its rich hangings and wooden furniture and fittings. With the building ablaze and surrounded by baying Nazis, four engines of the Dresden fire brigade arrived promptly. A specially equipped boat, anchored by the nearby banks of the Elbe, stood ready to supply water from the river for their hoses. The firefighters rushed toward the still-fresh conflagration—only to find themselves barred from quelling the fire by a ring of SS and SA thugs.

The fire-fighting professionals were permitted to save only the nearby buildings, including the community houses (Zeughausstrasse No. 1 and No. 3, later to feature as two of the infamous "Jew houses"), a warehouse, and the old youth hostel. Sometime in the small hours, the great wooden dome of the synagogue collapsed and sank into the sea of flame beneath it. By morning, what had once been the largest synagogue in Germany was a collection of smoking ruins.

The Nazi high burgomaster of Dresden, Dr. Kluge, announced that same day with open satisfaction: "the symbol of the hereditary racial enemy has finally been extinguished."

The painter Otto Griebel saw a ribbon of dark smoke curling into the air over by the river terrace, where the synagogue lay. Within minutes he was on his way to the scene of the atrocity. He saw firemen standing idly by beside their great, motorized fire engines, surrounded by a crowd of onlookers. It was an eerie scene. As Griebel watched, the scene turned even uglier:

> Uniformed SA people had hauled a group of totally distraught-looking and deathly pale Jewish teachers from the nearby Jewish community house. They forced crumpled top hats onto their heads and exhibited them to the baying crowd, to whom the unfortunate victims were forced, on command, to bow deeply and take off their hats.

> A well-dressed, gray-haired passerby, who looked like an actor, found this too much, and he called out, full of outrage: "Incredible, this is like the worst times of the Middle Ages!" But no sooner had he uttered these words than he was seized by Gestapo officials present among the crowd and taken away.

Transfixed with horror, Griebel wandered through the center of the city, picking out its best shopping streets—Seestrasse, Schlossstrasse, Prager Strasse. Every Jewish-owned business had been subjected to destruction and pillage. Perhaps he may have passed the young tram driver's son, Günter Jäckel, now eleven. That day Günter was late for school, because on the way to class he and his friends had made a detour to inspect the ruins of the synagogue. At lunchtime, when school was over:

> We went back via the Prager Strasse, passing Hirsch's fur store . . . A man in a white coat, with a ladder, was clearing up the shop window . . . and the people were standing there in front of it and looking on. It was a dead silence . . . I'll never forget . . . no provocation, no expressions of agreement or disagreement . . . just that dead silence. Nothing.

What the schoolboys didn't know, and perhaps few of the gawking, silent adults knew, was that during the night 151 Jewish men from Dresden had been taken from their homes and, amid scenes of casual brutality, shipped off to a special camp at Buchenwald, near Weimar. Most were eventually released, but from now on it was an unignorable fact that the only life possible for a Jew was one of escalating violence and increasingly perilous isolation. More than five hundred synagogues throughout Germany had been attacked, and in most cases destroyed in a similar fashion to the Semper synagogue in Dresden.

The embassies of the democracies and the aid offices of the Jewish Agency experienced a steep increase in the number of Jews in Dresden wanting to emigrate. However, the free nations had always been reluctant to "take" Jewish refugees, and little changed in that regard. A concession on the part of the British authorities meant that German-Jewish children could be sent on the *Kindertransporte* to London, where they were provided with foster homes and in many cases later sent to relatives and friends in North America and elsewhere. Henny Wolf's parents considered this route for her, but she refused to leave them.

It was typical of the regime, and especially of Gauleiter Mutschmann's even crueler local variant, that extra sadistic touches were added to the spiritual and physical damage wrought by the Kristallnacht mobs. Less than forty-eight hours after the synagogue's destruction, a circular went out from the Saxon Ministry of the Interior in Dresden to the mayors of all the state's major cities. Its author was Professor Dr. Martin Hammitzsch, among other things head of the Dresden Planning Authority. And also Adolf Hitler's brother-in-law. It read with charmless bureaucratic brutality:

> The synagogues that caught fire [*sic*] during the night of 9/10 November 1938 are a danger to public safety, spoil the immediate street scene and the wider urban landscape and are provoking public anger. In the light of this, and of early indications of their dilapidation, these ruins and any surviving sections of the buildings are to be cleared immediately, since the granting of permission for the reconstruction of the synagogues on these same sites is out of the question.

That same day, High Burgomaster Kluge and his colleagues met to discuss the circular. Since the ruins represented a public danger, it was the city's duty to demolish what was left of the Dresden synagogue, for which it selected the municipal Technische Nothilfe and the private demolition specialists Mätschke and Co. The cost would be borne by the Jewish community. Of course, everything must be done by the book. Representatives of the Jewish community were summoned to meet with the city authorities to receive their instructions.

As Construction Counselor Wolf observed dutifully in his notes: "The representatives of the Israelite Religious Community are not available." An inquiry at Gestapo headquarters evinced a formal confirmation of this. And the Gestapo should know—for the Jewish community representatives had been rounded up during the night and were either in prison in Dresden or being subjected to an educative roughing-up in Buchenwald concentration camp. There seems little doubt the good gentlemen of the municipality were also perfectly well aware of this fact. The "unavailability" of the Jewish community leaders meant that the city was entitled to clear the site without further consultation with its former owners.

On November 15 the district court in Dresden ordered that, in the absence of appropriate Jewish representatives, two Aryan lawyers could be appointed as administrators of the Jewish community, allowing them to authorize the wrecking teams of Mätschke and Co. to start their work. A large proportion of the bricks from the synagogue were dumped in the western suburbs of the city, where by a cruel irony Jewish forced laborers would later incorporate them into the new surface of the Meissner Landstrasse, the main highway leading northwest out along the Elbe.

The dynamiting of what remained of the famous Dresden synagogue was attended by a camera crew and the resulting footage used as an instructional film. Which, of course it was, though in ways those responsible could never have dreamed of.

By the middle of 1939 the Jewish population of Dresden was down to a little more than a quarter of the 1932 figure. Even after war broke out, some managed to escape. Among them were Günter Jäckel's Jewish neighbors in the Sedanstrasse, the Mattersdorfs, whose sons had been his playmates, and who for all their wealth had never discriminated against him as a tram driver's son. Herr Mattersdorf had been a banker, and until 1934—when forced to resign—president of the Dresden branch of the Goethe Society (as was Günter Jäckel himself, thirty years later). The Mattersdorfs found their way out through Italy and from there at great risk through Vichy-controlled French North Africa. Though it cost them what little of the family's wealth remained, they made it to America. There an immigration official looked at the Mattersdorf boys—both blond-haired, like Henny Brenner, despite their Jewish blood—and said, "Oh, you're a pair of fine Nazis!"

In 1945 Herr Mattersdorf, the former banker, was making a living in America as a humble bookkeeper. When he learned the fate of Dresden, no longer his home but an enemy city three thousand miles away, he wept.

But perhaps the last word should go to a grizzled Dresden street character named Franz Hackel. Acquainted with Hackel from their weekly encounters at the unemployment office, the artist Otto Griebel fondly called him "the Dresden Diogenes" for his habit of taking his ease on the bank of the Elbe and following the doings of the city's people with wry, philosophical interest. On November 9, 1938, Hackel spotted Griebel gazing in silent horror at the still-smoking ruins of the Dresden synagogue. He approached the painter, his tone conspiratorial, eyes blazing, and muttered: "This fire will return! It will make a long curve and then come back to us!"

And with that the old man melted into the crowd.

8

Laws of the Air

ON SEPTEMBER 27, 1939, the Polish capital, Warsaw, surrendered to the German army. The city had been under siege for more than two weeks, pounded by artillery and subject to repeated raids by bombers of the German Luftwaffe. On September 24 a thousand German planes devastated the city, the next day 420. By the end of that second day, ten thousand civilians lay dead within "Fortress Warsaw."

It is true that Warsaw had been declared a fortress by its defenders. It is also true that those defenders had been called on to surrender, and its civilian population to leave. The law of war, such as it is, stipulates that under those circumstances such attacks are legitimate. But that law, laid down in the Hague Convention of 1907 (just four years after the Wright Brothers' first flight) could not begin to imagine the devastation a sophisticated air force could wreak. The laws of siege remained, essentially, those established in the Middle Ages: Those within the walled city, if they refused to leave or surrender, could be attacked with whatever weapons the besiegers could wheel up to the walls and let loose.

The Luftwaffe's role in the Polish campaign was overwhelmingly tactical—designed to facilitate the army's advance rather than to damage the enemy's military and industrial infrastructure. In the early days of the war, attacks had been aimed against airfields and railway stations—in the first case to destroy the Polish air force on the ground, especially at airfields in the vicinity of Warsaw, in the latter case to hinder mobilization of the country's large conscript army. But there is no denying that civilian casualties were high; right from the beginning there was a whiff of total war and of willful lack of discrimination in target selection.

In the Luftwaffe's attack on Warsaw on September 13, Operation Coast (Wasserkante), a mix of half-and-half high-explosive and incendiary bombs, clearly indicated an attempt to destroy built-up areas by fire. The attack created "a sea of flame, so that accurate assessment of results was impossible." Heavy raids of this kind had originally been planned for September 1—the day Germany attacked Poland without declaration of war—but had to be postponed because of bad weather. With the capital surrounded and about to surrender, September 25 saw the crews of German Junkers 52s literally shoveling incendiary bombs out of their cargo doors onto the city below. Ten percent of Warsaw's buildings were destroyed, 40 percent damaged. Terrible destruction was inflicted on the Stare Miasto, the historic city center of Warsaw.

There was also widespread strafing and bombing of road traffic, intended to create maximum chaos and impede Polish troop movements, but also killing civilians and sowing terror among the endless columns of desperate refugees, Even the American ambassador to Poland, fleeing Warsaw in his embassy car, found himself under attack on the roads of southeastern Poland, far from the front. At a press conference after his escape, Ambassador Biddle told the world that the Luftwaffe's planes were everywhere. He had been repeatedly bombed and machine-gunned and finally forced to take refuge in a roadside ditch. His car had carried the Stars-and-Stripes flag fixed to the roof, but this proved no deterrent—rather to the contrary—and so the identification was hastily removed.

Meanwhile Dresden's last summer of peace was mellowing into the first autumn of war. The final, noblest grapes were being harvested from the south-facing slopes of the Elbe valley so that the winemakers of Pillnitz and Meissen could begin the process of creating a new vintage, just as they had done for eight hundred years. Of course, two hundred years earlier, the Warsaw that now lay in ruins had been the king of Saxony's refuge. This had been his twin capital, the safe place where he could wait out Frederick the Great's fury and the Prussian destruction of Dresden. Now troops and aircrew from that same Dresden were among those raining destruction on Poland's capital.

On September 24 German forces carried out the first mass murder of eight hundred Polish intellectuals and members of the local elite in the western Polish city of Bydgoszcz (Bromberg). The massacre

was at least an act of revenge—in the first days of the war, members of
the ethnic German minority had been executed as fifth columnists—
but worse, much worse, was to come. Within days of the fall of
Warsaw, the Germans began rounding up potential resistance members, political and social leaders, and—above all—Jews.

Sixty-year-old Victor Klemperer, expelled from his university post
five years earlier because he was a Jew, wrote on September 10 that the
atmosphere in Dresden was unnatural and strained.

> They are putting out no flags, even though Warsaw has already
> been reached in this first week. Nothing is said about the front in
> the West. The butchers' shops have to shut on the side facing the
> street: people must queue in the courtyard.—This is the view which
> has to be maintained: War only with Poland and the quickest possi-
> ble victory. But at the same time constantly intensified measures,
> which point to a long war . . .

On the day the Second World War broke out, Günter Jäckel, now
thirteen, had found himself staring up at the crowded heavens in adolescent fascination. Swarms of Junkers 52s, Heinkels, Messerschmitt
109s, and, of course, Junkers 87s—the "Stuka" dive-bombers that
would be the scourge of the Polish roads and railway stations—filled
the sky over Dresden as they headed eastward to Poland. "They were
flying low," he remarked, "probably as a show of power." But the next
day reality—if of a somewhat Chaplinesque kind—struck. He found
himself pressed into service by the painter and decorator from
upstairs, whom he refers to as "Hans U," a keen, uniform-wearing
member of the SA and a Hitler supporter. This man began to requisi-
tion potato sacks and fill them with earth.

> When I asked him the purpose of this activity, I got only the gruff
> military answer: "You'll soon see, you clown!" . . . The half-filled
> sacks would be placed in front of the round ventilation windows to
> protect us against gas or air pressure or whatever. (They stayed
> there until winter, when the earth had to be shoveled away; which
> happened without much fuss, for the final victory was of course
> within sight.)

• • •

DROPPING LETHAL OBJECTS onto enemies from a great height has been practiced as a military technique for thousands of years, whether from the battlements of castles onto the massed besiegers, or from the heights of mountain passes on unwelcome transient armies. There had been intimations of something more ambitious when, as Napoleon advanced on Moscow, some enterprising Russian patriots attempted to halt him with the aid of a balloon laden with explosives. Any deterrent effect on the French emperor has not found its way into the historical record. Balloons were occasionally used for rather eccentric localized bombing raids by, among others, the Austrians while putting down the Venetian rebellion of 1848–49. Nevertheless, until the twentieth century war remained essentially a two-dimensional affair. It could be fought on land or sea; with the aid of projectiles, certainly, but demanding that the enemy be somewhere reasonably close by on the earth's surface. Then in 1903 the Wright Brothers undertook the first manned powered flight in the Kill Devil Hills of North Carolina, and that event changed everything.

Just as Alfred Nobel had hoped that his invention of dynamite would make war inconceivable, so Orville Wright proclaimed many years after that first, dramatic trip through the air: "When my brother and I built the first man-carrying flying machine we thought that we were introducing into the world an invention which would make further wars practically impossible."

Not everyone shared their optimism. The British writer of futuristic fiction H. G. Wells took less than five years to produce a novel, *The War in the Air* (1908), that envisaged battles between massive airborne fleets. Within a handful of years the brothers were proved wrong in the real world also. In 1911–12, when Italy fought the Turks for possession of what is now Libya, bombs were tossed from the cockpits of biplanes onto recalcitrant Muslim tribesmen, the pilots having first pulled the pins with their teeth. It was the First World War, however, that really saw the beginning of aerial warfare directed against the enemy—and not just the enemy's soldiers but its citizens too.

Modern air warfare—one developed, militarized industrial state sowing destruction on another's population from the air—originated with the planners of the kaiser's Germany. At first aircraft were used

mainly for the vital task of reconnaissance, obviating the need for tele-
scope-wielding staff officers to seek out hillocks and church towers
from which to get a glimpse of the enemy. Now everything an army did
was spread out and visible to the soaring airplane, a fact that changed
ground warfare forever.

Other changes were quick to make themselves apparent. Not only
was the enemy's army naked, so was his country and the people in it.
From 1915, with the western front in bloody stalemate, German
Zeppelin airships crossed the English Channel to attack targets in
England. There was panic among British civilians, but the slow-moving,
massive, hydrogen-filled craft proved vulnerable to both the rapidly
assembled ground artillery shield and the agile biplanes of the Royal
Flying Corps, which were soon being produced in numbers by companies such as Sopwith. In November 1916, after suffering horrendous casualties, Zeppelins disappeared from the skies over England.

The real danger—and the real antecedents of later intruders—
were the sturdy, powerful Gotha airplanes—the first purpose-built,
powered-flight bombers, which the Germans developed midway
through World War I. These began regular sorties against English targets in May 1917. On the twenty-fifth of that month, twenty-one
Gothas bombed Folkestone (one of the major embarkation points for
British troops reinforcing the western front), killing 95 people and
injuring 195. In this case it could at least be argued that most victims
were soldiers. On June 15 a force of eighteen German aircraft attacked
London. A direct hit was scored on the busy commuter hub of
Liverpool Street Station. Altogether 162 people were killed in the
neighborhood, including 46 schoolchildren. Only eleven of the fatalities were service personnel. The German planes all returned safely to
their bases in France.

In July another force of twenty German bombers approached
London from the east, skirted north of the capital, then flew across
central London at ten thousand feet while shoppers in the streets
below incredulously watched. "The hostile air fleet," said a Press
Association account, "presented an unbelievable spectacle as in
stately procession it moved slowly, almost symmetrically spaced . . .
daringly low." Fifty-seven dead and 193 injured were the result of the
raid, in which only one Gotha was lost. There were anti-German riots,

with Lord Northcliffe's jingoistic *Daily Mail* comparing the incident to the humiliating Dutch naval incursion into the Thames estuary in 1667.

For a while an atmosphere of near panic infected the highest circles of British government. Both the rulers and the ruled recovered their nerve, but in January 1918 the Germans resumed night raids against London, using a larger variant of the Gotha, known as the Giant. At the same time they began to shell Paris with the huge long-range artillery pieces known by the collective name Big Bertha. The will was there to attack civilian targets, the technology was available, and whatever qualms the British harbored in the early years of the war rapidly disappeared. As *Flight* magazine commented in a tirade against the German High Command: "The psychological effect of 'frightfulness' on civilian populations is regarded by them as a weighty factor . . . We shall have to perpetrate quite a lot more frightfulness yet to bring them to a better frame of mind."

After a spring of bizarre political and strategic confusion in the newly established British Air Ministry, an independent British air force, to be known as the Royal Air Force, came into being in June 1918. Its commanding officer was the tough, aggressive Hugh Trenchard. The RAF's immediate task was to assist the Allied ground forces, still reeling from the German breakthrough of March 1918, which had brought Ludendorff and Hindenburg's resurgent armies within reach of Paris. Trenchard had no doubt that bombing was one of the answers.

Trenchard directed a new, intense campaign of coordinated strategic attacks in which the British, French, and newly arrived American air forces would all be involved. The aim was to disrupt enemy communications and industry, and not least enemy morale. "If I were you," the new air minister, Sir William Weir, suggested in September, "I would not be too exacting as regards accuracy in bombing railway stations in the middle of towns. The German is susceptible to bloodiness and I would not mind a few accidents due to inaccuracy." He added, "I would very much like if you could start up a really big fire in one of the German towns."

As regards the level of destruction that was possible with the existing technology, it may have appeared awesome to a population that had grown up under the peaceful norms of the "long nineteenth century," but compared with what was to come, the results of the bombing offensive were meager. Communiqués trumpeted Allied attacks on Bonn, Cologne, Koblenz, Frankfurt, Mainz, Saarbrücken, and Stuttgart, but the official British war history would reveal that in a total of 675 strategic raids mounted during the conflict by Allied air units, 746 German soldiers and civilians had died and a total of 1.2 million pounds' worth of damage had been inflicted. The new British air force had lost 352 aircraft, and 264 aircrew killed or missing. One dead airman for three dead Germans. Just over one dead German per air raid. The lost aircraft cost between 1,400 and 6,000 pounds each—very large sums at the time.

After the sudden collapse of the German army in November 1918, there was much discussion in victorious but impoverished Britain of whether air warfare represented value for money. The general view was that, while actual damage to enemy war production had been minimal, it forced the Germans to divert considerable resources to dealing with the physical and psychological effects of air attack. In the words of Weir after a postwar tour of inspection: "bombing has the immediate effect of causing the German to dig like the devil . . . this means a vast expenditure of man power."

Another politician, a rising star who first entered the cabinet in his thirties, and had both run ministries and fought on the western front as a commander, took a more jaundiced view:

> In our own case we have seen the combatant spirit of the people roused, and not quelled, by the German air raids. Nothing that we have learned of the capacity of the German population to endure suffering justifies us in assuming that they could be cowed into submission by such methods, or, indeed, that they would not be rendered more desperately resolved by them.

That prescient politician was the forty-three-year-old minister of munitions—soon secretary of state for air—Winston Churchill. Both he and Weir were to be proved right. But then who could imagine the sheer destructive power that within less than a generation would come pouring from the sky?

• • •

AFTER THE WAR ENDED, Sir Hugh (later Lord) Trenchard had to fight for the newly found RAF's independence with all the energy and cunning at his disposal.

Germany was soon largely disarmed. With the most important military threat to Britain removed, the RAF's political masters seriously considered dissolving its strength back into the army and navy whence it had come. Part of the reason for the RAF's survival, however, came from its potential usefulness in sorting out minor but irksome trouble elsewhere in the world. In December 1919 Churchill told the House of Commons, "The first duty of the Royal Air Force is to garrison the British Empire."

Sure enough, a three-week campaign crushed Muslim rebels in Somaliland, a victory gained by a single RAF squadron operating in tandem with the Camel Corps. The entire campaign cost 70,000 pounds. In comparison with the process of importing, feeding, and housing troops, and then stationing them wherever the trouble might be repeated, this was sensationally reasonably priced, and in the aftermath of the most expensive war in Britain's history, cheap was what the running of the empire needed to be. In March 1921, at a conference in Cairo, it was agreed that British control of Iraq would henceforth be based on the RAF squadrons stationed there. On the northeast frontier of India, where for the past hundred years British and loyal Indian armies had bled to death in incessant struggles with the fiercely independent tribesmen of the area, RAF planes could also enforce "order" at the drop of a leaflet or, if necessary, a string of bombs.

When the Geddes Commission was formed to enforce harsh savings on government departments in those bleak postwar years, Trenchard, a wily political operator, finally saved his baby by uttering magic words of parsimony in its defense, insisting that the RAF was "our cheapest form of defence." Further supported by Churchill against the machinations of the army and navy, both of which wanted their air arms back, by 1922 Trenchard and the RAF were safe both from the budget cutters and their rivals in other services. The era of independent air power had begun.

Trenchard's role as founder of the RAF, followed by his long reign as its commander, meant that the new force was inevitably his creature. The atmosphere was less stuffy than in the army or navy. Technical knowledge was all-important. This was a force made up not of cannon fodder but of technocrats. Trenchard also dictated another trend in the new air force. His experience in the First World War had convinced him that air power was the answer to victory in future wars, that through attacking the enemy's hinterland, the bloody deadlock that had cost millions of young men's lives between 1914 and 1918 could be broken. Trenchard wrote: "It is on the destruction of enemy industries and, above all, on the lowering of morale . . . caused by bombing that ultimate victory rests." When asked why Britain did not build up a purely defensive air force, he responded with typical conviction:

> If you play a game of football against an opposing team your objective is to win. If the opposing team start to attack, and the members of our team are told only to defend their own goal, they could not possibly win . . . Nothing is more annoying than to be attacked by a weapon that you have no means of hitting back at; but although it is necessary to have some defence to keep up the morale of your own people, it is infinitely more necessary to lower the morale of the people against you by attacking them wherever they may be.

These arguments, shorn of sporting metaphors, were the foundation of the notorious "moral bombing" doctrine, as well as of the conviction that nations could (contrary to what Churchill had written five years earlier) be forced into surrender by relentless air attacks alone. Trenchard was almost certainly thinking of the French when he expressed his forthright, sometimes chillingly practical opinions, for France was in the mid-1920s the only power with which Britain might conceivably engage in military conflict. "The nation that would stand being bombed the longest would win in the end," he said, adding: "The French in a bombing duel would probably squeal before we did." Trenchard included a prophetic rider in his observations about the primacy of bombing policy that would often be forgotten in the war to come. "In the future," he said, "increased means of defense may redress the balance."

Trenchard's doctrines took little account of the possibility of civilian casualties. The First World War had seen nations locked in life-or-death conflict. The war had witnessed unrestricted submarine warfare on the part of the Germans, leading to the deaths of thousands of civilians through the sinking of merchant shipping and passenger liners, and it had seen the beginnings of massed air attacks on enemy cities by both sides. Both sides had made attempts to starve each other out— treating the entire populations of their countries as if they were the inhabitants of medieval fortresses. In this the British, with an all-powerful surface fleet enforcing their blockade against the Central Powers (as Germany and Austria-Hungary were known), had been the more successful. Hundreds of thousands of enemy noncombatants died of malnutrition and disease brought on by shortages of food and other necessities. Even in Britain, massive sinking of merchant ships by the Germans' U-boat fleet had caused widespread shortages of imported foodstuffs. There was a sense of enemy civilian life being held cheap in a way that just a decade earlier would have been inconceivable.

As Trenchard put it when he addressed the Imperial Defence College in 1927:

> There may be many who, realising that this new form of warfare will extend to the whole community the horrors and sufferings hitherto confined to the battlefield, would urge that the air offensive should be restricted to the zone of the opposing armed forces. If this restriction were feasible, I should be the last to quarrel with it, but it is not feasible . . . whatever the views held as to the legality or the humanity or the military wisdom of such operations, there is not the slightest doubt that in the next war both sides will send their aircraft without scruple to bomb those objectives which they consider the most suitable.

Even more extreme were the views of the Italian army officer General Giulio Douhet. Douhet briefly (1913–14) served as director of the Italian army's aviation section before mysterious disciplinary hearings led to his transfer back to ground duties (some authorities have even suggested that he never learned to fly). He nevertheless caused a great stir with his book *The Command of the Air*, first published in 1921. The book attracted considerable attention, which led to the appointment of Douhet, an enthusiastic fascist, as undersecretary for air in Mussolini's first government.

Douhet argued that the invention of aircraft rendered all previous military thought irrelevant. Again, memories of the nightmare stalemate of the First World War, which haunted the minds of so many in the 1920s, could cause this to be viewed as something perversely like a blessing. No longer would millions of the nation's young men be doomed in inconclusive battles on land—the bomber would break the deadlock with its ability to deliver devastating attacks on enemy cities and industrial centers. Land forces would be needed principally to occupy and patrol enemy territory after the bombers had forced its surrender.

Douhet considered the large-scale slaughter of civilians to be perfectly justified. To this end, he was prepared to countenance the use not just of bombs, but of poison gas. Even more forcefully than Trenchard, he took his fundamental idea and carried it to its logical (and amoral) conclusion. Unlike Trenchard, Douhet appeared to rejoice in the apocalyptic visions his work induced in the minds of an appalled world. He also believed that the strategy could only work if the enemy's own offensive air capacity was first negated (by destruction of its air force on the ground) and then never allowed to rise again, thus leaving its cities permanently defenseless against the free-ranging bombers that would induce rapid surrender.

Douhet declared with characteristic ruthlessness that attacks from the air should be directed "against organs which are vulnerable both physically and morally, and which are . . . in no position to defend themselves through combat or counter-attack."

The gas attack that would achieve this total annihilation of the enemy's capacity to resist would be carried out with liquids that would "slowly emit gas, thus poisoning the atmosphere for weeks." Douhet spelled out the full horror of this in a final dramatic denouement:

> It has been calculated that it would be possible, using 80 to 100 tons of poison, to swathe great cities such as London, Berlin or Paris in a cloud of lethal gas, such that they could be annihilated by high explosive, incendiary and gas bombs, since the presence of the gas would prevent the fires from being extinguished. A system of attack has also been conceived which bears the name "cloak of gas." This consists of producing an invisible cloud of poisoned gas above the city, one heavier than air. As it slowly sinks to earth, it annihilates all things that it encounters, be they in the basements of dwellings or on the roof gardens of skyscrapers . . .

Douhet's book was not translated into English until 1942, and most of the Trenchard-style bombing doctrines were already in circulation before its first publication—after all, notions of breaking stalemates and destroying the enemy's hinterland were quite widespread during and immediately after the First World War. But his writings have a terrifying quality, which perhaps owes something to the aggressively antihumanistic foundations of fascist thought. Douhet's doctrine matches the machinelike, impersonal quality of high-altitude bombing like no other writer's. This is the bomber as weapon of terror, plainly expressed in word and thought. On the other hand, Trenchard's ideas, and even Douhet's, could be seen as having deterrence at their heart. The reasoning went: So terrible would be the damage the bombers would inflict that in the future no sane leader could envisage unleashing a European war.

The only alternative to deterrence was abolition, banning of the bomber weapon. But the conferences held to that end under the auspices of the League of Nations brought no practical results. A conference of jurists at The Hague in 1923, convened by the Western powers, proposed a ban on bombing purely as a means of terrorizing civilian populations. No government had or would ever publicly admit bombing enemy territory to that end alone. Far more subversive, therefore, was the conference's proposal that air raids against military targets should be permitted only if they avoided all damage against civilians. In practice, this would have permitted bombing only of battlefield targets or of military and industrial installations well outside towns or cities, thus nullifying most of the advantages of possessing a bomber fleet. The proposals were never ratified. For the next twenty years legal arguments about air power consisted of attempts to force pre-1900 rules of war into some kind of fit with the new realities of massive, geographically unlimited destruction. At the turn of the twenty-first century, the argument is still going on. As a modern air-war specialist writes:

What air power contributed to the development of a more total war than the Mongol conquests, or the destruction of the indigenous civilisations of the Americas, was the means with which to prosecute a greater degree of war, both in terms of destruction and in the perceptions of societies. Many civilisations have used whatever methods were available to prosecute war, often with few restraints, and air power was in reality nothing more than a further, if highly significant, step in this particular direction.

In other words, it is a law of war (in the scientific rather than the judicial sense) that unless overwhelming deterrence is in place, a power will use whatever weapons it possesses to achieve its objective. And the objective is victory, as cheaply bought (in terms of the belligerent's own reserves of lives and treasure) as can be managed.

ON APRIL 26, 1937, a town of seven thousand people in the Basque country of northern Spain was subjected to an air attack that would shock the world. It would remain notorious, despite all the horrors since witnessed, more than sixty years later. The name of the town was Guernica.

In July 1936 a reactionary Spanish army commander, General Francisco Franco, had raised the standard of rebellion against the country's democratically elected leftist government. Supported with arms and men by Nazi Germany and fascist Italy, Franco had nevertheless failed to win a swift victory, and as fighting dragged on into 1937, Spain remained locked in a bloody fratricidal struggle between Nationalists (Franco's forces) and Republicans (government supporters) that had already cost the lives of hundreds of thousands of its people.

Guernica lay ten miles from the front line in a war involving Spaniards against Spaniards, but the forty-three aircraft that took off that day were not Spanish. Their young pilots formed the core of the new German Luftwaffe, operating under the cover of an entity called the Condor Legion. This consisted of around one hundred aircraft— four bomber and four fighter squadrons—commanded by Major General Hugo Sperrle and Colonel Wolfram von Richthofen (a cousin of the great First World War fighter ace known as the Red Baron). Both were to play a major role in the Luftwaffe during the Second World War. There was also a tank unit, led by Colonel von Thoma, later commander of the Afrika Korps. Formally under the rebel General Franco's command (and justified as a response to the International Brigades fighting on the Republican side), the Condor Legion had the unadmitted function to "blood" the Luftwaffe's young pilots and to test out the Nazi state's new aircraft and weapons on real targets and live human animals.

Under most rules of war, the attack on Guernica was strictly speaking justified. It amounted to an act of interdiction—the forcible prevention of an army from retreating away from its victorious opponents and thus regrouping to fight another day. But Guernica was also a town, and a historic one at that, known since the Middle Ages as a center of Basque language and culture, where councils had traditionally met under the shade of an ancient oak. The Condor Legion's intervention was also spectacularly bloody and dramatic in a way that was rightly recognized at the time as setting a new, ghastly pattern for warfare, and that—scale apart—was to be followed with eerie similarity from Coventry to Dresden and on to Vietnam and Iraq.

Guernica was crowded, it is said, not just with fleeing Republican soldiers and civilians, but with shoppers, for it was market day. Even in the middle of the worst war, farmers must sell their produce, and people must eat, and with the Nationalist ground forces still miles away, there was no reason to believe that the town would be subjected to attack just yet. This was an assumption that no civilian population would ever be able to make again. The first German Ju-52s appeared over the town at about 4:30 in the afternoon. Bombs began to fall on the market square and the bridge leading into the city. The attack went on for some hours, with the German pilots using the existing fires as markers and therefore tending to drop even more bombs on the town center. After the bombers had finished, fighters came in and flew low to spray cannonfire through the streets.

The casualty figures of Guernica are still disputed more than sixty years later. The numbers of fatalities quoted still range between three hundred and sixteen hundred (if true, the latter representing almost a quarter of the town's population). There seems general agreement that around 70 percent of the town was destroyed. At the time Franco denied that the raid had occurred—or that, if it had, the Republican air force was to blame. Commenting in the 1950s, the German air ace Adolf Galland, who had taken part in the raid, described Guernica as a "regrettable mistake . . . of the kind that would be repeated countless times later during the Second World War." The world was learning that the air war—fallible human beings operating at high speed from high in the sky against targets that were hard to define or even find—

was especially prone to error, and that those errors could have truly terrible consequences.

Nor, as both reporters and historians would later realize, were the errors confined to the planners or the aircrew, or the weasel claims of "official spokesmen." If air attacks were the half-blind work of an instant for the perpetrators, for the human beings on the ground they amounted to an unimaginable melange of horror, fear, and loss: all the car crashes, house fires, and physical assaults that could be imagined, all rolled into one. Understandably, when questioned later, eyewitnesses offer accounts that vary wildly; the "facts" they quote (and that find their way into newspaper accounts, newsreels, and history books) are sometimes objectively disprovable, or their memories are found to be at fault or "tainted" by mixture with extraneous received wisdom.

A German historian from Dresden visited Guernica many years later, in the company of a Spanish colleague. The German was introduced to an elderly man who had been in Guernica when it was bombed, who proceeded to give a passionate, detailed, and heartrending account of the raid. As they walked away afterward the Spanish historian smiled and said that very little of what the eyewitness had told them was accurate. So was he lying? Of course not. He was telling the younger men *his* experience of April 26, 1937, *his* memories, *his* Guernica, and in that regard everything he said was the absolute truth for *him*. For the individual, memory is truth (for how else does a life recalled make any sense?), but for the historian it is just one factor, to be balanced and checked and seen as part of a whole.

Picasso's famous painting *Guernica* quotes no figures, gives no details, and yet says everything. It shows above all the horror, the fear, and the unnatural malevolent physicality of that day's mayhem. The mother/victim, wracked with anticipation, looks helplessly, in brutally distorted mimicry of a hundred painted pietàs, up to heaven, whence for countless generations her ancestors had expected not destruction but salvation. No longer. One moment a familiar, beloved town is there, just as it has been for hundreds of years, shaped by and shaping the land surrounding it. It is under threat, but the enemy troops are still, by traditional standards, quite distant. There is time for those planning to stay to buy and sell, and for those planning to escape to make their way across the river and continue on their journey. Then the planes come, and a few hours later the town is gone, a ruin, with many hundreds dead, injured, forever traumatized. This is war as it has never been experienced before. War at a remote distance, coolly impersonal, but more deadly than ever. Literally inescapable.

Franco's soldiers arrived two days after the raid, to occupy the ruins. Soon the Condor Legion would turn to bombing the city centers of Madrid and Barcelona in an attempt to wear down their citizens' stubborn resistance to Franco. Another experiment that would be repeated ad infinitum, over the cruel years to come. The Japanese had already used their increasingly powerful air force to bomb Chinese cities in their war against Chiang Kai-shek's regime, with devastating results. By the late 1930s it was clear that, whatever inhibitions politicians and voters in the democratic states might have about the use of air power to kill large numbers of civilians in towns and cities, it had already happened and was going to continue to happen for as long as war itself could not be prevented.

And it could not.

9

Call Me Meier

WITHIN HOURS of the German invasion of Poland on September 1, 1939, Franklin Delano Roosevelt, president of the neutral United States, made a plea to the major combatants to confine bombing to strictly military targets. The British and French signed on to the president's plan the next day (September 2). The Germans waited. Finally, after their forces reached Warsaw (which thereby became technically a "military target" and exempted from the agreement), they added their assent on September 18.

There were rather ineffectual British attacks on Kiel and Wilhelmshaven, important German naval bases, but by and large in that winter of the "Phony War," both sides stuck to their pledge. It suited them. The British could build up their productive strength, and supply the expeditionary force they now had in place in France, without the constant nagging worry of massive German air raids on London and other major cities. As late as 1938, despite rapid expansion of the RAF, whose strength tripled between 1934 and 1939, the development of promising new fighter aircraft, and above all the installation of the revolutionary new radar defensive system on the coasts facing the continent, the official view remained colored by former Prime Minister Baldwin's 1932 dictum that "the bomber will always get through." The authorities feared that something like 150,000 casualties could result from just the first week of Luftwaffe raids against London, the country's all-too-vulnerable capital. The British public's fear of extinction from the air was heightened not just by news reports of devastating bombing raids on towns in Spain, China, and

Poland, but also by screen fictions such as the film *Things to Come*, which featured an Armageddon-like war and the threat of worldwide destruction.

Nevertheless, the British public seemed to be of two minds. Following the attacks on the north German ports in the first week of September, the government ordered the RAF to concentrate on an operation code-named Nickel, which consisted entirely of dropping vast quantities of discouraging literature over the western provinces of Germany. If bombing techniques were still in their infancy, so were the RAF's leaflet distribution skills. The bricklike bundles each contained about fifteen hundred leaflets, held together by a rubber band. These were in turn combined in larger packs of twelve, done up with string. Over the "target," a crew member had to find the space to cut the string and feed the sections out through the flare chute. Once they emerged, the aircraft's slipstream was expected to free the leaflets from their restraining rubber bands and send them floating to earth over a suitably wide area. One of the early Nickel leaflets gives some of the flavor:

> We have no enmity against you, the German people . . . you have not the means to sustain protracted warfare. Despite crushing taxation, you are on the verge of bankruptcy. Our resources and those of our Allies, in men, arms and supplies are immense. We are too strong to break by blows and would wear you down inexorably. You, the German people, can, if you will, insist on peace at any time.

The notion that the "German people" had been able to "insist" on anything of any importance since the summer of 1933 was almost touching. Another leaflet, dropped over Berlin at the beginning of October 1939, went for the scandal sheet approach:

> Göring, whom Hitler has nominated his successor, has a fortune of not less than 30,030,000 Marks abroad.

> Goebbels possesses in Buenos Aires, Luxemburg and Osaka, in Japan, the handsome sum of 35,960,000 Marks.

> Ribbentrop is the richest of them all, since a sum of 38,960,000 Marks is invested for him in Holland and Switzerland . . .

Himmler (head of the Gestapo), who watches like a lynx that no German takes more than ten marks across the frontier, has himself smuggled abroad a sum of 10,555,000 Marks ...

Such are the men who are your leaders!

As the writer and cabaret artist Noël Coward remarked with typical acerbity, this was beginning to look like a campaign to *bore* the Germans to death. "But do we have the time?" he asked plaintively.

With reports coming in every day of devastated Polish towns and cities, there were murmurings against keeping such an expensive air force in near idleness, but the government still did not order major bombing attacks. Perhaps the idea was, *pace* Coward, to buy time, for every month that passed enabled Britain's factories to produce more fighters and bombers and thus lessen the gap between the Luftwaffe's seeming invincibility and the RAF. The government duly started emphasizing that these were not merely leafleting trips but reconnaissance missions, spying out the enemy's weaknesses.

The war against Poland might have been the "Blitzkrieg," but the winter of 1939–40 witnessed the "Sitzkrieg" (sit-down war). The Anglo-French forces, dug in behind the Rhine defenses and the Maginot Line, spent a freezing winter staring at their victorious German opponents and perhaps asking themselves if the promise to aid Poland had been such a good idea after all. This was when British minister Sir Kingsley Wood, asked to approve a scheme to bomb the Black Forest and set it on fire, is said (possibly apocryphally) to have replied: "Are you aware it is private property? Why, you'll be asking us to bomb Essen next!" It was also when Marshal Göring, World War I ace and creator-commander of the Luftwaffe, made one of his most notorious and probably most unfortunate remarks: "If a single enemy plane reaches Reich territory, you can call me Meier"—meaning "I'm a monkey's uncle."

The "Phony War" ended in April 1940 with the successful German attack on Norway, where the Luftwaffe's command of the air over the Baltic enabled use of airborne troops and also provided almost faultless cover for the forces being poured in by sea. Another key element was the revolutionary degree of coordination between ground and air forces through radio contact (techniques way ahead of the Anglo-French capability, developed above all by Wolfram von Richthofen during his time as chief of staff in Spain and Poland).

Come the attack against France and the Low Countries just over a month later, and the Luftwaffe seemed to be transforming itself into an invincible legend, the airborne equivalent of Frederick the Great's famous Prussian grenadiers.

Skillful propaganda also enhanced the German air force's prestige. Two major propaganda films had toured the world during the first winter of the conflict, both of them dealing with the Polish campaign, and both building up the Luftwaffe into a glorious, unbeatable, and above all terrifyingly adept instrument of Germany's national will. The first film, *Feuertaufe* (*Baptism of Fire*), showed the role of the air force in the fall of Warsaw, and spared little effort in showing just how destructive German bombers could be. In the second, *Kampfgeschwader Lützow* (*Combat Squadron Lützow*), the crew of a low-flying Heinkel 111 swoops down over a Polish road, recognizes not just a column of kidnapped ethnic Germans but also manages to distinguish between them and their cowardly but bombastic Polish guards—successfully picking off the latter in true movie-hero style and leaving the innocent Germans unscathed. The film managed to present one of the standard propaganda justifications of Hitler's invasion of Poland (ethnic Germans had to be saved from brutal Poles). At the same time, like *Feuertaufe*, it also emphasized in wildly exaggerated style the superhuman skills of Luftwaffe aircrew, thereby presenting a warning of what would happen to those hapless nations that opposed the Third Reich in its march to power.

The second great example of the woes that would befall those enemies came after Hitler sent his forces, without declaration of war, into wealthy, complacent, and traditionally neutral Holland. Many key defenses were captured by German paratroopers during the first few hours of the surprise attack. When the garrison of the Netherlands' second city and largest port, Rotterdam, refused to surrender, the Wehrmacht sent in the bombers, destroying large areas of the old city and killing hundreds and possibly thousands of civilians as well as soldiers. Since the Netherlands never expected to be involved in a European war, air raid defenses and shelters were minimal, and this contributed to a higher-than-necessary death toll.

Rotterdam became notorious, as Guernica had been, and as Coventry was to become. The Germans argued that it was a mistake (again), resulting from, among other things, glitches in radio communications and the fact that Holland at that stage set its clocks twenty minutes ahead of Greenwich mean time. The British and their allies, to whom Rotterdam was a genuine horror but also a propaganda boon, pointed out that incendiary bombs had been used, that excessive force had been brought to bear against a city that was about to surrender in any case, and that Holland was traditionally a peaceful country (at least as far as its relation with European countries was concerned—those absorbed into its extensive and profitable Far Eastern empire might beg to differ). The time had come, evidently, for the restrictions adopted at Roosevelt's behest to be reconsidered.

Attempts by British bombers to intervene in the ground conflict, as the Luftwaffe had done so brilliantly in Poland and Norway, had been disastrous. On May 10, the day the Germans broke through the Allied lines, thirty-two near-obsolete aircraft, Fairey Battles, were sent in to attack the southernmost German armored columns, which were racing through Luxembourg toward the French border. The Battles flew at low level—250 feet—to avoid German fighters, but instead were greeted with withering antiaircraft fire. Thirteen of the thirty-two were shot down, the rest damaged. The next day an early air raid by German fighters caught one of the two squadrons of the slightly more up-to-date Blenheim bombers on the ground. That same day eight of the hapless Battles tried again against the German armor. Seven were shot down, and the only surviving aircraft crashed when landing back at base. On May 12 nine Blenheims of 139th Squadron sought out German armor on the Maastricht-Tongres road, this time keeping to six thousand feet to avoid the fate of the low-flying Battles. Predatory Luftwaffe fighters ambushed them. Only two Blenheims survived. Heroic failure became the pattern, accompanied by unacceptable loss rates of both aircrew and machines. Raids on the bridges over the Albert Canal in Belgium (four out of five planes destroyed) and against road junctions and bridges at Maastricht (fourteen out of twenty-four lost) continued the sorry tale.

Within forty-eight hours the RAF bomber force operating on the continent had lost almost half its aircraft. On May 14 the remaining planes were called out by a desperate High Command, at the request of the French, to help close off the German bridgeheads just established across the Meuse. A further forty Battles and Blenheims were lost to enemy fighters and antiaircraft fire. This reduced the force to a quarter of its original strength. One of those who survived wrote of his comrades' vomiting from sheer terror and exhaustion before climbing into their aircraft and flying off to do battle with an enemy whose equipment was superior to theirs in every way.

The unequal fight could not go on. A few days later, after more disastrous losses, the RAF policy switched to night bombing. Less accurate, but much safer. Losses dropped dramatically. The failure of the bombers in a tactical role played a part in other decisions made during these grim days. The Air Staff had considered that if the Germans invaded the Low Countries, it would be time to "take the gloves off." Even when that happened, the politicians still hesitated. Then came Rotterdam. The next day the War Cabinet approved the plan long supported by the Air Staff and Bomber Command. Attacks against German targets.

Bomber Command had already been carrying out night raids against marshaling yards and communications inside Germany but only west of the Rhine (that is, immediately behind the front), reckoning that these would count as legitimate attacks by any international standards. The decision of May 15 permitted British aircraft to cross the Rhine into the industrial powerhouse of the Ruhr area and bomb oil plants, prominent, self-illuminating industrial targets such as blast furnaces and coke ovens, plus the usual transport and communications facilities. The RAF's ambitions were dual: to show they could hit vulnerable enemy assets, while at the same time (they hoped) drawing back into a defensive struggle inside Germany the Luftwaffe forces that were creating such mayhem in France. In doing this, the RAF dropped bombs not on troop concentrations or defended fortresses but on quasi-civilian targets—the kind of targets, now and later, whose legitimacy would be endlessly argued over.

The raid on the night of May 15–16, 1940, involved a large-for-the-time force of ninety-six Whitney, Wellington, and Hampden bombers. It was not especially successful. Seventy-eight were given oil plants to attack, of which only twenty-three reported actually finding the targets. Sixteen aircraft—one-sixth of the force—failed to make any kind of attack at all. On the other hand, only one aircraft failed to return, having crashed into a hillside in France on the return trip. So was completed the first operation in RAF Bomber Command's campaign against the interior of the Third Reich. The attacks may have contributed to hesitation by the Germans in the period before Dunkirk, but the effect was not major.

From these ineffectual but low-cost beginnings would follow five years' expenditure of blood, toil, and treasure by Britain and its allies, and untold suffering for the civilian population of Germany.

WITH THE UNITED KINGDOM standing alone and bracing itself for a possible German invasion, the primary aims of most British military plans during the fateful summer of 1940 were defensive ones.

Just a week after Rotterdam's immolation and surrender to the Germans, British bombers attacked the city's oil refineries in an attempt to deny their use to the enemy. London's propaganda might bewail the "barbaric" Luftwaffe bombing of Holland's second city, but war was war. In June Blenheims hammered German airfields at Rouen on the channel coast, Amiens, and Schipol in Holland (now the site of Amsterdam airport). Factories, especially those connected with aircraft manufacture and supply, were attacked at Bremen (Focke-Wulf), Gotha (Messerschmitt 110), Deichshausen (Junkers 52) and the Rhineland, including Cologne. Plants to produce artificial oil from coal, built at enormous expense by the Reich to make its armies and air force independent of vulnerable foreign oilfields, were also targeted, including those at Gelsenkirchen in the Ruhr and the giant works at Leuna, near Halle in Central Germany, which was troubled by ten visits from the RAF from mid-June to mid-August 1940.

Nevertheless, at that time Bomber Command was definitely the junior partner to Fighter Command. Political pressures in the late thirties had led to a switch in purchasing policy. The government had invested massively in new fighters, leading the bomber arm to suffer relative neglect. The unfortunate results had shown themselves clearly in the early days of the battle for France. However, during the subsequent battle for Britain the government's decision seems to have paid off. The defending Hurricanes and Spitfires of Sir Hugh Dowding's Fighter Command—newer models, closer in quality to the opposing German aircraft than their bomber equivalents—fought off the Luftwaffe's attempts to establish air superiority and thus prevented a German cross-channel invasion.

During the high summer of 1940 the Germans concentrated on

attacking British airfields, an effective tactic that came close—though the enemy did not know this at the time—to inflicting a fatal blow on the island's fighter defenses. Then, smarting from the humiliation of a tiny RAF raid on Berlin and, it is said, hoping to lure the stubbornly resisting Spitfires and Hurricanes into the air in defense of their own capital, the Nazi leadership ordered its forces to attack targets in the heart of London. This decision is generally agreed to have been a great strategic blunder on Hitler's and Göring's part, far-reaching in its consequences. The spectacular results of the air raids on British cities, in terms of damage and casualties, may have bolstered morale in Germany, temporarily alarmed the Reich's enemies, and impressed neutral onlookers, but in direct military terms they achieved little except to infuriate the British people and provoke calls for revenge attacks by the RAF. Arguably the continued bombing campaign was actually an admission of defeat, a desperate alternative to the direct, annihilating contest with Fighter Command that, had the Luftwaffe emerged victorious, would have enabled it to rule the skies over England. Fighter Command survived. Its losses in the Battle of Britain were quickly replaced by the British aircraft industry, which was surprisingly little affected by the German raids that persisted through the autumn and winter.

FOR THE GERMAN PEOPLE, the war it had embarked on rather reluctantly—if we are to believe other witnesses, including Klemperer—had apparently ended in triumph. Hitler's aggression seemed to have been vindicated. He was unquestionably hugely popular, even with Germans who had hitherto counted themselves skeptics.

The Third Reich held victorious sway from the Spanish border to the western Ukraine. If the British could not be conquered, they could be confined to their island and eventually made to sue for peace. Germany had defeated the traditional enemy, France, and driven the British back to their tight little island. This was 1918 in reverse, with the Germans on the winning side.

The war was over. This was the widely held belief in Germany. After all, what else could Churchill and his commanders do now except make peace? The Reich's civilian population would one day find out, but not yet.

Dresden and other Wehrmacht command centers in central and eastern Germany had been important jumping-off points for the absorption of the Czech lands and the invasion of Poland, but they had not had to suffer air raids, unlike the western German industrial cities. To the happy citizens of the German Florence, basking in the summer sun, for the moment all that mattered was that the war had been won—so it seemed—and won at minimal expense.

By the end of July 1940 units of the all-conquering Wehrmacht were coming home to heroes' welcomes in towns and cities that were as if at peace. On August 9, 1940, Dresden turned out in blazing summer weather to watch the returning Fourth Saxony-Dresden Infantry Regiment march through the streets and form ranks in the great cobbled square of the Altmarkt. The parade was, as a historian records, "an absolute high point in the life of the city. . . . Hundreds of thousands of onlookers—young and old—stood cheering on the pavements. They, like the soldiers, hoped that the war was now as good as over."

Two weeks later, on the night of August 25, 1940, the RAF bombed Berlin. Despite minimal damage, Goebbels claimed that there was "colossal anger against the English." To the Reich's propaganda chief this was healthy, since "now Berlin is also right in the middle of the war, and that is a good thing."

More British raids on Berlin were to follow during the tail end of summer and into the autumn, though they were in truth largely symbolic. On September 11 Goebbels wrote in his diary mainly to gloat over German air raids against London:

> The reports from London are horrendous. An inferno of unimaginable extent. The city is coming to resemble a hell. It is already possible to discern small indications of deteriorating morale. How long will this city of eight million people hold out? We have no examples we can judge that by . . . the question is: can London be brought to its knees in this way? I would assume, yes. But we must wait things out and attack, attack!

By contrast, the same morning's entry deals with an RAF raid on Berlin the previous night, the latest of several since August 25. It is a masterpiece of cynical opportunism.

Attack on the government quarter. Brandenburg Gate, Academy of Arts and Reichstag hit. Nothing serious, but I organize for the matter to be given a little extra help. Through fake incendiary bombs. Wodarg [one of Goebbels's aides] has this photographed immediately. A splendid propaganda device.

So feeble were the British efforts against German soil in the autumn of 1940 that Goebbels had to resort to faking British "atrocities" to rouse the German public. It was a time his fellow countrymen would look back on with nostalgia as months turned into years and it became clear that the war was not over.

10

Blitz

THE GERMAN LAND FORCES might parade through the homeland, giving their civilian compatriots the illusion of peace, but the Luftwaffe's air campaign against Britain continued with little respite through the winter of 1940–41.

On the evening of November 14, 1940, some 515 German bombers crossed the English coast, heading inland toward the country's industrial Midlands. Two-thirds of the attackers belonged to the Luftwaffe's Third Air Fleet, the rest to the elite pathfinder group, Kampfgruppe 100. It was a clear night with a full moon. Codename of the raid: Operation Moonlight Sonata.

The aircraft of KG 100, whose Heinkel 111s had been designated to lead the assault, were following the beam of a newly developed German direction-finding device known as the X-Gerät. This worked by a system of staged, intersecting beams sent out from two coastal transmission stations on the French side of the channel, one in Cherbourg and the other in Calais. The device not only enabled the aircraft to pinpoint their target to within a few score yards' accuracy over hundreds of miles, but also, by linking the operations of a special clock in each aircraft to a series of automatically triggered signals, provided a basic system of automated bomb aiming.

The X-Gerät was a very clever invention, though like all inventions in the air war, the enemy had quickly begun to find ways of countering it. That night the British (perhaps helped by their ability to read the enemy's coded signals traffic, or perhaps just benefiting from a lucky interception) had gotten a fix on the beams and were preparing to jam them, thus throwing the Germans off course. Unfortunately, owing to human error, they were not close enough to the X-Gerät's frequency to fully realize that ambition. There was still a sufficiently clear distinguishing signal for the German navigators to follow it, so dooming their target to a fate that, like those of Rotterdam, Warsaw, and Guernica, would brand itself on the collective memory of the world.

The city of Coventry, just south of Birmingham, had been an important town for more than a thousand years. Lady Godiva, wife of the Anglo-Saxon Earl Leofric of Mercia, had in legend ridden naked on a white horse through its streets in protest against his imposition of crippling taxes on the common folk. In response, Leofric is said to have lifted all taxes except the one on horses. He also founded a Benedictine monastery on the site of a convent sacked by the Danes, whose church, dedicated to St. Mary, became a cathedral in the Middle Ages as the town became seat of the bishopric of Coventry and Lichfield. The town itself became prosperous from the wool trade that the Benedictines had established, and later from the silk-weaving trade, though the old cathedral fell into decay after the Reformation, when the monastery was dissolved and the bishop's seat transferred to Lichfield. Only in 1918 was the bishopric revived, and the historic Gothic church of St. Michael—a building recognized as one of the finest examples of such architecture in Europe—designated as its cathedral, making Coventry by British rules once more able to call itself a city. By this time the silk trade was dead, as was the watch-making industry that had grown up in the eighteenth century. In common with most of the British Midlands, Coventry had become a center for light manufacturing and engineering, including bicycles, cars, airplane engines, and—fatally—after 1900, munitions.

Coventry, a city of more than 320,000 inhabitants, was therefore, in terms of what little law existed on the subject, a legitimate target for aerial bombing. It was also true that, in contrast to many industrial towns, but typical enough of the English Midlands, a lot of Coventry's factories and workshops were of a small to middling size and located in the ancient heart of the city, tucked in among and behind the half-timbered houses and winding lanes. In July and August the city had been subjected to fleeting air raids, which had killed a few dozen of its citizens, most memorably destroying a splendid new cinema, which had been finished just before the war. The next day it had been due to start showing *Gone With the Wind.*

Like other wartime air raids that seemed to show extraordinary results, and thereby stick in the common memory, Coventry illustrated the use of a novel method—the X-Gerät. However, in this case there appeared not just one novelty but three. The second new procedure came when, at 7:20 P.M., thirteen Heinkel 111 aircraft dropped a combination of special incendiary canisters and parachute flares onto the designated target area. As the canisters, which were filled with a phosphorus mix, fell to earth, they emitted a shower of sparks, almost like fairy lights on a tree. This method of delineating the target was to be copied and used by all sides throughout the rest of the war—notably by the RAF in February 1945. The final, lethal innovation was the use, along with conventional high-explosive bombs and a few so-called air mines, of masses of incendiary bombs. There was no mistaking the intention of the Luftwaffe on this moonlit November night: Its plan was to set Coventry ablaze.

At 7:30 P.M., once the phosphorus canisters had been dropped, marking the target, the first wave of the main bombing fleet arrived. Soon it was time to add the high-explosive bombs to the mix. These were calculated to knock out the water supply, the electrical network, and the telephones. Then the railway. The great craters in the roads and streets would also make it hard for fire trucks and engines to reach the city center when the next stage came. For an hour streams of bombers droned overhead, relentlessly pouring incendiary bombs (regular magnesium and petroleum ones) onto the vast conflagration that was consuming Coventry.

At around 8 o'clock the first bomb had penetrated the roof of the cathedral, where Provost Howard of the cathedral; Jock Forbes, the master mason of the church; and two young men were keeping watch. Then came a second and a third bomb. The fire brigade was summoned, but they seemed to take their time. Two of the incendiary bombs were extinguished before they could catch, but the other had lodged above the great organ, in the space between the roof and the high ceiling, where all kinds of detritus, including dried-out birds' nests, provided excellent fuel. It was the first really serious blaze in the cathedral. Astonishingly, the little group of firefighters managed to bring it under control, using hand pumps and buckets.

Other bombs were falling all the time, however, and other fires starting. Provost Howard described their despair:

Another shower of incendiaries fell, four of them appearing to strike the roof of the Girdlers' Chapel above its east end. From below a fire was seen blazing in the cellar. Above on the roof smoke was pouring from three holes and a fire was blazing through. These were tackled by all of us at once, but, with the failing of our supplies of sand, water and physical strength we were unable to make an impression; the fire gained ground and finally we had to give in.

Almost every street in central Coventry was now on fire. The fire brigade headquarters itself had been badly damaged by a direct hit and was ablaze, which accounted for the delay. As were landmarks such as the Warwickshire Hospital, which had been hurriedly evacuated, all except for fifteen women in the gynecological ward who were not judged fit to be moved, and in another wing a dozen fracture cases who lay helpless in bed with legs in plaster, suspended from the ceiling. Through a great hole in the roof (a bomb had hit the head nurse's office), the patients gazed up at the sinister crimson glow of the burning city, and above it the German planes still circling overhead.

Without help from the fire brigade, the tiny band of amateurs trying to save the cathedral were bound to fail. When finally the professionals arrived, because of shattered water mains, they found themselves unable to access adequate water for its hoses—just as the Luftwaffe planners had hoped. Eerily, after the main body of the church had collapsed, for a few hours the spire stayed upright, its bell striking each remaining hour. Then, after midnight, the spire too disintegrated and tumbled to the ground. The greater fire that was consuming the city could also not be controlled. Almost sixty thousand buildings were destroyed or damaged that night, a hundred acres of Coventry's built-up area. In all, the Germans had dropped five hundred tons of high explosives, thirty thousand incendiary bombs, fifty parachute landmines (a large metal box that would drift slowly and silently to earth and explode above ground level) and twenty incendiary petroleum mines. This was a new level of annihilation. The Germans invented a proud, jokey expression for what had been done; thereafter, any town that suffered a similar degree of destruction was said to have been *coventriert*—"Coventrated."

There were, nevertheless, limitations to what the Luftwaffe could achieve. The German aircraft, two- or three-engined and designed to be used in a ground support role as well as for strategic bombing, carried relatively light loads. The Coventry raid, like most major German raids in England, was, in essence, shuttle bombing. The enemy planes had to fly two hundred miles back from Coventry to their French bases to be reloaded with bombs. This led to lulls during which, despite the chaos, much could be achieved by those on the ground; fires fought, civilians evacuated. The raid did not actually reach its climax until almost midnight—four and a half hours after the Pathfinders of KG 100 had first marked the target. The bombers kept coming until the small hours, with the final all clear at 6:15 A.M., more than eleven hours after the first warning.

So, despite the massive destruction of buildings and manufacturing capacity, a total of 568 civilians died in Coventry that night—blown apart, asphyxiated, above all incinerated, so that many of the bodies were unidentifiable.

Appalling as the effects of the fire had been, these casualty figures were relatively modest compared with Warsaw or (depending whose figures one believes) even Rotterdam, and as nothing compared with what would be achieved much later in the war. Cold comfort for the victims' relatives and the outraged British public, but the message about the feasibility of terrible mass killing by fire was chillingly clear, as was the identity of the message's sender: the German Luftwaffe.

Not that the Germans had done a "perfect" job in this raid or other attacks on British cities. As a cool and supremely professional observer would write after the war:

> The Germans again and again missed their chance, as they did in the London Blitz . . . of setting our cities ablaze by a concentrated attack. Coventry was adequately concentrated in point of space, but all the same there was little concentration in point of time . . .

The writer was Arthur (later Sir Arthur) Travers Harris, also known as "Bomber" and "Butcher." By then he knew exactly what he was talking about. Until earlier that autumn he had been commander of Fifth Group, Bomber Command, based in Lincolnshire. At the time of the Coventry raid Harris was deputy chief of the Air Staff, a desk job, based at the Air Ministry. Just over a month later, he was to witness another major German fire raid, this time on London.

At about seven o'clock on the moonless evening of Sunday, December 29, 1940, Luftwaffe bombers arrived over the "City" of London—the historic heart, where much of its finest architecture and most venerable buildings were to be found. This was marked on the aircrew's maps as target area "Otto." Guidance beams previously locked onto the docks, farther east along the river, had been redirected to cross exactly there. The Pathfinders of KG 100 started dropping their special incendiaries—more than ten thousand even in that first, marking phase—and by the time the main force arrived the area was already well on fire. A total of 136 aircraft were involved, and because they were operating on short hops from their French bases—much shorter than in the case of Coventry—they could save fuel and devote yet more cargo space to bombs, the vast majority of which were incendiaries.

The authorities had so far failed to set up a systematic network of firewatchers. The waters of the Thames were at their lowest point. Firefighters had to wallow through a chill, deep quagmire of mud as they attempted to connect the water for their hoses. Above all, it being a Sunday, the City was almost empty of the worker bees who filled its office buildings during the week. There were also few local residents to help with the struggle against the hundreds and even thousands of small fires that began to break out as soon as the first incendiary bombs fell. Individual incendiaries were generally fairly easy to douse, with the aid of a bucket of sand, and fires could be tackled with a water pump; it was the quantity of them, in big raids, that made it impossible to reach all of them before it was too late and larger fires formed. In the City there was also the added problem that many empty buildings were locked.

That night, London lost eight churches built by Sir Christopher Wren, as well as its exquisite fifteenth-century Guildhall. Fires also raged through the narrow streets and alleys in and around London's ancient center of printing and the book trade, Paternoster Row, hard by St. Paul's Cathedral. There the works of Shakespeare and many other famous British authors had first been set in type. All the area's fine old buildings were destroyed, along with thousands and thousands of precious books and other printed matter. Because of the confined, warrenlike layout of this part of the City, it was especially difficult to maneuver engines and fire-fighting equipment within it. The result was that the smaller fires combined to form a single fire covering half a square mile, which was officially named a "conflagration." It was the nearest London came to what later would be called a firestorm.

A published wartime account described that night, considered by many to be the worst of the entire London Blitz:

> But though the flames licked its very walls, as buildings on each side of the Churchyard blazed, a southerly wind and the Fire Brigade saved the [St. Paul's] Cathedral. Fifteen firebombs that fell on the historic Guildhall were dealt with promptly by the ARP [Air Raid Precautions] staff. But an unchecked fire in Gresham Street spread to the church of St Lawrence Jewry, which was locked and unattended, and from the belfry of which sparks were carried to the Guildhall roof. Among the famous buildings gutted were the churches of St Bride's Fleet Street; Christ Church, Newgate Street, and six other Wren churches; Girdlers' and Barbers' Halls; the Cathedral Chapter House; Dr Johnson's house in Gough Square; Trinity House on Tower Hill . . . when the City returned to work on Monday, the whole area north from St Paul's, including Paternoster Row, Amen Corner, long stretches of Newgate Street and Cheapside and northwards along Wood Street, were smoking ruins.

The big attack on the City of London had been planned as a nine-hour operation, which might have caused incalculably more widespread damage, but in fact it was abandoned after three hours. Dense cloud had developed over the bombers' home bases in northern France, making the intended shuttle operation impossible. The weather worsened further overnight, eventually turning to snow and grounding the entire German bomber fleet. Fate had intervened, as it had done before and would so often again. The consequences of a firestorm in London—many thousands dead from blast, fire, and suffocation, and the heart of a huge city wiped from the earth—would have been incalculable. The Luftwaffe would never come so close again. Now it was the RAF's turn.

According to Albert Speer, Hitler's personal architect and later chief of war production, if in 1940–41 the Führer failed to destroy the capital of the British Empire, it was not through lack of desire. After dinner one night in 1940 he "worked himself up to a frenzy of destructiveness" and demanded of his guests:

Have you ever looked at a map of London? It is so closely built up that one source of fire alone would suffice to destroy the whole city, as happened once before, two hundred years ago. Göring wants to use innumerable incendiary bombs of an altogether new type to create sources of fire in all parts of London. Fires everywhere. Thousands of them. Then they'll unite in one gigantic area conflagration. Göring has the right idea. Explosive bombs don't work, but it can be done with incendiary bombs. What use will their fire department be once that really starts!

London proved itself, at least at the tail end of 1940, a lucky city. Due to the curtailment of the raid, and the fact that, because it was a weekend, the City was virtually uninhabited, "only" 160 civilians died. Twenty-five firefighters also lost their lives. Unscientific as the concept of good or ill fortune may be, it was to be proved in raid after raid that chance events, and the operation of providence in the shape of the weather, were often mightier arbiters of a bombed city's fate than its defenses, the enemy planners' intentions, the numbers of attacking aircraft, or even the skills of their bomber crews.

Arthur Harris wrote afterward:

I watched the old city in flames from the roof of the Air Ministry, with St Paul's standing out in the midst of an ocean of fire—an incredible sight. One could hear the German bombers arriving in a stream and the swish of the incendiaries falling into the fire below. This was a well-concentrated attack . . . the Blitz seemed to me a fantastic sight and I went downstairs and fetched Portal [chief of the Air Staff] up from his office to have a look at it. Although I have often been accused of being vengeful during our subsequent destruction of German cities, this was the one occasion and the only one when I did feel vengeful . . . Having in mind what was being done at that time to produce heavy bombers in Britain I said out loud as we turned away from the scene: "Well, they are sowing the wind." Portal also made some comment to the same effect as mine, that the enemy would get the same and more of it.

The future chief of Bomber Command was certainly learning lessons. From September 1940 to March 1941 the Luftwaffe's bombers would launch raids on targets in Britain as far apart as London and

Liverpool, Gosport and Glasgow, killing more than forty thousand British civilians. Between September 7, 1940, and New Year's Day, 1941, London was attacked on fifty-seven nights. Fourteen thousand of its inhabitants died, a rate of around 250 fatalities for each day of bombing. These are impressively grim figures, even compared with the later British bombing of German cities. In 1944 the average daily death toll caused by Allied bombing within the Reich's borders was 127. This was to rise dramatically only during the final terrible endgame of 1944–45, when German military decline and massive increases in the frequency and efficiency of Bomber Command's operations drove death rates in the Reich, at nearly ten times that figure, to truly apocalyptic levels.

More than a year would pass before Arthur Harris could begin to turn his own crisp prophecy into fierce reality, but in the meantime he had forgotten nothing of what he had heard, seen, or felt during the winter England's cities burned. Especially after the Coventry raid, the British government successfully appealed to the court of world opinion. The seemingly wanton destruction of a fine medieval city center, and above all a great house of God, appalled neutral opinion, in the United States and elsewhere, and did much to win over doubters to the British cause. There was little discussion of the heavy concentration of war industries in Coventry.

As for senior RAF officers, they left outrage to the journalists and got on with the cold, quiet business of examining how the Germans had managed what was in strictly military terms a signal success. As Harris explained, "It would have taken Bomber Command much longer to learn how to attack Germany if it had not been for the lessons of the German attack on Britain."

FROM THE SPRING OF 1941, Hitler's bombers would ease off on their raids over Britain, for despite the hopes of many Germans that their country, dominant in Europe, would now bask in peace and prosperity, the Führer had other uses for them.

Hitler was planning further conquests, and for them he needed his air force. There would be new work for the Luftwaffe to do in the conquests of Greece and Yugoslavia, and from late June 1941 in the vast, treacherous spaces of the Soviet Union, where battle swallowed men, tanks, and aircraft alike at a rate unknown in modern history. The Luftwaffe's bombers did not cease their London raids until May 1941 (the last attack included a direct hit on the House of Commons), and would return to Britain from time to time, but never in the same strength or with the same devastating effect as during the winter they had set England's cities ablaze.

From now on, the roles gradually reversed. Britain moved onto the offensive in the air. In consequence, the story of the RAF's part in the war against the Third Reich became largely the story of Bomber Command; of the production and use of Wellingtons and Halifaxes and above all the powerful Lancaster four-engined bombers against German targets. As for the Luftwaffe, very soon its fighters, hitherto cast in an invasion-support role, would have to take over major and previously unaccustomed responsibility for dealing with that increasingly destructive British bomber offensive.

11

Fire and the Sword

"FIRE AND THE SWORD," as the grim phrase has it, have always belonged together. Whether to drive the enemy from his stronghold, or deprive him of shelter, or to destroy his morale, organized burning was always one of the nastier methods of waging war; indiscriminate, impersonal, killing and maiming its victims in terrible ways. It was, nevertheless, permitted under the rules of conflict, though arousing almost universal horror. Who could forget the burning of Atlanta, the destructive climax of the American Civil War? Or the manmade fire that in 1631 consumed the ancient, treasure-laden German city of Magdeburg after its sack by the Imperial army, leaving six thousand charred corpses in its ruins? Or the thousands of rebellious Catholic peasants and their families who perished in the towns and villages of the Vendée during the 1790s as the armies of revolutionary France systematically burned and pillaged the entire province into submission?

Centuries before the Second World War, filled projectiles had already been invented that released oil or petroleum to set fires. As witness the bomb-throwing Prussian grenadiers of the siege of Dresden in 1760, by the eighteenth century they were in fact commonplace. Once aircraft began to be used for military purposes, it was no great imaginative leap to adapt this kind of simple but cruelly effective weaponry for dropping from a great height onto human habitations, defensive emplacements, or workplaces. A bomb filled with high explosives would blow up, causing damage and death through the explosion and the violent distribution of shattered metal in the immediate vicinity, but by and large that was it. The great advantage with the much smaller incendiary device was that it could, if dropped in sufficient quantities and in a sufficient density, create limitless damage over an ever-expanding area. This could be achieved through the calculated spreading of individual fires and their eventual joining together into a burning mass, so extremely hot and so wide in extent that no fire brigade could bring it under control.

The chief challenges were twofold: first, to devise a way in which the bomb could be made to start burning well away from the aircraft; and second, once it had, to ensure that after impact with the target it continued burning until the fire could catch and spread.

The Germans came up with two basic designs. One, a fairly thin-skinned oil-filled bomb (*Flammenbombe*) was constructed simply so as to explode on impact with buildings or other objects on the ground. The other, the *Phosphorbrandbombe*, contained a mix of phosphorus and petroleum, which (because of the phosphorus element) would burst into flames after being exposed to contact with the air. German civilians tended to refer to all incendiaries as phosphor bombs, although only in a few cases was the terrifying, blister-inducing substance used as any more than an aid to initial ignition. The mixes and host casings were so conceived as to ensure that most incendiaries would burn for between eight and thirty minutes once ignited.

Later in the war the incendiaries were combined in "clusters," that, with the aid of a small explosive charge, would break apart on or shortly before impact, thus scattering a number of burning bombs over the immediate area and increasing the chances of their combining quickly to make a larger fire. There were many modifications in the delivery methods, and science found alternative choices of fire-inducing fillers, but these were the two basic principles that dominated the production of the incendiaries used in air attacks by both sides throughout the Second World War.

The British, who had begun the war with a modus operandi almost exclusively based on the use of high-explosive (HE) bombs, aiming at specific industrial, fuel, and transport targets, quickly understood the potential of incendiaries, and caught up with the Germans over time.

On the night of December 16–17, 1940, the British mounted a bombing raid on the German city of Mannheim. Perhaps deliberately, the operation was officially designated as "revenge" for the immolation of Coventry and other English provincial cities, the planes carried what for Bomber Command at that stage of the war were unusually large quantities of incendiary bombs—the standard British four-pound magnesium "stick," eighteen inches long and weighted at one end to direct its fall. It was somewhat unreliable in both aim and flammability, but en masse it did its job. Mannheim was attacked with a force of 134 bombers, the largest number yet sent against an individual target. This was also the first time Bomber Command had sent its forces against the heart of a specific urban area rather than an individual factory or airfield or other military installation. Another ethical threshold crossed, this time as a result of Coventry. Almost five hundred buildings were destroyed, 47 people were killed, and 1,266 people were made homeless.

Research and development proceeded rapidly. A young Cambridge University chemistry graduate, Vaughan Southam, was one of those recruited straight into the Ministry of Aircraft Production to accelerate and facilitate the development and production of incendiary bombs for the RAF. These youthful tweed-jacketed progress chasers (called "ginger groups") were given powers to overrule, where necessary, both factory managers and regular civil servants, which caused a predictable tension in an industrial and bureaucratic culture where age and experience conferred authority.

As Mr. Southam explained, "We had to make sure the bombs didn't take too long to burn through, because the RAF didn't want the Germans to have enough time to pick them up and throw them out of the window." In fact, the basic designs weren't considered in themselves sufficiently nasty, by either side in the contest. He added:

> Of course, the Germans had a nasty habit of putting a fuse in the nose of their bombs, so you never knew when you were brave and picked up an incendiary bomb and threw it out of the window whether you had a hand grenade that was going to go off in your hand . . . We did the same. We had these nasty things with an explosive device in the steel, and they were indistinguishable from a normal incendiary bomb. And these four-pound incendiary bombs were just showered down from clusters, and you hoped that by the time they arrived at the roof of a German house they would have enough velocity to punch a hole through the slates and go in either into the roof space, where they might lodge, or into a room where they might lodge.

The bombs Mr. Southam worked on were both manufactured and tested in the lowlands of Scotland. There lay the testing ranges, and also the factories where the incendiary bombs themselves were assembled. "Young Scots lassies" did the work.

Special tests were held just north of London in Watford, at the Building Research Department. They often created mock-ups of typical German dwellings, with German furniture—the backdrops to enemy civilians' everyday lives—enabling the researchers to tailor the incendiaries as closely as possible to the buildings they would be destroying.

The Americans, after they entered the bombing war against German territory in 1943, went even further. They constructed several German-style buildings at Dugway Proving Ground in Utah, where chemical and biological weapons were developed, even going to the great trouble and expense of re-creating a "German apartment block" designed by a German refugee architect, Erich Mendelssohn. On these buildings, incendiary bombs were systematically tested.

In what became known as fire raids, the role of the high-explosive "general purpose" bomb, following the example of Coventry, was to knock out the utilities and block or crater the access roads to the point of impassibility. The conventional bomb would penetrate the roofs of buildings, thus facilitating both the access of the incendiaries that followed and also, by creating openings to the outside air, ensuring the drafts that would enable the fires so caused to grow and spread as quickly as possible. The last was very important. Thermite (a heat-producing mixture of powdered iron oxide and aluminum) created its own oxygen, so did not need an air supply to start burning, but after a minute or two, having activated the magnesium, the Thermite would go out. If the incendiary was to earn its keep, it had then to ignite the area around itself. For this, drafts were essential.

But so, as the RAF began to realize, was a revision of its entire modus operandi.

AIR MARSHAL ARTHUR HARRIS, late deputy chief of Air Staff and (from June 1941) head of the British air mission in Washington, was appointed commander in chief of RAF Bomber Command on February 22, 1942. It was a post he retained until four months after the end of the war in Europe. His doughty person became so closely associated with the theory and practice of "area bombing" that it seems almost a pity to point out that this policy had already been adopted months before Harris started the job that made him one of the most famous—or notorious—figures of the Second World War.

Harris took over an organization that possessed, according to him, only 378 aircraft, of which 68 were "heavy" four-engined bombers. Moreover, though Bomber Command was extremely popular with the general public (to whom it represented the only British force properly "hitting back" at the Germans), within government circles serious questions were being asked about Bomber Command's effectiveness. Under pressure from the prime minister's scientific adviser, Frederick Lindemann (later Lord Cherwell), a civil servant belonging to the War Cabinet Secretariat, by the name of Mr. Butt, had been authorized to carry out a searching inquiry into the real effectiveness of Bomber Command's highly publicized night-bombing sorties over Germany and occupied Europe.

Butt examined hundreds of photographs captured by on-board cameras during bombers' attacking approaches in June and July 1941. For comparison, Mr. Butt also combed through summaries of operations, navigational reports, and so on. His conclusions horrified his masters. The most shocking was that, of the aircraft recorded as hitting the target, only one in three had actually gotten within three miles. The news got worse: This figure included the French ports, where two out of three bombs dropped fulfilled this rather generous requirement. Over Germany as a whole, it was only one in four, and over the vitally important Ruhr industrial area, *one in ten*. The success rate was reasonable at full moon, but plunged when the moon was new. Clear conditions enabled half of the bombs to get within distance of the target, but haze reduced the number again to one in ten. Intense antiaircraft fire further reduced the number of "on-target" sorties.

Even more discouraging was that these figures applied only to those 30 percent of aircraft that came within five miles of the target.

In short, a huge proportion even of the aircraft previously credited with successful attacks had in fact dropped their loads in open country.

Not surprisingly, Butt's figures were dismissed as too pessimistic, especially by Bomber Command. Then headed by Sir Richard Peirse, Bomber Command already found itself faced with difficult changes in its role. For most of the first half of 1941, on Churchill's orders, it had turned to tasks connected with the Battle of the Atlantic—bombing ports on the French and German coasts, U-boat pens, dockyards, and, where they could be found, German surface raiders.

On July 9 a new government directive had sent Bomber Command's aircraft deeper into Germany and revealed yet more cruelly its lack of capacity for on-target, destructive—and therefore cost-effective—night bombing. Not only were the raids inaccurate and ineffective to a proven degree, but casualty rates among aircrew had started to rise.

In 1940 the Germans had a very underdeveloped air defense, with only forty defensive fighters at its disposal. By the late summer of 1941 the system had been dramatically augmented, especially around key areas such as the Ruhr. In the first eighteen nights of August 1941, some 107 British aircraft were lost; in September, a total of 138 (62 crashed inside England); and in October, 108. A shocking 12.5 percent of those sent to Berlin, 13 percent of those sent to Mannheim, and 21 percent of those bombing the Ruhr did not come home. Added to these figures was the implication, from the Butt report, that these raids were not just costly in money, men, and aircraft, but near enough futile. They just weren't hitting much. On November 13, 1941, Bomber Command was ordered to halt long-range operations. This was, for the moment, a serious admission of failure.

However, with no progress on any land front, and the public still shocked and embittered by the effects of the Blitz, there was no question that, in the longer term, bombing Germany—"hitting back"—remained a key component in the machinery of morale maintenance. As a result, whatever the concern felt in government circles about the effectiveness of Bomber Command's raids, this was not the message put out by the Ministry of Information. Newsreels and skillfully made documentary films such as *Target for Tonight* (released in August 1941) had served to convince most British civilians, and many Americans also, that the RAF was doing an excellent job over Germany. They did not know that so far more RAF aircrew had been lost over Germany than enemy civilians killed on the ground.

Meanwhile, the bombers could not be left inactive. Bomber Command's major operations in that midwinter period were against the elusive German warships, the *Scharnhorst* and the *Gneisenau* in their secure anchorage at Brest, and the publicity spotlight was accordingly turned on those. Through the winter there were discussions about the future role of the strategic bomber.

Churchill might instinctively want to keep the bombers in the air, but there were those in government, and in the navy and Coastal Command in particular, who felt that the vast sums expended on building and crewing heavy bombers simply represented money better spent elsewhere. In North Africa, British forces were struggling against a new and cunning adversary, the German general Rommel. After Pearl Harbor, they faced a rampant Japanese enemy in Asia. The second-in-command of the Far Eastern Fleet, Admiral Willis, wrote following the fall of Singapore:

> It certainly gives us furiously to think when we see that over two hundred heavy bombers attacked one town in Germany. If only some of the hundreds of bombers who fly over Germany (and often fail to do anything because of the weather) had been torpedo aircraft and dive-bombers the old empire would be in better condition than it is now . . .

By New Year 1942 it was clear Willis could think as furiously as he liked—it would do him no good. Bomber Command was too important a propaganda symbol, and its promise as a weapon, against the German war machine and against the German homeland, too attractive. Despite the bombers' failures, the Air Staff's long-expressed desire for a much larger force of heavy bombers had begun to find a more sympathetic ear in some government circles. After all, what alternative was there but to press on with the bomber offensive against Germany and occupied Europe?

On February 14, 1942, St. Valentine's Day, the Air Ministry issued a new directive to Bomber Command. Intensive bombing by night against Germany was to recommence, to be carried out whenever weather conditions permitted. The important aspect of the directive, however, was the new principle according to which the campaign would be waged. The key phrase was "area bombing." From now on, rather than carrying out "precision" attacks (bombing a specific oil plant or munitions factory or transport center) and accepting the unfortunate inevitability of civilian casualties as a by-product, raids were to be mounted on cities or areas of cities, on the assumption that damage to the German war effort would be done as a result—whether to factories and railway yards, or to power and water supplies, or to postal and telephone services. Like the Luftwaffe's switch from bombing British fighter bases to bombing London in the summer of 1940, this was actually a continued admission of failure, arising from Bomber Command's proven inability to bomb with sufficient accuracy for the notion of precision targeting to have much meaning. However, the directive went further:

> It has been decided that the primary object of your operations should now be focussed on the morale of the enemy civilian population and, in particular, of the industrial workers. With this aim in view, a list of selected area targets . . . is attached.

Peirse was not, as it happened, the officer to whom it fell to put the new principles into action. He was sent out as commander of air forces to the Far East, where he would have his hands full. A little more than a week after the area bombing directive had landed on the AOC [Air Operations Center]'s desk at Bomber Command, Peirse's successor arrived. The new man's name was Air Marshal Arthur Harris. He was stubborn, combative, opinionated, a fearsome perfectionist determined to fight his corner, and under the circumstances almost certainly the very leader Bomber Command needed.

ARTHUR HARRIS WAS BORN in 1892 at Cheltenham, the youngest among the six children of an official in the Indian civil service, George Steel Travers Harris, and Mrs. Caroline Maria Harris. Arthur was a true child of the British Empire. Later that year the family returned to India, where the boy was baptized at Gwalior and his christening recorded at St. George's Church, Agra, in the diocese and archdeaconry of Calcutta. At five he was sent "home," first to what they called a "baby farm," run by gentlefolk who provided care in loco parentis for the offspring of such respectable but relatively impecunious pillars of the British Empire as Harris's parents. Later he attended a mediocre but kindly English prep school. The red-haired Arthur was not an academic, and by the time he reached twelve, there was in any case not enough money in the kitty for him to follow his elder brothers to Sherbourne School and Cambridge. He ended up at an obscure boarding school in Devon.

Although his father had been an official with the Indian Public Works Department, the background of the family was strongly military. It therefore came as a surprise to his parents when young Arthur announced that, rather than apply to Woolwich or Sandhurst for officer training, he was determined to go to the British colony of Rhodesia (now Zimbabwe) in southern Africa and become a farm manager. This he did. His plan, after gaining sufficient experience and saving some money, was to take up one of the two-thousand-acre allocations granted to new settlers, and thereby to become independent.

"He had learned a whole range of practical skills," in the words of his most recent biographer, "and could turn his hand to almost anything. He could shoot, improvise, live rough, cook, and cope with the unexpected. He was able to organise, to run a small business, and to direct the men and women who worked for him. Above all he had acquired the self-confidence necessary to launch out on his own."

Then came war. In 1914 Germany and Britain clashed not just in Europe, but also in Africa. The Germans threatened to launch an invasion of British Southern Africa from their own colony of South-West Africa (now Namibia). Already a group of pro-German Boers had risen in rebellion. A call went out for the young colonialists to come to the aid of the empire. Harris joined the Rhodesia Regiment, a roughrider outfit, which tackled the enemy in the Veldt and the Kalahari Desert. Within months the German territories had fallen into British hands, but there was grave news coming from Europe, where the conflict had settled into the bloody stalemate of trench warfare. Harris decided that, for the moment, his duty lay back in the old world. He returned to London and started pestering the War Office for a service commission—not with the ground forces but with the army's Royal Flying Corps.

Harris later put this choice down to his weariness with marching around after the rigors of the South-West Africa campaign. As a consequence, he said he was "determined to find some way of going to war in a sitting posture." By January 1916 Harris had qualified as a pilot in the RFC, still insisting that he would return to his beloved African bush just as soon as the Germans had been vanquished. He was to stay with the RFC and its successor, the RAF, for thirty years, until he retired. Only then did he return to Southern Africa, though under very different circumstances.

Harris served as a pilot and later as a commander of fighters detailed to protect London against the enemy Zeppelins and Gothas. Then, in September 1916, he was sent to France. He was almost immediately wounded and forced to crash land, suffering further injuries in the process. A spell in the hospital followed, then a period on light patrol duties in eastern England before he was deemed fit to return to the front. Promoted to captain, Harris led a flight of Sopwith Camels on "offensive patrols" over the Flanders trenches. He shot down five enemy aircraft during the rest of 1917, making him formally an "ace."

The battle going on below him as he flew his missions was Field Marshal Haig's bloody quest to take Passchendaele Ridge and break through to the Belgian coast at Ostende. A hundred thousand British and Commonwealth soldiers would die for that stretch of mud, and about the same number of German defenders. Harris witnessed the full horror of what was happening on the ground, the wholesale slaughter of millions of young men in an apparently endless struggle between equally matched military and industrial powers. From his experiences came the opinions he would hold later, to the effect that air power, and more especially the use of bombers, would be able to break the stalemate and—however many casualties bombing caused among aircrew and people on the ground—buy the end of any new war more cheaply than had been possible in 1914–18. After all, how could the price be any higher?

The armistice found Harris back in England, preparing a new unit, 44 Squadron, for night fighter duties. Outspoken, often to the point of rudeness, confident, a keen disciplinarian, Major Harris (as he now was) also had a reputation for caring about his men, pilots and ground crew alike, and for leaving nothing to chance.

That he did not go back to Rhodesia was due to two factors. First, the new air force showed him a satisfying respect (he was awarded the Air Force Cross), and second, Harris had married in 1916, and it soon became clear that his culture-loving, upper-middle-class English spouse would find life hard in the African bush. Harris made a decision that would prove fateful not just for himself and the Royal Air

Force, but arguably also for Germans born and yet unborn. On August 1, 1919, he accepted a permanent commission as a squadron leader.

In tune with the RAF's postwar role as the British Empire's policeman, Harris commanded squadrons first on the northwest frontier of India, where there was yet another Afghan war in progress and British forts under siege. Transferred to the rebellious former Ottoman province of Iraq, now controlled by the British, he led bombing raids, first against alleged Turkish infiltrators on the northern border, and then against an attempt to establish an independent Kurdish state.

The dropping of warning leaflets over rebel villages before raids sometimes, but not always, saved lives. It is possible, more than eighty years later, to see history repeating itself not just in similar ways but also in the same places. Nevertheless, Harris made a reputation for himself and his men, despite what he described as "the appalling climate, the filthy food, and the ghastly lack of every sort of amenity." Though he was an obedient servant of the empire, his political judgments could be surprising. He showed sympathy with the Iraqis. They had, he agreed, "been led to expect complete independence and got instead British Army occupation and a horde of Jack-in-Office officials."

Despite his abrasive character traits, Harris steadily gained promotions and experience. The outbreak of the European war saw him an air vice marshal and air officer commanding in Palestine and Trans-Jordan. He traveled straight back to England to take over at 5 Group, Bomber Command, and was soon back in the thick of action once more.

There were similarities between the roles and characters of Harris and Churchill, who came to know each other well over the next few years. What each of them had to do when he reach his peak—Harris as AOC Bomber Command, Churchill as prime minister—was already clear, even predestined. What each brought to his job, and which transformed the situation in each case, was personality, energy, and determination.

Bomber Command headquarters was in an underground bunker near High Wycombe, a town in Buckinghamshire about thirty miles north of London, best known for the furniture produced there from local beechwood. In February 1942, when Harris walked through the reinforced doors of Bomber Command headquarters for his first operations meeting, it was as if a typhoon had hit the place.

• • •

THE CHANGE IN ATTITUDE at Bomber Command was not due just to new, more energetic leadership, or a more ruthless bombing policy. The other developments that gave the proponents of large-scale bombing, chief among them Harris, confidence in its future were scientific.

The relative lack of success Bomber Command had experienced in the first two years of the war was due partly to inferior aircraft, partly to inexperienced aircrew, but equally crucial was the lack of aids to accurate navigation and target location. Here the Germans had held the edge during this time, to some extent due to Britain's understandable concentration on defensive rather than offensive technology. The British might have invented centimetric radar, enabling night interception of enemy aircraft, but they had nothing to match the X-Gerät used at Coventry or its more advanced successor, the Y-Gerät (also known as Wotan). Within weeks of Harris's assumption of command, however, the first major British breakthrough in this area of research was ready to be used in a real attack on Germany. This device was named Gee.

Gee was not a single beam to be followed, like the German devices, but a web of such beams covering enemy territory. This was created by a "master" and two "slave" stations. A cathode-ray tube receiver carried in each aircraft picked up these signals and from them created a grid on a screen. Hence the name Gee for "G," which stood for grid. This device enabled a navigator to plot his plane's position with accuracy far superior to any previous guidance system. Gee further enabled aircraft to follow one another along a prearranged course in much better order, and to maintain or retain that course much more easily.

Gee's great advantage was that the role of the equipment on board the bomber remained completely passive; it was simply a receiver, emitting no signals that the Germans could track down. And without such signals, the network of beams had no meaning as far as the enemy was concerned. At least for a while, this would have the German radar experts perplexed. It was expected that they would start considering countermeasures only when a Gee-equipped aircraft was shot down and they got a look at the device. This could be a matter of weeks or months, but until then aircrew had an extremely useful aid that enabled them to navigate to the target and all the way back.

The main drawbacks with Gee were twofold. First, because limited by the curvature of the earth, its range only extended over around 400–450 miles (not as far as Berlin or central Germany). Second, although it brought the aircraft to within two to three miles of the target (remembering that five miles had previously been considered impressive), it was not a pinpoint system. This was shown all too clearly in the first Gee operation against Essen, heavily defended home of Krupp and Holy Grail of British bombers, on March 8–9, 1942.

Harris had optimistically believed that Gee's introduction would "be equivalent to multiplying my force of 300 aircraft by seven . . . I should be able to destroy completely Essen and three other towns in the Ruhr within three months." The 211 aircraft got there, but the city was so fogged with industrial pollution that they couldn't find it. Some houses and a restaurant were damaged. Eight aircraft were lost. The same thing happened the following two nights, over Essen and other Ruhr industrial towns. Most of the Krupp manufacturing facilities lay right in the middle of Essen. This was rare; in most industrial towns they were in the suburbs and therefore harder to find. If you couldn't hit Krupp, what *could* you hit?

The failures over Essen and its neighboring industrial centers were blamed on the haze and a lack of clear landmarks, old problems that impeded the final and most important stage of the attack, the actual dropping of the bombs. All the same, the Ruhr, as Germany's heavy-industrial powerhouse, was Bomber Command's most important single destination, to which had it no choice but to return again and again. "We had to accept the fact that the Ruhr was for the time being impregnable," Harris admitted, "because in this area industrial haze made visual identification almost impossible."

Harris's opinion of Gee somewhat improved when better results were observed in smaller raids over other parts of the continent. These bombers were getting much closer to their targets than before. Further experiments, in Germany and especially against the Renault works at Boulogne-Billancourt, had shown that devoting more aircraft to each attack, operating in a much more bunched-together sequence, led to great improvements in the quantity and concentration of damage on the ground. Those who predicted a disastrous rate of collisions because of this were proved wrong. Bomber Command was still progressing by trial and error, but with new, more dynamic leadership, things were also starting to look up.

• • •

HARRIS HAD DECIDED to start his reign at Bomber Command with a bang. Real "blind" bombing aids, which would enable his aircraft to find targets even in poor visibility, were still in development. So Harris sought out targets that might be less vital to the enemy war effort but were easy to identify. No target was completely problem-free, but coastal towns were definitely at an advantage in this regard (or disadvantage, depending whether you were at ground level). After some discussion Harris fixed on the picturesque old Baltic port of Lübeck, known from the novels of the Nobel Prize–winning German novelist Thomas Mann, who grew up there. It lay beyond the range of the Gee transmitters, but Gee-equipped "leaders" could help the aircraft most of the way out and then most of the way home.

With a population of about 150,000, Lübeck was also an industrial center and a training center for U-boat crews. Its docks were the chief route through which iron ore, vital for Germany's military industries, flowed from Sweden. Above all, it boasted a medieval Altstadt filled with highly combustible buildings, a fact of which Bomber Command's planners were well aware. Harris himself described it as "more like a firelighter than a place of human habitation."

On the moonlit night of March 28–29, 1942, Palm Sunday, 234 British bombers were dispatched to Lübeck. The first arrived over the city at 11:16 P.M. Going in low in accordance with their briefing, around three-quarters later claimed to have found the target, dropping 160 tons of high-explosive bombs and 144 tons of incendiaries. The latter included both conventional "sticks" and a thirty-pound bomb containing a mixture of Benzol and rubber, these calculated to cause fires over a distance of ten meters from the impact point. Defenses over the city were modest. The resulting blaze could be seen by aircrew a hundred miles away, and the level of devastation was as awesome as that at Coventry. Almost a third of Lübeck's built-up area was burned to the ground, sixteen thousand people made homeless, and a great deal of the city's infrastructure wrecked, including its main power station and numerous factories. The cathedral, dating from 1173, was wholly destroyed, along with another grand medieval church, the Church of Mary (Marienkirche). A total of 320 people were killed, the largest number of fatalities in a single raid since the

RAF began its sorties over Germany. A dozen British aircraft (around 5 percent of the total) were lost.

Harris wrote with satisfaction: "On the night of the 28/29 March, the first German city went up in flames." He added: "However, the main object of the attack was to learn to what extent a first wave of aircraft could guide a second wave to the aiming point by starting a conflagration: I ordered a half an hour interval between the two waves in order to allow the fires to get a good hold before the second wave arrived."

This was the first admitted instance, not just of a major incendiary attack, but of the so-called double-blow technique that became much more widespread during the latter part of the war. It led not only to target-marking fires, as Harris noted, but also to far greater chaos and destruction in the target city as the local fire brigades—and forces called in from neighboring cities—struggled in vain to fight the fires and at the same time avoid being destroyed on the ground by new waves of attackers. Dresden was to be its apogee.

Thomas Mann, living in political exile in California, broadcast shortly afterward on the BBC. Speaking in his native language, he declared with a stern certainty that many of his fellow Germans must have found both chilling and infuriating that, much as he regretted the destruction wrought on his native city, which included the house where was born, "I think of Coventry, and have no objection to the lesson that everything must be paid for. Did Germany believe that she would never have to pay for the atrocities that her leap into barbarism seemed to allow?"

Goebbels, who less than eighteen months earlier had searched for ways of playing *up* the level of destruction from British raids, was shown newsreels of the destruction. He betrayed something approaching panic, describing the damage as "really enormous" and, clutching at straws, commenting, "Thank God, it is a North German population."

The alleged phlegm of North Germans was to be much tested over the coming months and years. The Baltic city of Rostock, another Hanseatic town even farther east than Lübeck, was bombed over three nights in the third week of April 1942, causing huge damage, and forcing thousands of citizens to flee the city for the countryside. Rostock had the same combination of close-packed merchants' houses and dockyards, plus a Heinkel aircraft factory and a U-boat production yard. Almost 70 percent of the town's area was destroyed, six thousand of its inhabitants killed or seriously wounded. Both the damage inflicted and the casualty figures were climbing alarmingly—or satisfyingly from Harris's point of view.

The German leadership, hitherto preoccupied with preparations for a spring offensive in Russia, was stung into revenge. Planning for the retaliatory strike began after the Lübeck raid. On the night of April 25–26 the Georgian spa town of Bath in the west of England was bombed, once just before midnight, then again in the small hours (the bombers having rearmed and refueled at their French bases). A German double blow. Four hundred high-explosive bombs had fallen on Bath, and over four thousand incendiaries. Four hundred civilians had been killed. Casualties might have been higher, but unlike many other old towns in Germany and England, Bath's predominantly eighteenth-century architecture favored wide thoroughfares and crescents, and the buildings were of Cotswold stone. It was an environment much less suited to firebombing than narrow, winding streets and timbered buildings.

> The Führer declares that he now intends to repeat such attacks, night after night. He completely shares my opinion that we must now attack cultural centers, spas, and towns where the middle classes live; there the psychological effect will be much stronger, and at the moment the psychological effect is the most important thing of all . . .

Goebbels's comments are revealing. Both sides now admitted, privately at least, the relative feebleness of their attempts to cripple each other's war industries, and were essentially beginning to concentrate on "morale bombing."

In the period April to June 1942, the Germans also bombed Exeter, Canterbury, Norwich, and York—all ancient cities and principally noteworthy as tourist attractions. After the first raid, the German Foreign Ministry's spokesman, Baron Gustav Braun von Stum, was said to have told the press that the Luftwaffe would "bomb every building in England marked with three stars in the Baedeker Guide." This flippant remark led to the campaign being dubbed the "Baedeker

Raids." Hitler's order for attacks against the British cultural centers had stated cold-bloodedly that "preference is to be given to those where attacks are likely to have the greatest possible effect on civilian life . . . terror attacks of a retaliatory nature are to be carried out against towns other than London." Large quantities of incendiaries were used in the raids.

Harris, meanwhile, was planning another "spectacular," the largest one so far. He gained Churchill's approval for a huge raid on a German city, code-named Millennium, which, despite the fact that Bomber Command's actual available strength was under four hundred, would involve a thousand aircraft. Coastal Command, asked to contribute, refused, but the two Operational Training Groups contributed 368 aircraft, some manned by instructors, as did the Heavy Conversion Units, whose task was to prepare qualified but inexperienced aircrew to handle the big four-engined bombers. Eventually the number did indeed exceed a thousand. As at Lübeck, the plan was for a first wave (No. 1 Group) to drop incendiaries to set the city alight. A second (No. 3 and No. 5 Groups) would follow up an hour later, when the fires had taken hold, to sow further mayhem. All the leading aircraft were equipped with Gee for the first time, which placed limitations on the choice of target. Both the proposed cities were within range: Hamburg (first choice) and Cologne.

The raid was postponed on successive nights because of unfavorable meteorological reports until, as May 30 dawned, the planners were faced with their final opportunity. It was the last night of the full moon, and also the last in which interrupted training schedules could remain that way. When Harris entered the ops room at Bomber Command, the updated weather forecast for Germany, prepared by his punctilious meteorological officer Magnus Spence, was a little better for the Rhineland than the northwest coast. Group Captain Dudley Saward was there at the decisive moment. Harris stared at the charts, moved his forefinger across Europe to a town in western Germany, and then pressed on it until the blood drained from his fingernail. Finally he turned to Saward and the other staff officers and said calmly: "The Thousand Plan tonight."

The city selected was Cologne. Hamburg had been saved, at least for a year, by the vagaries of the weather.

The effects on the ancient city by the Rhine were horrifying.

Almost the entire public utility and transportation system out of action, damage to factories measured in months of lost production, fifteen hundred commercial and industrial premises destroyed, thirty-six major factories wrecked and another three hundred damaged. All in all, forty-five thousand people had been made homeless, thirteen thousand dwellings destroyed. Many of historic buildings, including most of what was left of Cologne's Roman remains, were pulverized. Only the death toll was astonishingly light: fewer than five hundred dead, with just over five thousand injured. These scarcely exceeded those from the Luftwaffe's recent attack on Bath, perhaps showing the difference between the undefended British town, with its relatively poor provision of shelters and air raid protection, and a city like Cologne that had always known it would be heavily attacked and had prepared accordingly. All the same, the damage to Cologne was widespread and serious. It showed not just a new level of British capability, but a change of policy to more or less indiscriminate bombing of urban areas. The raid was, from the RAF's point of view, a great success. Only 2 percent of the planes were lost. It proved to be a tremendous morale booster for the home front—and made Harris a public relations star.

The effect on local morale was so serious that citizens of Cologne leaving the city were required to sign a chilling pledge of secrecy:

> I am aware that one individual alone can form no comprehensive idea of the events in Cologne. One usually exaggerates one's own experiences and the judgement of those who have been bombed is impaired. I am therefore aware that reports of individual suffering can only do harm, and I will keep silence. I know what the consequences of breaking this undertaking will be.

On the day of the thousand-bomber raid, Goebbels wrote with habitual febrile cynicism in his diary:

> We have stationed sufficient squadrons of bombers in the West that we are in a position to respond in kind to each blow, if necessary with twofold force. We shall leave no attack by the English unanswered, and since attacks so far on military and economic targets have scarcely been worthwhile, we shall now, as previously, attack cultural

centers, which is what the English are also doing, if without acknowledging the fact. For our part, we also don't need to say anything on the matter; we just need to do it. In connection with this, the Führer has once more charged me with the responsibility for ensuring that all the Reich's valuable artistic treasures are made safe.

As for Arthur Harris, on June 11, 1942, it was announced in the *Times* of London that he had been appointed a Knight Commander of the Bath.

From the other side of the channel, the signals were equally clear. All targets were "legitimate" targets.

ON OCTOBER 8, 1871, the most devastating forest fire in American history swept through the northeastern part of the state of Wisconsin. Probably started by railroad workers clearing brush, the fire consumed 1.2 million acres and claimed between twelve hundred and fifteen hundred lives, including around eight hundred in the thriving logging village of Peshtigo, almost half its population. Many witnesses spoke in awe of the approach of fire—"so fast" or "like a tornado." A few outran it, while most survived by taking refuge in the Peshtigo River, just outside the town, or in the wells of farms and houses, where they near-submerged themselves or wrapped wet clothing around their heads and their exposed parts. The fire crossed the river with ease, starting blazes on the eastern side, trapping many who had thought to escape by crossing the wooden bridge. For some hours the desperate "swimmers" found themselves surrounded by flames. Fireballs and burning objects rained down on them, and the air became almost unbreathable. Some suffocated, some were burned by the flying debris; some became exhausted and just sank beneath the water.

What happened at Peshtigo by accident, by a freak of nature, was a firestorm. This phenomenon arises when such an intense heat is created—eight hundred, nine hundred, a thousand degrees—that hot air's natural tendency to rise sucks all the oxygen out of the air at ground level. So there exists at some height above the ground a pillar of flame, but beneath this a vacuum that, searching for oxygen to fill it, rushes horizontally along the ground, enveloping everything in its path. There can be a moment of stillness before the storm arrives, as before a natural tornado. Then comes a rush of searing air that roasts the lungs, leaving the few survivors with only poisonous carbon monoxide to breathe. At Peshtigo, when bodies were found, there were often "no visible marks of fire near by, with not a trace of burning upon their bodies or clothing."

There had been several weeks of hot, dry weather all over the Great Lakes states during that fatal autumn of 1871. Seventy-two years later and three thousand miles away, at the end of July 1943, Hamburg was suffering from a heat wave of similar proportions. Daytime temperatures had reached 27 degrees Celsius (over 80 degrees Fahrenheit).

On Tuesday night, July 27, 1943, a total of 787 aircraft took off from Bomber Command bases in eastern England. This was the second in a sequence of raids aimed at Germany's second largest city, the port of Hamburg. The RAF had first attacked on July 24 by night, the Americans twice by day, without spectacular success. With almost two million inhabitants, the largest of the medieval Hanseatic ports was an important center for shipbuilding and U-boat construction, manufacturing, and the merchant marine. It was heavily defended.

What would provide the Hamburg-bound bomber fleet with protection, though, was something quite novel and sensational: a recently invented device—first used just days previously in the first RAF attack—whose purpose was to throw the German radar defenses into total confusion. Its name was window.

More than two thousand strips of tinfoil fluttered to the ground from the first of the aircraft that had been loaded with window. Each bundle of metal fragments created an echo that, to the German radar stations, was indistinguishable from that of a British bomber. Since the rate of fall was slow, it continued to give that impression for fifteen minutes. Every bomber in the stream was trained to drop one bundle, every minute, until it reached that point on the return trip.

Window caused complete chaos for the German defenses. The German night-fighter defense system had groups of aircraft, each assigned an area or "box" to patrol, of which the most westerly ones, where first contact with the RAF could be expected, were the most crucial. The Würzburg stations that guided the night fighters suddenly "saw" literally thousands of "enemy aircraft," and initially sent the fighters chasing all over the sky. A Luftwaffe control officer described the chaos: "It was just like trying to find a glass marble in a barrel of peas." The on-board radar in the German fighters was also susceptible, again leading to utter confusion. "With one stroke, the whole defense was blinded!" declared General Walter Kammhuber, the commander of the German air defenses in the west.

Later the Germans would learn how to cope with the worst effects of window, but not yet. On that night the bombers still faced a severely disadvantaged enemy.

The RAF planners' idea on the night of July 27 was to exploit this situation by saturation-bombing Hamburg. With almost eight hundred aircraft involved, for the time this was a very big raid. Five group captains and two air commodores traveled as passengers to check the exciting effects of window.

A total of 2,326 tons of bombs, a high proportion of them incendiaries, was dropped on Hamburg. The main concentration was in the working-class areas of Billwärder, Borgfelde, and Hamm, which had cheaply constructed tenements lining dense, narrow streets, with few parks or open spaces. The district was actually attacked in error—the planned aiming point was about two miles away—but it was also, in many terrible ways, the perfect environment for what happened next. That night, the word "firestorm" appears in the records of the Hamburg Fire Department.

With small incendiaries raining down in their thousands and lodging inside roof spaces, and the smaller quantities of big thirty-pound "phosphor" bombs penetrating several stories to start fires inside the buildings, blazes had soon started all over the area. The bulk of the Hamburg Fire Department, meanwhile, had been concentrating on dealing with the last of the fires from the night of Saturday to Sunday, on the far side of the city. Those emergency teams that did try to reach the more recently affected districts were inhibited by craters and rubble blocking the approach roads—as had happened in Coventry almost three years before.

The big blast bombs mixed in with the incendiaries shattered walls, blew out countless doors and windows, creating drafts that fed the fires inside the buildings. The resulting updraft further caused burning sparks and debris to float over and scatter over wide areas of the neighboring parts of the city, starting fires there also. Within twenty minutes there was a massive conflagration a square mile in extent, centered on a blazing sawmill and timber yard. By 3 A.M. sixteen thousand apartment houses were reckoned to be on fire, along more than 130 miles of close-packed narrow streets. Trees were torn from their roots by the air rushing in to fill the vacuum over the center of the blaze, and people were blown off their feet and pulled into the flames by the same invisible force.

Like the Peshtigo fire, the great firestorm of Hamburg, on the night of July 27–28, 1943, did not only burn and char its victims. Thousands died in basement shelters from carbon monoxide poisoning. They were found perfectly preserved and still in the attitude of death, turned cherry pink by the gas but otherwise seemingly unharmed. It was two days before some of the worst hit streets had cooled down sufficiently for the rescue teams to go in to recover what remained of the victims.

At the heart of the apocalyptic fire there were no survivors found, none at all.

Four mass graves were excavated in the Ohlsdorf cemetery, on the outskirts of Hamburg, each containing up to ten thousand bodies. The postwar United States Strategic Bombing Survey conceded that the estimate of forty thousand victims might still not do the carnage of Hamburg justice. This was the worst of Douhet and Trenchard's imaginings made burning flesh. By the morning after the raid, tens of thousands had begun to flee the city, carrying what essential belongings they could, some still in their nightclothes. The authorities set up evacuation points at assembly areas such as the racecourses.

Albert Speer, Hitler's war production chief and personal architect, wrote that "Hamburg had put the fear of God in me." He recalls informing Hitler that "a series of attacks of this sort, extended to six more major cities, would bring Germany's armaments production to a total halt."

Goebbels, who since October 1942 had also been charged by Hitler with arranging "aid measures arising from war damage," reacted quickly to the Hamburg disaster. He was immediately in telephone contact with the Gauleiter of Hamburg, Kaufmann:

> He speaks of a catastrophe . . . Here we must envision the destruction of a metropolis on a scale without parallel in history. In this

context, problems arise that are almost impossible to master. The inhabitants of this huge city have to be fed, housed, and where possible evacuated, provided with clothing and bedding. In short, here we have to deal with tasks that a few weeks ago would have been unimaginable . . .

Goebbels was so affected by Hamburg that, in a rare loss of nerve, he ordered the partial evacuation of Berlin, in case the British now turned their attention to the capital. He and the other Gauleiters had to spend the rest of the summer damping down the panic that resulted. From now on he had a twin aim in his propaganda: to label the Anglo-Americans as barbarians, but at the same time to minimize the morale effect on the German public of such increasingly frequent and severe devastation. It was a perilous tightrope walk even for the wily propaganda minister, and to keep his balance he wielded a balancing pole crafted from the very finest lies.

But the triumphant Harris could not press home his advantage by bombing other German cities, as Speer feared. For political reasons—Mussolini had fallen and the British government was trying to tip the Italians into surrender—the next night, Bomber Command was ordered to attack big-city targets in northern Italy. That raid was then canceled, but for the next two nights there was bad weather over both England and Germany, and the moment was lost.

After the huge, seemingly low-cost destruction of the firestorm night, Bomber Command had become fallible once more. Situation normal.

Dresden, a long way off, in an area as yet untroubled by air raids, had already begun to take refugees from the Ruhr and Rhine. Now came Hamburg—at least some of the evacuees ended up in Saxony. Dresden, beautiful and undamaged, only occasionally disturbed by air raid warnings, must have felt like a paradise to such people, who had become bomb damaged in body and mind.

The refugees carried with them to Saxony, at the same time, the bacillus of fear. But for the moment, the "cultural city" of Dresden seemed like a sanctuary. A sanctuary about to become crowded with pilgrims.

12

The Reich's Air Raid Shelter

ALMOST A YEAR after the beginning of the Second World War, the first air raid warning sounded in Dresden.

The sirens began to wail at 2:15 A.M. during the night of August 28–29, 1940. The all-clear was given forty-five minutes later. There would be another eleven warnings in the city before the end of the year, mostly because of British night raids on Berlin, a hundred miles to the north. On October 20 at around 10:45 P.M., three high-explosive bombs actually fell in a field at Bühlau, just southeast of the city, as the Gauleiter's headquarters in Dresden reported. They left two craters, eight meters across by two meters deep, and a hole that was thought to contain an unexploded bomb.

In the months to come, leaflet drops and the occasional stray stick of bombs disturbed the peace of the nearby Saxon countryside, but none fell within the Dresden city limits. In the all of 1941, there were seven air raid warnings. In 1942, four. For most Dresdeners, the war must have seemed a long way away. And all the news seemed good. Victory imminent in Stalingrad. Rommel surely about to take Cairo. Admittedly, there were the occasional funereal-bordered newspaper notices—more frequent after the summer of 1941—that carried a black Iron Cross, a message from a Dresden family, and the formula "a hero's death" or "fallen for the Fatherland."

The otherwise cheerful advertisement section of *Der Freiheitskampf* on September 30, 1942, carried job offers by the dozen. There were lists of what was playing at the theater and the cinemas, which numbered almost fifty. The clown Charly Rivels was appearing at the Circus Sarrasani. "Dance amusements" (*Tanzlustbarkeiten*) had been banned since April 1941; otherwise, these pages didn't feel as if they belonged to a city at war, in a country beginning to bleed to death on the eastern front. But many of the employment advertisements sought women to do traditionally male tasks—from draftspeople to dentists, laboratory assistants to butchers. And all but three of the thirteen lonely-hearts requests were from women for men.

At this time, in Dresden, almost no special precautions had been taken in case of air raids. Citizens were encouraged to keep buckets of sand and water at hand to deal with fires. Cellars and basements were fitted out, by the obedient or the careful, with emergency supplies and gas-proof doors. In factories and schools, there were the usual demonstrations of what to do in an air raid and how to extinguish an incendiary bomb, and of course there were fire alarms. Many public buildings had cellars or stores converted into shelters, though rarely were the kinds of modifications and additions undertaken that would have provided real protection. Dresden had been excluded from the Führer-Order of October 1940 (following the British air raids on Berlin), which decreed that eighty-one German towns and cities would begin the construction of certifiably bombproof shelters with immediate effect. In central Germany, Leipzig, Halle, and the nearby Leuna hydrogenation (synthetic oil) plant had been the main beneficiaries of this centrally directed building program.

The big Zeiss-Ikon complex, by this time employing over ten thousand workers in Dresden on war contracts, contained the only large private buildings to be properly secured. The Goehle-Works in the Grossenhainer Strasse, on the northern edge of the city (later notorious for its use of forced labor), had a bunkerlike appearance and was supplied with sloping slats over its windows, designed to ward off incendiary bombs. The Ernemann factory, also a Zeiss-Ikon plant, had its central stairwell and the landings on each story reinforced to withstand the impact of a thousand-pound bomb. These were factories that had been constructed since the early 1930s, when a law had been promulgated making it compulsory for all new buildings to comply with air raid regulations. Such precautions were rare, except in new factories and housing developments.

One exception was the state-of-the-art air raid shelter built beneath the Bramsch distillery in the Friedrichstadt industrial area, complete with strengthening girders, gas filters, sealed doors and windows, and emergency exits. Dresdeners smiled incredulously when given the tour by the manager responsible, Dr. Bergander. This was air raid security gone crazy! Why would anyone ever need such a thing?

Meanwhile, from 1942 onward, the RAF had exacted a severe toll on western Germany. In March 1943 began the Battle of the Ruhr, when the RAF eventually began to cause some real damage to the industrial heartlands of the Reich. Essen, Dortmund, Düsseldorf, and the rest were battered relentlessly for the next four months. To Gee was soon added Oboe, which allowed basic blind bombing in poor visibility. Since its transmitter stations were limited in number, and it could therefore handle only a few aircraft at a time, it was fitted only to Pathfinder planes, which used it to mark the target for the following bomber fleet.

A little later came H2S, a radar set that, by registering varying echoes on a cathode-ray screen known as a "Plan Position Indicator," enabled first Pathfinders and then bombers in general to make out the rough shape of the town or below even through cloud. The technology, perfect for following the outlines of coastlines, had at first been confined to Coastal Command, but its adaptation and transfer to Bomber Command began early in 1943, boosted by the confident assertion that "the accuracy of bombing with H2S in blind conditions will produce a concentration of bombs about the aiming-point comparable with the best results that can be achieved at present by crews in perfect visibility." By this time Air Marshal Harris also had almost sixty squadrons fully operational, more than half of them equipped with the new Lancaster heavy bomber, which could fly faster and higher and carry more bombs than the old Wellingtons and Halifaxes. Soon the U.S. Army Eighth Air Force would be joining them, and daylight raids would once more begin.

The development of the advanced British "heavies" (as the big long-distance bombers were called) should have been a warning signal to eastern Germany that the area was no longer beyond the RAF's normal bombing range. But through most of 1943 the Rhine and Ruhr areas (plus, notoriously, Hamburg) had been on the receiving end of Bomber Command's growing offensive power. Increasing numbers of women and children, evacuated from Düsseldorf and Dortmund,

Krefeld and Cologne, were sent to Saxony, including Dresden, for temporary resettlement. The locals wondered at these tough survivors of Allied bombing and their hair-raising stories, their cynicism and defiance. Soon Saxony came to be known as "the Reich's air raid shelter" (*Reichsluftschutzkeller*), partly because of its role as an evacuation area, and partly because the local population thought itself safe.

Without clear instructions from Berlin, the air raid protection situation in Dresden depended on the local authorities. Even after major Allied air raids elsewhere in Saxony (at the end of 1943), little progress was made in improving provision for the general public. A Construction Office for Air Raid Protection was established, but with increasingly scarce concrete, steel, and labor already earmarked for larger military projects elsewhere in the Reich, it was by now much too late to start major works in Dresden.

It was a matter of public scandal, meanwhile, that the really solid, technically advanced air raid protection in Dresden was available only to those in official positions. The "Local Air Raid Leadership of the NSDAP" shared with the office of the chief of police a reinforced bunker two stories under the surface of Albertinum, the massive neoclassical building on the Brühl Terrace that housed what had been the royal archives and art galleries. The deeply dug, well-constructed "command posts" of the *Gauleitung,* at the Lockwitzgrund, on the far southeastern edge of the city, and of the police and SS leadership at the Mordgrundbrücke, in woods on the north bank of Elbe just a brief drive from the city center, could at least justify themselves in terms of their official function.

When, in the middle of 1943, Gauleiter Mutschmann had a secure bunker constructed in the garden of his villa in the Comeniusstrasse (close to the Grosser Garten and confiscated from a Jewish businessman), he employed SS engineers from the Dresden garrison. The SS commander in the city was moved to protest to no less a person than Reichsführer Himmler:

> I do not dispute that such a bunker is necessary. I even believe that
> it is based on an order from the Führer. However, I do not think it
> right that such a bunker be installed in the Gauleiter's garden, of all
> places, because the greatest part of the population still has no access
> even to a properly constituted air raid shelter . . .

Nor would it have been unknown to many Dresdeners that in the Ruhr, Hamburg, and Berlin, experience had proved that the safest places to take refuge were the big, government-built bunkers, which could hold hundreds or even thousands of civilians. After all, plenty of survivors from the air raids on these places were now living among them in Dresden as refugees, and one had only to ask.

THE ACCOMMODATION of thousands of refugees, many of them children, from the heavily bombed areas was the responsibility of Gauleiter Mutschmann, and a highly problematic one. In early July 1943 he circulated a secret instruction to the burgomasters of the major Saxon cities ordering them to organize the enrollment of schools and entire classes (with teachers) composed of evacuees from the west.

The aim in "importing" the school groups whole was both to provide a familiar environment for the new arrivals and to make supervision through the Hitler Youth and other state organizations easier. With many school buildings already being used for military or industrial purposes, the result was vastly increased school rolls with even less accommodation. The average new intake into the school system in Dresden in 1940 was fifty-five hundred; in 1943 it was eighty-five hundred to nine thousand. Many schools were already running double shifts for their students. A worthy "Law for the Protection of School-Aged Youth" of January 1943, which provided for all children to take refuge in school shelters in case of air raids, was in fact bound to be ineffective from the outset. By 1943 almost all the schools in Dresden were devoted to the military or war-related uses, whether as stores, offices, or accommodation. This made the provision of adequate teaching space, let alone easily accessible air raid protection, difficult or impossible.

As the year went on, the children from the west were joined by similar arrivals from Hamburg and Bremen. The children were often with their school classes rather than their families. At Christmas, rather than visit their home cities, for safety reasons they would stay in their accommodation in Dresden and make toys and Christmas decorations. The older pupils took part in the Hitler Youth's war work, delivering mail and so on, to free up adults for the military and for industry.

By the end of 1943 Dresden considered itself full. It could take no more children from outside.

IN THE SMALL HOURS of December 4, 1943, the industrial and mercantile city of Leipzig, famous since the Middle Ages for its musicians and its trade fairs, and more recently as the center of German language book printing and publishing, was bombed by the RAF.

There had been a raid on Leipzig in October, of modest extent and even more modest effect, aimed at the Erla factories in the city (which made one in three of the Luftwaffe's Me 109 fighters), but the December attack was, as everyone immediately realized, something else. In sixteen minutes 432 Lancasters, Halifaxes, and Mosquitoes dropped a mass of air mines and high-explosive bombs, plus 313 large and 12,550 small incendiary canisters, and just over 280,000 standard four-pound incendiaries. This was a major firebombing raid.

At first the number of dead had been estimated at the incredibly low figure of ninety-five, but by February 1944, when the clearing of the rubble and the searching of the cellars was completed, almost two thousand of Leipzig's civilians were registered killed or missing and four thousand injured. What was worrying about the attack were two things: the complete insufficiency of the fire-fighting measures, and the firestorm effect—limited but chillingly clear to those who knew how to judge such things—that Bomber Command was able to unleash that night.

The great Hamburg firestorm of July 1943 had shown Harris's planners what could be done if everything went "right," and from then on, his planners were eager to promote other firestorms wherever possible. On October 22, 1943, another firestorm occurred in Kassel, a much smaller city of around a quarter-million inhabitants. A stream of bombers almost a hundred miles long dropped more than four hundred thousand incendiary devices, mainly on the historic center. As in Hamburg, high-explosive bombs and mines blew open doorways and shattered windows to feed the draft.

Forty-five minutes after the first bombs had fallen on Kassel, the firestorm had already reached its climax. Ten thousand civilians died, including one in ten of the inhabitants of the Altstadt and 4.2 percent of the city's total population (Hamburg lost 2.73 percent). The weather was not hot, as it had been in Hamburg, but conditions were dry, and the thousand-year-old city's streets were narrow and many buildings half-timbered. The nearest good-sized city was almost a hundred miles from Kassel, so no substantial fire-fighting help was available until the fires had long raged out of control.

Leipzig did not lack outside help, for such assistance was quickly at hand. The problem was that the hoses of the neighboring fire brigades that rushed to the city's aid needed adaptors to fit the local fire hydrants. These were supposed to be available at each police station, but none could be found. The outside fire-fighting teams were forced to stand by helplessly as buildings burned, house fronts collapsed, and attic fires (the curse of incendiary attacks) spread laterally from house to house.

Add to all this the fact that more than half of Leipzig's own fire brigades had been sent to help Berlin. There the Anglo-American air forces had been conducting a systematic campaign of destruction, which would continue into the spring of 1944. The long-planned upgrading of Leipzig's venerable water system had been repeatedly postponed, meaning that, even when the hoses could be connected to the mains, the water pressure was inadequate. It was a fiasco for which the authorities—and ultimately Gauleiter Mutschmann—were wholly responsible. Most of Leipzig's Altstadt was destroyed, plus wide areas of the inner suburbs, including many historic buildings, 80 percent of the Trade Fair Center (which had been converted into workshops and factories), and a great deal of the university complex.

With regard to Leipzig, an official report made grim reading:

> The rise of several big firestorms of an extent and with consequences similar to those previously seen only in Hamburg and Kassel. For example, strong trees were ripped from their roots, automobiles thrown around, petrol pumps torn from their supply pipes and hurled through the air, fire hoses whipped up against trees and power lines, officers and men of the fire-fighting forces whirled across streets and squares, and in the process killed or injured.

These words were written three weeks later by someone who had happened to be in Leipzig at the time of the raid, but whose special knowledge was by no means haphazardly acquired: Major General Hans Rumpf, inspector of fire-fighting forces at the main office of the Order Police in Berlin. He continued:

> From an urban construction point of view, Leipzig was worse rather than better off compared with other historic cultural centers. The Altstadt, as a central feature of the fair, is crammed with an extraordinary large number of buildings dedicated to storage and display, which lie alongside closely built up residential areas. Even in the case of a weaker attack, one would have had to reckon with a considerable loss of such buildings.

By Rumpf's expert calculations, each fire-fighting team of eight men was faced with eighty to one hundred fires—at least ten per fireman. Add to this the incompatible fire hydrants and the ancient water supply . . .

The actual scale of the raid's effects approached that of Kassel and Hamburg, but the death toll was much lower. The economic warfare officer for Leipzig, attached to the staff of Defense District IV (of which Gauleiter Mutschmann was the commissioner) found a reason why: The residents helped themselves.

> The inhabitants of Leipzig . . . in courageous fashion extinguished the fires that arose, even while the raid was continuing. They were able, in many cases, to save their homes and property—although in this they acted against air raid regulations. According to instructions from the air protection authorities, they were actually obliged not to leave their shelters before the conclusion of the enemy raid.

In other words, the Leipzigers disobeyed the authorities, emerged from their shelters to fight the fires, and for the most part survived. Whether, just over a year later in Dresden, it might have been possible to save more buildings and property if this course had been followed can never be known for certain, but the Leipzig experience seemed to teach that it was best to leave the shelter as early as possible and explore the situation in the open. Both buildings and lives were saved as a result. In the street there was the danger of fire and falling masonry, but in the enclosed world of the cellars perhaps the even greater peril of being buried alive or suffering the insidious effects of carbon monoxide poisoning. One thing was clear: to stay in the shelters substantially beyond the end of the raid may have seemed safer, but—counterintuitively—it was actually the most dangerous course of action.

There were many more raids against Leipzig in 1944. In the course of the war, the city suffered thirty-eight attacks, eleven of them major. Leipzig became one of the favored targets for "precision" bombing by the Eighth Air Force, especially against the Erla, ATG, and Junkers works. These plants employed thousands of workers directly producing aircraft and parts for the Luftwaffe, and attacks on them were therefore considered essential in the campaign to destroy as much of the German aircraft industry as possible and thereby undermine the Reich's air defenses. In this the Leipzigers were fortunate. Especially in cloudy conditions, where H2S was used, the practical results of the USAAF's precision bombing could be hard to distinguish from British area attacks, but at least the Americans did try to aim for the factories (which were in the suburbs), and their preferred mix was much lower in incendiaries than Bomber Command's. There is no doubt that civilian casualties from American attacks were much lower.

All the same, nothing that happened in Leipzig boded well for Dresden, its sister Saxon city.

DRESDEN AND LEIPZIG were sixty miles apart, traditional rivals in the way that only two roughly equal-sized cities vying for prominence in a small country can be. At the beginning of the war, both were thought beyond the RAF's reach. Leipzig was the first to be proved wrong, and the shock was considerable.

Not that it should have come as a complete surprise. The RAF had been creeping farther east for the past few months. In Dresden, the number of air raid alarms rose to 52 during the course of 1943, an average of one each week, and would reach 151 in 1944—an alarm every other day.

During the latter part of 1943, in response to concerns about possible air raids on Saxony, the authorities in Dresden had been making confidential preparations for a mass evacuation of the city's children—its future human capital—to places of safety in the countryside.

Accommodation for individual school classes was arranged, teachers selected to take charge of the children when the time came, and plans made with the railway authorities regarding special trains. By late November everything was in place. "Highly confidential" information packages were sent out to head teachers all over the city. The plan remained strictly secret to avoid causing unnecessary panic. Mutschmann, in his role as governor, would make the final decision: "Probably the commencement of such an operation will come in question only after the occurrence of a major air attack."

Two days after the head teachers received the information, Leipzig was hit by Air Marshal Harris's first eastern firestorm. On December 6, 1943, they were authorized to inform their staffs, though they were forbidden to refer to "evacuation." The approved phrase was "country vacation." The following day Mutschmann's office produced an "Address to Dresden Parents," which was printed and sent to the school authorities, arriving on December 9. Its tone was characteristically bland:

> The danger exists that after his attack on Leipzig, the enemy will extend his aerial terror to other cities. In order to keep losses as low as possible, it is planned to transfer schoolchildren from the most at-risk cities to less threatened places.

With Christmas just around the corner and nerves on edge from reports of the raid on Leipzig, this mealymouthed official missive was not calculated to inspire a sense of enhanced security among Dresdeners. Nor did it. A series of new announcements and orders rapidly followed. Parents were assured that any move would be voluntary. "Transfers" would in any case take place only after the Christmas holidays. Head teachers were chided for not having "prepared" the parents sufficiently for this news. The Ministry of Education specifically forbade further mention of the new instruction in the press, so as not to stoke the potential panic up any higher.

Parents' meetings were held at Dresden schools on December 11 and 12, to give the authorities a chance to explain their position, and to encourage parents who could not make arrangements with friends or relatives in the countryside to sign their children up for the government-sponsored schemes.

Unfortunately, sign-ups remained disappointing. Only four thousand children were enrolled by their parents. Almost thirteen thousand other children were either already accommodated with relatives outside Dresden, or their parents planned to arrange this in the immediate future. That left 70 percent of Dresden's children still living in the city, and therefore under threat from an Allied air attack.

Clearly, parents and children were reluctant to be parted. In many cases, fathers were away fighting at the front. The situation was not helped by a further order stipulating that mothers could not travel with their children to their new homes to "settle them in." The authorities cited shortage of accommodation, but the real reason was that they wanted to keep control of the children and minimize disruptive parental influences.

Anita Kurz, then twelve years old and an only child ("Yes, my parents were . . . perhaps fixated on me"), had her father at home, but only because he had been invalided out of the army in Poland after being kicked by a horse. A sales manager for a bank before the war, he now found it hard to hold down a job. Anita's mother kept the family together by working at a shop in the Schlossstrasse. They were a close little family to begin with, and the misfortunes of war had brought them even closer together. When they received the evacuation order, there was a family conference, Anita recalled :

> Toward the end of the war all children were supposed to leave the city and sleep away from Dresden. I had a cousin who lived out there . . . Anyway, we all sat down and discussed it. And I said, because this was my wish: "I want to be where you are." And that was accepted. Of course, the circumstances turned out quite different to the way we had imagined . . .

On January 18 Gauleiter Mutschmann issued a new order, now attempting to put indirect pressure on the parents. "If children stay behind in the endangered urban areas at the express demand of the parents, they will not be entitled to schooling." This order was delivered to parents, who had to register their receipt of it, and expressly give or refuse permission for their children to join their school friends on the "country vacation." The increased pressure had little effect. February brought reports that many of the children already in the country were unhappy, missing their parents and family life. The Gauleiter wrote a confidential circular to Saxony's burgomasters and district officials, warning them about the problem and encouraging them to do their best to enforce the program.

The struggle continued throughout the year. Parents showed amazing ingenuity. They would register their children as resident with relatives in the outer suburbs or even in the country, while the children would actually be coming into the school in the inner city as normal. Even when the authorities began closing down these schools and turning them over to the military for use as hospitals, convalescent homes, and antiaircraft installations, parents still would not send their children to the country, where schools were still open. The drift back into the city continued. A Nazi Party report from August 1944 noted with an element of resignation:

> Whatever the parents' reasons for bringing their children back, that is what is happening. Many are making efforts to ensure their children receive schooling, many are not. But all are still living in Dresden and are not to be moved from there.

Nine days later, on August 24, 1944, a daylight American raid caused the first casualties within Dresden itself. Seventy-eight Flying Fortresses of the USAAF's 486th and 487th Bomber Groups attacked the town of Freital, southwest of the Dresden city limits, with bombs falling on Gittersee and in considerable quantities on the suburb of Alt-Coschütz, which was administratively part of Dresden.

The Rhenania Ossag hydrogenation works was the target, and 241 people died. Many fatalities were from the factory, which produced a special oil used in the Wehrmacht's tanks. One casualty was, according to a comrade's account, a British POW who worked at Rhenania Ossag. He was given a soldier's funeral at the English cemetery in Dresden, with German guards firing a salvo over the grave. These British POWs, billeted at a school in Freital, were also put on corpse recovery and burial detail. One of them, Robert Lee, commented that the "shelters" in the town were basically just dugouts, which collapsed inward at the slightest impact. The British, some of whom had been working in the town for years, were on familiar, even friendly, terms with many of the victims, who were mainly adults but also included some children.

A police report from the nearby suburb of Gittersee noted gloomily: "Trust in the leadership is diminishing steadily."

The Freital raid was the signal for a final push on the part of the authorities. More school meetings were called. The Ministry of Education resumed its pressure, drawing up lists according to which children who refused to leave the city or were known to have left and then returned would be compelled into "useful activity" through the medium of the Hitler Youth. Teachers were instructed to make home visits to recalcitrant families. The local Nazi Party machinery was drawn into the campaign. Ortsgruppenleiter (local group commanders) were encouraged to contribute to the new publicity drive. Then, on September 15, 1944, came a brusque communication from the National Socialist People's Welfare headquarters in Berlin, which changed everything. It informed the Dresden authorities that new definitions of "total war" (Goebbels's favorite phrase) meant that there were no longer financial or human resources to spare for the protection of the civilian population. Everything was to go into the war effort; civilians would have to fend for themselves.

So the story of the city's forlorn attempt to save its children ended in failure, due to the seemingly insoluble bonds between parents and children. Would it have been different if the "country vacation" plan had not been so thoroughly mismanaged?

A few hundred Dresden schoolchildren remained safely in the country, but later in 1944 thousands of refugee children began pouring in from the endangered eastern provinces of the Reich, seeking the deceptive security of the only major German city to remain pristine and undamaged.

So, by the beginning of 1945, there were certainly more, not fewer, children in the city than a year earlier, when the vain attempts at evacuation had begun.

But surely Dresden, of all places, must be safe. It was a *Kulturstadt*, famous for nothing except its beautiful buildings and treasures. Why would anyone want to attack Dresden?

13

A City of No Military or Industrial Importance?

ACCORDING TO THE 1944 handbook of the German Army High Command's Weapon Office, the city of Dresden contained 127 factories that had been assigned their own three-letter manufacturing codes, by which they were always referred to (for example, Zeiss-Ikon = dpv; Sachsenwerk = edr; Universelle = akb). This assured secrecy, while at the same time allowing the military authorities to identify individual weapons, munitions, and military equipment back to their manufacturing sources. An authority at the Dresden City Museum describes the handbook's code list as "very incomplete," and it did not include smaller suppliers or workshops that were not assigned any codes. Even by this measure, however, Dresden was ranked high among the Reich's wartime industrial centers. As the 1942 *Dresdner Jahrbuch* (*Dresden Yearbook*) boasted:

> Anyone who knows Dresden only as a cultural city, with its immortal architectural monuments and unique landscape environment, would rightly be very surprised to be made aware of the extensive and versatile industrial activity, with all its varied ramifications, that make Dresden . . . one of the foremost industrial locations of the Reich.

The "industrial activity" had always included a surprisingly large number of precision engineering companies, but before the war Dresden remained known to the outside world almost exclusively for its leisure-related and luxury industries. This was an important factor in the pervasive postwar myth that at the time of the bombing raids it was a "city without industry" and "of no military or industrial importance." Even many prosperous Dresdeners seem to have been only half aware of what went on in their city, away from the pleasant suburbs where they lived.

However, when it became clear that the war was going to be a long one, Dresden quickly followed the rest of Germany into the integrated war economy, a process accelerated after the campaign against Russia began. Most of the factories formerly making consumer goods or luxury items had, by 1944, been turned over almost entirely to war work.

The Wehrmacht High Command's list of company codes includes firms that any casual browser would never conceive of as war-related. It contains machine and engineering companies, which are more obvious, but also manufacturers of leather goods, wooden furniture (including a company making "school benches"), curtains and lingerie, pianos, and towels.

A typical example of the how Dresden's economy, after a slow start, became a war economy is Seidel and Naumann, the well-known typewriter and sewing machine manufacturer, whose factory was situated close to the Friedrichstadt marshaling yards. An employee reported:

> From 1923 I worked at the company Seidel & Naumann-AG, which before the war produced typewriters, sewing machines and bicycles . . . bicycle production was halted in 1937. Sewing machines and typewriters are now made only in small quantities. Production has been overwhelmingly switched to armaments. Parts are made under cover names. Only a few employees know what the end product looks like or what it will be used for.

The same was true of Richard Gäbel & Co., a smaller company with factories in the Pirnaische Strasse and the Caspar-David-Friedrich-Strasse. It had been founded almost sixty years earlier to produce machines for making waffles and marzipan, and later packaging for the candy industry. However, in March 1944, the company's own report to the regional Rüstungskommando (Armaments Command) in Dresden stated that 96 percent of its production was for the Wehrmacht High Command, including torpedo parts for the navy. There reigned a similarly strict regime of secrecy, as at Seidel and

Naumann. Wartime instructions included an order to employees that they must, when referring to items made at the factory, at all times use codes ("list attached") to identify them ("That is, not cartridge cases but KG31/630").

Another larger firm was the J. C. Müller Universelle-Werke in the Zwickauer Strasse. Founded in 1898 as a maker of cigarette-making and -packaging machines, Universelle was bought during the First World War by J. C. Müller, who rapidly diversified into making shells for the army. With the peace, the factory returned to its original business. After 1933, when Germany began to rearm, new assembly sheds were added and the firm took on armaments contracts again (including, in 1936–37, one producing aircraft parts for the Spanish Nationalists). By 1944 Universelle was employing four thousand workers, many of them foreigners from the occupied countries. In the latter half of 1944 there arrived seven hundred women, many Jews, from Ravensbrück concentration camp. Everything the company now made was for the war, including machine guns, searchlights, aircraft parts, directional guidance equipment, torpedo tails, and much more. In that year J. C. Müller was able to chalk up a record turnover of around 40 million Reich marks.

Ilana Turner, one of several hundred women concentration camp prisoners brought to Dresden from Liztmannstadt (Lodz) in Poland via Auschwitz in October 1944, tells of filling bullets in a former cigarette factory on a machine adapted from making cigarettes. The factory was known by the harmless name of Bernsdorf & Co. She and her coprisoners were working for the Deutsche Waffen- und Munitionswerke (German Weapon and Munitions Works), part of the complex of companies owned by the fabulously rich Quandt family.

Born in Lodz to a middle-class Jewish family, Ilana had been set to go to high school when the Germans invaded Poland. Soon the family had to move from their comfortable home to the ghetto; 230,000 Jews were crammed into this area before it was sealed. In 1942 the schools were closed and all the children over ten put to work. "Otherwise you were sent away. We didn't know where to, but not to a good place."

So for more than two years, from the age of thirteen to fifteen, Ilana worked in a military uniform factory making hats. Sometime a few months before the dissolution of the Lodz ghetto in the early summer of 1944, the authorities decided she was capable of handling heavier work. She was set to making military saddles, which was indeed demanding and hard. Then came the evacuation of the ghetto, which by 1944 had been emptied of the sick, the young and the old and had become almost entirely a center for forced labor.

Ilana Turner went from Lodz to Auschwitz, but after a few days was transferred along with her fellow workers to Stutthof, a small but notorious camp on the Baltic coast near Danzig (Gdansk). It was not an extermination camp, but conditions were appalling, the German guards brutal, and sickness prevalent. Luckily for the girl and her comrades, the Germans now desperately needed workers. Even Jews.

It was sometime in the second week of October 1944 when they were put into boxcars and transferred by train from Stutthof to Dresden. The beauty of the city's streets was stunning. They were marched through the streets from the station to the factory in their rags and their shaved heads, but still they looked around in wonder. A pedestrian bridge connected the factory with the big Zeiss-Ikon plant on the Schandauer Strasse.

> Work was done on the first floor, in the cellar, and in the two floors above. We slept on the third floor. So in this we were lucky. We had bunks. Three-story bunks. But it was warm, and that was the main thing. It was a very cold winter. And I had a rag for a dress. Nothing under it and nothing over it. In Auschwitz they took away everything. We remained without anything and without hair because they shaved our heads. At least it was warm, but we were hungry. And there was no medical treatment if anything happened. I mean, they wanted to help us but they didn't have any way of treating us. So many people died there, but they died of diseases that they contracted in Stutthof. They brought those things with them to Dresden. They were so weak and so hungry . . .

Since the slave laborers also lived under guard there, the factory amounted to a camp. It was nevertheless the least bad place the Jews from Lodz had been. They even encountered kindness, but the work itself was hard—twelve hours a day, seven days a week, making cartridges. It was repetitive, demanding labor. Ilana had to coordinate pressing a foot pedal with pushing each cartridge through a small hole. One thousand per hour.

The list of factories turned over to war production goes on—camera factories, cigarette makers, confectionery manufacturers, furniture, electrical goods. Even arts and crafts centers.

The Deutsche Werkstätte (German Workshops) at Hellerau, in the leafy northern reaches of Dresden, were founded in 1898. They represented an idealistic attempt to make fine German furniture and crafts while at the same time exploiting modern machines, with designers and craftspeople living in an idyllic, purpose-built garden-suburb setting based on the English model.

An artists' colony grew up around the original workshops and dwellings, attracting famous visitors from all over Europe, including Rilke, Kafka, George Bernard Shaw, Upton Sinclair, Diaghilev, Le Corbusier, and Rachmaninoff. A. S. Neill, later founder and headmaster of Summerhill progressive school in England, started his first school at Hellerau. In 1911 a magnificent art school–cum–festival hall was built, providing space for an educational institute as well as for performances of progressive dance and drama. This international bohemian flowering was sadly cut short by the First World War, and though the workshops continued, the artists' colony never really recovered.

By 1944, the Festival Hall was an SS barracks. As for the workshops, these too were pressed into working for the Wehrmacht. As metal became scarce toward the end of the war, craftsmen at Hellerau made wooden tail assemblies for planes, and possibly even wooden parts for V-1 flying bombs and V-2 rockets.

In the Albertstadt industrial area, the war years also saw a huge expansion by a company that had formerly represented the brave new world of communications and entertainment. Radio-Mende, founded by Otto Hermann Mende in 1923, had begun with modest manufacturing premises in the former ordnance shops, but its embrace of new technology (in this case the radio receiver) brought rapid growth. Soon Mende's radios were being sold all over Germany.

Mende was a right-wing nationalist who welcomed the Nazi regime. By the late 1930s, after a disappointing involvement with the manufacture and sale of the Nazi-encouraged "people's receiver" (*Volksempfänger*, colloquially known as *die Goebbelsschnauze*—the "Goebbels Gob"), the company decided to ride the wave of rearmament and began to exploit its in-house expertise (and underemployed workforce) in pursuit of Wehrmacht contracts.

The move was successful. By July 1943 Radio-Mende was employing about twenty-five hundred workers, more than half of them women. The company, like most in Germany, was crying out for labor. During the next year and a half, more women were recruited from forced labor sources—Russians and Poles, and from the concentration camps of Flossenbürg (three hundred women) and Bergen-Belsen (six hundred women). They manufactured field telephones, radios that fit into knapsacks, and two-way radios. They had also expanded into other forms of communications equipment for the Wehrmacht, including teleprinters, artillery observation devices, and—in the tens of thousands—electrical fuses for the Luftwaffe.

In the same Albertstadt industrial area, the other larger firms, including the box and packaging firm AG für Cartonnagenindustrie; the Infesto-Works, which made steering elements for torpedoes, aircraft, and U-boats; Gläserkarrosserie GmbH Works III (parts for Messerschmitt planes); and many of the others, were according to the Albertstadt industrial area's chronicler, almost all working for the armed forces in some way or another. There was a navy-approved testing station for specialized turbines, which were also produced there by Brückner, Kanis & Co.

The largest employer in Dresden by far, though, was Zeiss-Ikon. And it was a long time since that distinguished company had produced anything as innocent as a snapshot camera.

NO. 2 SPORERGASSE, in the Dresden Altstadt. The date is November 23, 1942. (We know this because this is a scene from a film, in which titles have been inserted to explain the date and context of the events.) In front of the house stand overfilled, unemptied garbage bins. This is one of Dresden's "Jew houses" (*Judenhäuser*). Young men with "Jew stars" on their clothing are carrying household objects out of the house, while uninvolved passersby hurry past on the pavement. The same scene plays out in front of another house, the Jewish old people's home at 24 Güntzerstrasse. Individual pieces of luggage come repeatedly into the shot, with the names of their Jewish owners clearly legible.

The next sequence of pictures is introduced with the title: "City Disinfecting Institution."

The delousing of these respectable middle class citizens is a deliberate humiliation, especially for the women. The camera dwells implacably on them as white-coated, dutiful city employees pick through their hair for vermin. Some of the women drop their gaze, embarrassed by the camera's presence, others stare with an edge of defiance.

Four men watch from the yard of the disinfecting institution. Two are uniformed Dresden Gestapo officers, one a police officer in civilian clothes, and the fourth is a civilian, in well-tailored coat and Homburg hat, smoking a cigar. The Gestapo men are grinning and chatting in animated fashion. One is SS-Scharführer Martin Petri, the other SS-Untersturmführer Henry Schmidt, head of the Jewish Department in Dresden, the notorious Judenreferat. Schmidt, hands clasped behind his greatcoated back in an attitude of relaxed command, is almost laughing as he addresses a remark to the plainclothes policeman. Between them is the enigmatic, cigar-smoking civilian, Dr. Johannes Hasdenteufel, a powerful executive of Zeiss-Ikon AG, prewar giants of the camera and lens industry and Dresden's largest single private employer.

The scenes at the disinfecting institution conclude. The Jews fetch their overcoats and suitcases, which have been separately handled. Then, in the rain and temperatures only just above freezing, they must set off on the five-kilometer walk to their destination. This is the *Judenlager* (Jew camp) Hellerberg, on the northern outskirts of the city, near the old aerodrome. In the barrack huts there they will be housed until the Gestapo decides what to do with them. They pause and stare into the camera—obviously by order—as, protected by a scattering of umbrellas, some in rain capes foraged from luggage, they trudge through the rain toward the entrance to the camp. The camp lies in a sandy depression bordering the St. Pauli cemetery and not far from the entrance to Weinbergstrasse. At the camp they are "greeted" by the Gestapo men, who have traveled there by official car.

Some 293 of the surviving Jews in Dresden were moved into the camp on the Hellerberg on November 23–24, 1942. The film, shot by an employee of Zeiss-Ikon for unknown reasons and discovered after his death in the early 1990s, provides a record of how one group were "processed."

The Jews concerned were still alive, for the moment, because they already had jobs in the armaments industry. At that stage this still protected them from following the other Dresden Jews to the east, to where "transports" had been dispatched since January of that year. The Hellerberg Jews did not know that Gauleiter Mutschmann, ever keen to preempt Berlin in the pursuit of an anti-Semitic plus-point, had vainly tried to have even the so-called *Rüstungsjuden*—armaments Jews—from Dresden sent to the concentration camps with the others. But Mutschmann had suffered a temporary defeat in this particular bureaucratic battle.

Victor Klemperer, saved by his marriage to an Aryan, wrote in his diary about the fate of his friends and neighbors:

> It is quite deplorable that this imprisonment is already considered to be halfway good fortune. It is not Poland; it is not a concentration camp! One does not quite eat one's fill, but one does not starve. One has not yet been beaten. Etc. etc.

The "lucky" Hellerberg Jews would stay that way for less than four months. On March 3, 1943, all the Jews from the Hellerberg were marched to the goods station in Dresden-Neustadt and loaded into goods trucks. Their destination was Auschwitz-Birkenau. Most were murdered soon after their arrival.

But in the meantime, they were to continue to work for Zeiss-Ikon AG, whose cigar-smoking representative, Dr. Hasdenteufel, had stood shoulder-to-shoulder with the Gestapo and watched the entire procedure, from the disinfecting institution to their secure arrival in the *Judenlager*.

The Jews lived in six out of seven huts, the seventh a communal building with a dining hall and washrooms, the others consisting of three rooms, each containing sixteen people. Couples were together, singles divided according to sex. Children from the age of four lived separately, again according to sex.

There was no wire around the camp, though a guard watched over the camp entrance. These guards were supplied by one of the local private security companies, the cost being charged along with the inmates' rent. Relations with the guards were described as quite friendly. The inmates would leave at regular shift times without formal inspection, though to leave the camp at other times they needed special papers. "We could have put up quite well with life in the camp," wrote one of the few inmates who survived the war. "I wish they would have left us there . . . Everyone would still be alive."

The Hellerberg camp also had the advantage of being just outside the city limits, thus satisfying the objectives of the Nazi authorities (and Goebbels particularly) of making Germany's cities "Jew-free." The camp inmates were spared the five-mile march from the "Jew houses" in the center of the city, for their place of employment, the Goehle-Works, now lay less than a quarter of that distance away:

> After a twenty-minute journey on foot, they would reach the factory . . . The company specialized in the development and production of optical and precision instruments. Since 1940 the Dresden plant had mostly assembled timed fuses with built-in precision delays. These were a standard part of the torpedo armament in German submarines. A second product, equally important for the war effort, was the development of bomb-aiming apparatuses for the Luftwaffe. Dresden's Jews were engaged in both lines of production, for the most part carrying out the precision work of assembling the equipment. The atmosphere in which they worked was therefore very quiet and concentrated . . . The charge hands and foremen were usually only interested in turning out a high quantity of items per shift, and in keeping the numbers of rejected products low. The work was therefore physically light, but because of the necessary concentration, extremely exhausting.

What the inmates did not know was that in Berlin, others were already considering whether even this limited, institutionalized existence would be allowed to continue. Should Jews be permitted to survive because of their use to the war effort? The Jews from the Hellerberg camp had been saved from Mutschmann's private vendetta the previous year, but now Berlin was catching up with the Saxon Gauleiter's prescient malice. Goebbels, also a violent anti-Semite, knew that the balance was swinging against the *Rüstungsjuden* (armaments Jews). He insisted that the Jews were not "indispensable"— adding that the Führer was also adamant on this point. Even before the Dresden Jews were transferred to the Hellerberg camp, a lethal struggle was being carried on within the Nazi military-industrial complex between those who wished to continue to use the Jews for slave labor and those who espoused simple extermination.

The debate continued through the winter and into the New Year. As Goebbels had predicted, with the Führer's support, the ideologists were winning their struggle against the pragmatists. Finally, on February 20, 1943, the RSHA (Reich Main Security Office) issued new guidelines for the "technical accomplishment of the evacuation of Jews to the East (Concentration Camp Auschwitz)." The guidelines were the same as before, except for one small change. Jews involved in war industry were no longer exempt.

A "factory action" was planned for a week later. In the early hours of February 27, 1943, before they were due to go to work, all inmates of the Hellerberg camp were placed under arrest. The camp was declared a "police detention camp," surrounded with barbed wire, and police guards placed on the perimeter. The remaining leaders of the Jewish community, who had been permitted to remain living in the city, were arrested and brought to the Hellerberg, to be followed in the next few days by Jews from Erfurt, Halle, Leipzig, Plauen, and Chemnitz. This was now a general holding camp for Saxon Jews, and for those few last days, filled to bursting.

On March 3 the "deportation" began.

Nearly three hundred Dresden Jews were loaded into boxcars in Dresden-Neustadt that day. The destination was Auschwitz. Eight survived the war. The great majority of the rest are marked as "disappeared" (their names were never entered in the camp register, which usually indicates they were gassed immediately after arrival.) A few are traceable as "died" or "murdered" in the camp itself.

Zeiss-Ikon continued to produce for the war effort, using "eastern workers" or foreign labor from elsewhere in the Nazi empire to fill the gaps that could not be plugged with that precious, ever-dwindling commodity: Aryan German employees.

AFTER MARCH 1943 the only full Jews left in Dresden were those married to Aryans, like Victor Klemperer, or the children of mixed marriages, like Henny Wolf. They knew each other slightly—though they were very different in personality and background. Henny Wolf had been working at Zeiss-Ikon's Goehre-Works since July 1941, when, at sixteen, she was barred from further education and contracted to forced labor.

While for the Hellerberg Jews, during their brief stay at the camp, it was a twenty-minute walk, for Fräulein Wolf it was still four miles on foot from the family's flat in the southeast of the city, near the Grosser Garten. She did twelve-hour shifts, working on the same timed fuses and clocks for U-boats that many of the Hellerberg Jews were assembling. They had one small stroke of good luck: the Aryan foreman was a "fine human being, who took no notice of the fact that we were Jews." All the same, this was strict piece-work with a magnifying glass and tweezers, hour after hour under artificial light, which in turn was limited because of wartime power-saving regulations. It caused a deterioration in Henny Wolf's eyesight from which she never fully recovered.

After the Hellerberg's inmates were deported, the camp was closed, and with it the "Jewish department" at Zeiss-Ikon. Henny Wolf, many years later, can still recall the anxiety of those left alive—all married to "Aryans" or *Mischlinge*—until they were assigned to other, smaller companies in Dresden, apparently not associated with the armament industry. She thought that the "Jewish department" had been shut down because it wasn't worth the effort with most of the Jews gone, but the real reason was more sinister. The authorities planned for the moment to stick to the letter of the law and allow married Jews and the *Mischling* children of such liaisons to live. However, should it also be decided to "deport" such people at a future date, the action would be easier and attract less attention if the victims were dispersed and working in selected low-profile concerns not associated with the war effort.

The spring of 1943 saw Henny Wolf begin work at the Adolf Bauer cardboard packaging company, which lay just east of the Altstadt. It was a middling-sized enterprise, owned and managed by Herr Bauer, a man rumored to be a Nazi Party member, but who nonetheless behaved decently to his Jewish employees.

Henny Wolf stayed at Bauer's for most of the rest of the war. Not that Jews were completely safe there. On several occasions the Gestapo snatched colleagues, who were never seen again, or sent friends of hers to punishment assignments at factories known to be much harder than the box factory. The few remaining Jews in Dresden were subject to the whims of Henry Schmidt and his henchmen at the Judenreferat, and they knew it. "The fear was worse than anything," wrote Henny Wolf, "worse than the hunger or even having to wear the star . . ."

There were occasional kindnesses. The young woman who slipped her some ration coupons, the occasional passerby who would whisper, "Chin up!" or the butcher who would hide some extra meat in with the meager allowance permitted on a Jewish "J" ration card. Then again, there were the youngsters who taunted her as a Jew, the men who followed her home on the long trek back from the factory after the nightshift, the young fanatic who picked her up when she fell off her bicycle, then saw her yellow star and let her drop bruisingly to the cobbled ground once more.

Most Dresdeners, though, just looked away or through her, like the nervous passersby on the Sporergasse, caught on film watching the young Jews loading the furniture onto the trucks headed for the Hellerberg camp.

BETWEEN 1940 AND 1945 more than thirteen hundred human beings were killed within the confines of a building in Dresden's comfortable southern suburbs (known as the Südvorstadt). The building, the Justizgebäude, housed the central courts and central remand prison for the whole of Saxony. At the time it was built, in 1907, it was considered an advanced, model institution—with the offices of the clerks and prosecutors, the courtrooms, even the cells where the accused were held awaiting trial, seen as spacious, light, and airy. The facilities, even down to a prison library and the visiting rooms, which separated prisoners from visitors by a wide table rather than a set of bars, were absolutely modern.

Despite its potentially solemn, even grim purpose, the Justice Building, which vaguely resembled the tastefully fortified residence of a middling-ranked princely family, was reckoned by a contemporary critic to be "not wholly foreign to a sense of benevolent humanity." The prison part of the complex was hidden away around the back behind a fifteen-foot wall. This was not just because its originators and architects intended the appearance of the Justice Building to represent the new, modern reforming ideals of that still-hopeful time, though it was a factor. The complex also needed to fit in with the surrounding suburbs, which were inhabited by the kind of educated, respectable middle-class folk who frankly would be reluctant to live next door to a prison—especially if it looked like the traditional idea of one. Under

Hitler, however, the prison wing became a place where draconian punishments were meted out.

There was a "death row" in the prison. These were the small cells where condemned prisoners spent their final days, each with a little bench and desk. There was the nearby guardroom to which they would be led and kept while the guillotine was prepared, and then the courtyard where the advanced instrument of death was set up. The rate of executions reached its apogee after the annexation of Bohemia and Moravia, the so-called protectorate that Hitler hacked out of the rump of Czechoslovakia. Eight hundred Czech dissidents and resistance workers were executed here with the aid of the new electrically operated guillotine. Some Poles, mainly from the Posen (Poznan) province, which had been annexed in 1940, were also brought here to be killed. German victims included members of resistance groups, Social Democrats, and Communists, among them former members of the Reichstag, and individuals such as the idealistic doctor Margarete Blank, denounced in 1944 by a patient for making "demoralizing" remarks.

Dr. Blank died under the blade early on the morning of February 8, 1945, a few days before bombs brought salvation to some of her fellow prisoners.

AT THE END OF 1940 Dr. Walter Schmidt, president of the Directorate of the German State Railways in Dresden, described the rail industry as "at the core of its being, so closely connected to the Wehrmacht." The railway's slogan was "Wheels Roll for Victory" (*Räder Rollen für den Sieg*). The entire development of the German railways in the middle of the nineteenth century had been designed to facilitate military mobilization and to move troops and equipment rapidly to meet the enemy threat at whichever front, wherever it might arise.

In this system, Dresden was not only one of the largest regional directorates but a key junction, through which ran both the north-south and east-west axes of the German railways.

After the German-speaking Sudetenland was ceded to Germany by Czechoslovakia in October 1938, its railways were absorbed into the Reich's railway system. In the summer of 1939, just before the outbreak of the Second World War, the Dresden directorate had a payroll of eighty-eight thousand employees and controlled movements of rolling stock over more than three thousand miles of track. By the end of 1943 the Dresden directorate employed a total of 128,000 workers of all kinds.

Almost all the various stations and yards in Dresden were within a few hundred meters of the river Elbe. The magnificent main station, or Hauptbahnhof, was completed in 1898 and ambitiously refurbished and extended in the mid-1930s. It lay on the southern edge of the city center, the long nineteenth-century shopping street of the Prager Strasse linking it with the Altstadt to the northeast.

Adjacent to the Hauptbahnhof on the Wiener Strasse, another splendid building from the time of the kaiser housed the railway directorate itself. From here the traffic between eastern Saxony, parts of the Sudetenland and Bohemia (now the Czech Republic), and on east into Silesia and Poland, was managed, regulated and timetabled. On the left hand, as the tracks ran northwest toward the Elbe, lay the directorate's huge repair sheds. Then came the vital Friedrichstadt marshaling yards, which served an old-established industrial area and also acted as transfer facility for the barge and river traffic that moved in and out of the nearby Alberthafen (Albert Harbor). From there the tracks ran across the river to Dresden-Neustadt, passing first the goods station (where the Hellerberg Jews had begun their journey to Auschwitz) and continuing to the main Neustadt station, at which most passenger trains stopped on their way to and from Dresden Hauptbahnhof.

This was the heart of the railway system in Dresden. It was important to the city, to the region, and to the war in the east.

For the Polish campaign in August-September 1939, the Dresden directorate laid on fifteen thousand extra trains. This meant drastic reductions in normal passenger and commercial usage, but the traffic for the first battles of the war ran for the most part smoothly and on time.

By the last year of the war, with Dresden's industries running at full capacity the railways had become a key factor in the city's importance. Over the past few years, special tracks and platforms had been installed to expedite supplies to and from the major armaments and war-related factories in the city. Germany's—and Dresden's—war-industrial production tripled between 1940 and 1944. The Dresden Chamber for Industry and Trade declared at the end of 1941, "the work rhythm of Dresden is determined by the needs of our army."

The most serious problem the railways faced in 1943–44 was that a great many younger members of the trained staff were conscripted to the Wehrmacht. Remaining skilled railwaymen were often moved around to keep trains running in other parts of the Nazi empire, or to help repair the increasing numbers of railway installations and tracks destroyed or damaged by Allied air raids. Much of the basic work was now done by women, foreign labor, or prisoners of war, who were extensively used in track repair gangs and for loading and unloading freight.

By the end of 1943 about 12,500 foreign workers lived in camps in and around Dresden, working exclusively for the railway directorate. Their managers were exhorted to "get as much out of them as possible," though they were to be "correctly and humanely treated." The same stipulation was less emphasized in the rules regarding prisoners of war. It was missing altogether from the conditions of the five hundred men from the concentration camp at Flossenbürg, near Weiden in northeastern Bavaria, who were employed as forced labor at the repair shops from September 1944 onward. Even the railway's established German employees worked a sixty-hour week, with no increase in wages since 1939.

The deportations of Dresden's Jews began on the morning of January 21, 1942, with the freighting of 224 human beings into unheated boxcars. They traveled as third-class single-journey passengers, with a group discount. Only the guards had return tickets. All journeys to the concentration camps were classified as "special trains" because of the group bookings, but were otherwise treated as normal passenger journeys.

And it was not only Jews from Dresden who passed through the station. This was an important junction. A great many deportation trains, heading for the extermination camps at Belzec or Auschwitz, and to the ghetto at Theresienstadt, passed through Dresden:

> The Dresden Department . . . was therefore responsible not just for the transports to Theresienstadt but also for the journeys from there to the extermination camps. So on January 16, 1943, in Dresden they received the "round-trip plan for multiple utilizable goods trains, round-trip 125," originated in the General Management Authority (East). Between January 20 and February 2, 1943, a train

would shuttle five times between Theresienstadt and Auschwitz. "Number of passengers: two thousand each trip."

If there were technical problems, such trains could remain at Dresden and changes might be necessary. A prisoner en route from Dachau, near Munich, to Auschwitz, described an enforced stop and change of trains at Dresden:

> We were led over busy platforms. In our prisoners' clothing we are conspicuous to everyone. I look at the faces of passersby, try to read their thoughts. I find no sympathetic gaze. They look at us as if we are war criminals. Are all Germans on the side of the SS?

With the conquest of vast areas of Eastern Europe and Russia after 1941, Dresden became part of a huge, overstretched rail network.

The importance of Dresden as a transit point for military traffic can be seen from the figures for October 1944, when the western Allies' advance from Normandy was starting to slow down, but the fronts in the east and southeast were coming perilously close, and east-west movements of forces were heavy. A total of twenty-eight military trains, altogether carrying almost twenty thousand officers and men, were in transit through Dresden-Neustadt *each day*.

There is no reason to believe that three months later, in the first weeks of 1945, with Russian offensives on the Oder and in the region of Budapest, and the Ardennes offensive against the Anglo-Americans going into reverse, the frantic yo-yoing movement between eastern and western fronts would have decreased substantially from the level in October of the previous year. That was part of the Allies' calculation when they started to place Dresden in their sights. After the war, an American former prisoner of war wrote:

> The night before the RAF/USAAF raids on February 13.14, we were shunted into the Dresden marshaling yard, where for nearly twelve hours German troops and equipment rolled into and out of Dresden. I saw with my own eyes that Dresden was an armed camp: thousands of German troops, tanks and artillery and miles of freight cars loaded with supplies supporting and transporting German logistics towards the East to meet the Russians.

Both the railway directorate and the Dresden authorities were also aware that the Hauptbahnhof and its surrounding area, to the south and northwest lined with industrial sites and warehousing, would be obvious targets for any Anglo-American bombing attack on the city. One railway official noted that "by their location and size they represented a good target." The target maps in the possession of the Allies showed this configuration clearly.

There was toward the end of the war some consideration by the local authorities of how to provide shelter of a kind for any military personnel or civilians using the station in case of air raid. Unfortunately, as generally happened in such matters in Dresden (unless dedicated to the protection of the authorities), the preparations that were actually made lacked either thoroughness or urgency. At the end of 1944 an inspection of the air raid shelter arrangements at the Dresden stations, including the Hauptbahnhof, showed serious failings. At the Hauptbahnhof, they had mostly converted existing underground storage cellars.

> The air raid shelters of the Hauptbahnhof could hold two thousand persons. The shelters lacked . . . gas filters or ventilation . . . the anti–air raid precautions in the surrounding areas were also extremely inadequate: the antishrapnel trenches had been only partially constructed by February 1945, and there was a lack of properly constructed air raid bunkers in the neighbourhood.

The teenage Götz Bergander was downright dismissive:

> By the last winter of the war, there existed no more possibility of the construction of effective anti-air raid precautions. So as to be seen to do something, antishrapnel trenches were dug in various places. In the Bismarckplatz by the Hauptbahnhof, for instance, there was this zigzag arrangement, a pathetic refuge, just in case the basement rooms beneath the station should be overfilled. But not even those existing underground spaces were secure against bombs. The corridors were only partly reinforced, there was no ventilation system, and no emergency exits in this air raid shelter for 2,000 people.

His view of the authorities in Dresden was even more withering:

Only a few hundred or thousand privileged individuals belonging to the party, the organs of the state, the Wehrmacht, SS, or police and city administration could take a relaxed attitude towards air raids so far as their personal safety was concerned: in the deep cellar of the New Town Hall, in the command post under the Albertinum, and in the escape bunkers cut deep into the cliffs of the Lockwitzgrund for the *Gau* leadership, the SS leadership at the Mordgrundbrücke, or in Mutschmann's garden bunker.

All this time, the eastern front was moving relentlessly closer. Although Dresden kept the deceptive appearance of peace beneath the surface, it was working harder than ever for the victory the Führer had promised in those sunlit days of parades and jubilation, back in 1940–41.

But now the eyes of the enemy were finally fixed upon the city.

PART TWO

TOTAL WAR

14

Ardennes and After

AS CHRISTMAS 1944 turned to New Year 1945, only a fool or a fanatic could have believed that Germany would now emerge triumphant from this war. The imponderable factor was, how long would it take for the Reich to be defeated? And at what continuing cost to the Allied forces and the still-captive peoples of Europe? These were the hard questions that planners in the British and American camps were asking themselves during these weeks. And their mood was in many ways far from optimistic.

After the "breakout" from the Normandy beachhead in late July 1944, quickly followed by Anglo-American landings in the South of France, the western Allies had appeared to sweep all before them. They took Paris on August 24 and Brussels on September 3, pushing on toward the Dutch and German borders, in a blitzkrieg comparable to the German triumph of 1940. Then came the disastrous failure of British forces to seize the Rhine crossings in a bold combined land and airborne operation known as Market Garden. The Allies had advanced too quickly for their supply lines. The only major French port in their hands, Cherbourg, provided too limited a conduit for the food and supplies needed by an expeditionary force now totaling around two million men.

The vital Belgian port of Antwerp had been captured on September 4. However, it took almost three months in total to clear the nearby Scheldt estuary of enemy forces and the harbor approaches of German-laid mines. Meanwhile, every day the German V-1 launch sites in Holland sent flying bombs not just against London and southern England but against Antwerp. More than six thousand of Antwerp's citizens were to die under this rain of V-1s and then V-2s—twice the number of Londoners killed by these German "wonder weapons"—as the western Allies struggled to exploit fully their possession of the Belgian port. The Anglo-American advance ground to an almost complete halt short of the westernmost German city, Aachen, which was taken only after costly house-to-house fighting amid the ruins of the historic city.

Meanwhile, with their supply lines much shorter than the Allies', the Wehrmacht's forces in the west were able to regroup. They had their backs to the natural barrier of the Rhine and the man-made defenses of the German fortified line known as the Westwall, and they now started to put up a much fiercer resistance. The Hürtgenwald, southeast of Aachen, formed a dense, twenty-square-kilometer triangle of forest between the west Rhenish towns of Eschweiler and Düren and the village of Schmidt. This area—impassable for armored vehicles—proved nightmarishly difficult for General Courtney H. Hodges's U.S. Seventh Corps to wrest from its stubborn, skilled German defenders, and until it was taken, no secure progress could be made across the Rur River (not to be confused with the Ruhr, farther east) and toward the western bank of the Rhine. This small area would cost the Americans ninety days of close-contact fighting between mid-September and mid-December 1944, resulting in twenty-four thousand men killed, wounded, or captured.

In the eastern French province of Lorraine, Patton's army's advance was snaillike. The flamboyant hero of Sicily and Avranches took Metz at the end of November at the cost of 2,190 soldiers' lives. In three months, his Third Army had advanced a little over twenty miles, and was still some miles short of the Westwall. Since landing in Normandy Patton had lost forty-seven thousand men.

In the German homeland, the screw of total war was, in the meantime, being turned even tighter. From October 1944 the draft age had been lowered to sixteen and raised to fifty, putting an additional three-quarters of a million male Germans under arms. The new Home Guard, the Volkssturm, was sworn in, boys and old men tossed together and paraded in town squares all over Germany. Short of weapons and uniforms alike, they were a symbol of insane defiance, of the Nazi elite's determination to resist to the end, whatever the cost to the German people.

Any hopes that Hitler would concentrate his defensive efforts against the hated Russian enemy in the east, or that the Wehrmacht would somehow "prefer" to give in to the Anglo-Americans, had been shown to be vain. Moreover, the Führer was planning a great counteroffensive—not against the Soviets, who had now paused just east of Warsaw, but against the Americans in the Ardennes.

In the early hours of December 16, 1944, Operation Autumn Mist was launched. A massive force of two hundred thousand German troops, with six hundred tanks, attacked the eighty thousand Americans holding this hilly part of eastern Belgium and northern Luxembourg. The object was to smash their way through and recapture Antwerp, some 120 miles distant.

Despite having long since lost control of the air, the Germans were protected from Allied aircraft by a spell of cloudy, overcast weather—a situation that was to persist for just over a week. The northern part of the offensive, led by Sepp Dietrich's SS Panzer Division, made slow progress, but Field Marshal Mantueffel's Fifth Panzer Division managed to advance sixty-five miles and encircle the transport center of Bastogne, just south of the river Meuse. Here, after demanding the American forces' surrender, the Germans received from the commander of the 101st Airborne Division, General Anthony C. McAuliffe, the famous one-word refusal: "Nuts!"

Bastogne was about as far as the Germans got. Already short of fuel for their tanks and other vehicles (they had been told to supply themselves from captured Allied fuel dumps), and forced to advance along narrow, twisting roads over vulnerable bridges, the Germans found their offensive grinding to a halt. On December 24, the weather lifted and thousands of Allied aircraft took to the skies. German supply routes, airfields, and forces were now subjected to all-day bombing and strafing. Men and materiel could be moved only at night along frozen roads. On December 26 Patton broke through and relieved the American First Airborne Division in Bastogne. The game was over. Hitler's desperate last gamble had failed.

All the same, the Allies—and public opinion back in Britain and America and the Dominions—had been given a serious shock. Part of the Allied front had caved in. The Germans had fought fiercely for every inch of the contested ground. The V-1s and V-2s continued to fall on Antwerp and London. There might be no chance of further German advances, but the fighting in the Ardennes was to drag on for almost six weeks. The Americans suffered eighty-one thousand casualties of all kinds, including a shocking nineteen thousand killed (most in the first three days after the German surprise attack), and the British fourteen hundred (two hundred dead). German casualties totalled almost one hundred thousand, and these were men Hitler definitely could not afford to lose. The Ardennes offensive would be reckoned a catastrophe in the longer term for Germany, but in the meantime morale had been bolstered and the invincibility of the western Allies cast into question.

This was the difficult, ambivalent, and fast-changing situation in which the purpose of the Anglo-American bombing campaign was being reconsidered during the particularly hard winter of 1944–45. One thing was certain: Anyone bold enough to say that the war was all but over would have received pretty short shrift from soldiers and public alike.

THE SPRING OF 1944 had seen Bomber Command's rapidly growing strength given over mostly to preparations for the Allied landings on the continent planned for late May or early June. French factories, railways, and bridges had been bombed relentlessly in a precision campaign that contrasted with the "area bombing" visited on German towns and cities during the previous two years.

Understandably, there was reluctance to cause more casualties among the French population than strictly necessary, and in most cases the targets were quite specific, selected to prevent the Germans from deploying their forces efficiently once the Anglo-American invasion started.

The bombing in support of the D-Day landings was successful, though, despite all the preinvasion qualms, quite costly in terms of French dead. The Allied air forces continued in support of the ground troops as they first consolidated their position in Normandy and then engaged in vicious, costly fighting to expand it. Only after the "breakout" at the end of July (in which precision bombing again played a key role) could the highest echelons of Allied military power turn to consideration of what they might now do with Bomber Command and the USAAF.

In the case of Bomber Command, there ensued a drawn-out struggle between Sir Charles Portal, the RAF chief of staff, and his longtime subordinate, Sir Arthur Harris. It was to prove crucial for the fate of Dresden. Harris had put his force at the behest of Supreme Headquarters Allied Expeditionary Force in the spring of 1944 with reluctance, but there can be no doubt that Bomber Command did its job. The problem came in the early autumn when Harris wished to return to what he had seen since 1942 as his chief mission in life: destroying Germany's cities.

On December 7, 1943, three days after the British incendiary attack on Leipzig, Sir Arthur Harris had submitted a lengthy analysis to the British Air Ministry, assessing his view of the destruction caused so far in Germany. He wanted a great expansion of the Lancaster bomber fleet, improved countermeasures, and radio aids. If he got those, he argued, Bomber Command could force Germany to surrender by April 1, 1944.

This plea was, for the moment, Harris's last attempt to pursue the "bomber dream." It soon became clear that nothing of the kind would be granted him, and that Bomber Command's role in the New Year would be subordinate to the needs of Overlord, the cross-Channel invasion of Europe. Harris, since the end of December 1943 in discussions with Air Chief Marshal Leigh-Mallory, commander of AEAF (Allied Expeditionary Air Force), knuckled under. However, he also warned in his inimitable blunt style that if his forces were made to concentrate on tactical support for too long, the German homeland, which Bomber Command had been battering so assiduously for the past two years—with, he considered, increasingly devastating results—would be given an invaluable respite:

> The effects of strategic bombing are cumulative. The more that productive resources are put out of action, the harder it is to maintain output in those that survive. It is easy to forget, however, that the process of rehabilitation if the offensive stops or weakens is similarly cumulative. To put it shortly, the bomber offensive is sound policy only if the rate of *destruction* is greater than the rate of *repair*. It is hard to estimate the extent to which Germany could recoup industrially in say a six months' break in bombing.

Harris would also have been grimly aware of the role of Germany's fighter force and its capacity to regroup and recover if given the opportunity. His last great project before he placed Bomber Command directly at

SHAEF's disposal—and probably his greatest single failure—was the sustained campaign against Berlin in the winter of 1943–44. Much had been expected, perhaps even the dreamed-of "knockout" blow. But the distant capital of the Reich, vast, spread-out, with few clear topographical distinguishing features (and many of those easily concealable through camouflage measures), subject to extreme cold and filthy weather conditions, had proved an extraordinarily tough nut to crack. Berlin's flak defenses became legendary, including the two massive concrete "flak towers" in the Tiergarten, whose lower stories also doubled as secure air raid shelters for thousands of Berliners.

The lengthy flights without fighter escort exposed the British bombers in merciless fashion to the depredations of the well-organized German fighter defenses. Aircrew spoke of missions against "the big city" with mingled pride and apprehension. As spring 1944 approached, Harris abandoned the attacks on Berlin. More than ten thousand Germans had been killed in the city itself. On the British side, 2,690 bomber aircrew died and almost a thousand were captured. Losses over Berlin between August 1943 and March 1944 averaged 5.8 percent, a total of 625 aircraft. In Martin Middlebrook's words, "The Luftwaffe hurt Bomber Command more than Bomber Command hurt Berlin." This painful fact was, perhaps, to influence later decisions about where decisive blows might be delivered against Germany at relatively low cost.

In the autumn of 1944 Harris was eager to return to the "city-busting" fray. On September 30, 1944, Harris wrote to Churchill, who had passed on some ULTRA information about Germany's current prospects, agreeing that "the Boche would fight his damnedest when driven back to his own frontiers." The Germans were clearly, Harris said, attempting to regain the initiative in the air, aided by the respite in attacks on their aircraft factories. Full advantage must now be taken of the vast Allied air superiority to "knock Germany finally flat."

Churchill replied quickly:

> I agreed with your very good letter, except that I do not think you did it all or you can do it all. I recognise however that this is a becoming view for you to take. I am all for cracking everything in now on to Germany that can be spared from the battlefields.

Bomber Command formations began to go back into Germany during the first part of August, on what at first glance have seemed like little more than harassment raids, often using Mosquitoes to attack from a much higher, and therefore safer, altitude than Lancasters. From mid-August, however, there were more serious attacks. Stettin was hit with 461 aircraft on August 16–17, Rüsselsheim on August 25–26 (targeting the Opel auto factory), and the port of Kiel was attacked the next night with nearly five hundred Lancasters, followed by similar assaults on Königsberg in East Prussia (twice)—a very long flight. On September 11–12, 1944, some 262 British aircraft of 5 Group started a firestorm in the middling city of Darmstadt, killing between eight thousand and twelve thousand of its inhabitants—roughly 10 percent of the total population—and proving that Bomber Command did not need massive quantities of planes to wreak ultimate mayhem.

A few days later Harris was formally released from the direct command of SHAEF. If that was what he had been able to do in August–September, what could he do now, free to choose his targets and with the longer winter nights providing a safer operational environment for his bombers?

But there were problems between Harris and his superiors during the period October 1944 to January 1945. They began with a memorandum that Air Chief Marshal Tedder, Eisenhower's British deputy at SHAEF, sent to Portal on October 25, 1944. This criticized the patchwork pattern of Harris's resumed strategic offensive and strongly advised a higher priority for attacks on the German transportation system, including the synthetic oil plants. (The Soviet conquest of the Romanian oil fields had just cost Germany its remaining access to conventional fuel supplies.) A copy was sent to Harris. So was another directive giving oil top priority. Harris's reaction was fierce. He argued about the decisive effects on target selection of weather and tactical factors (including the need to keep the enemy guessing). He repeated his principle that "bombing anything in Germany was better than bombing nothing." While he resented the continuing pressure from the "panacea merchants," he was attacking the Ruhr and the oil plants whenever possible, but remained worried lest the fifteen major German cities that remained unbombed (one of them Dresden) were to be left intact.

To and fro the correspondence went, with Harris's position stiff-ening somewhat as December arrived. The contemptuous word "panacea" cropped up once again in relation to attacks on oil targets (though at the same time he insisted that he was "missing no worth-while opportunity" to attack them).

On January 8 Portal replied wearily that what he really wanted was "*your* determination" in regard to the oil targets. This implied pretty strongly that Harris was merely going through the motions when it came to these operations. Portal's letter evinced a reply from Harris in which the AOC Bomber Command used his ultimate weapon:

> I will not willingly lay myself open to the charge that the lack of suc-cess of a policy which I have declared at the outset—or when it first came to my knowledge—not to contain the seeds of success is, after the event, due to my personal failure in not having really tried. That situation is simply one of heads I lose tails you win, and is an intol-erable situation. I therefore ask you to consider whether it is best for the prosecution of the war and the success of our arms, which alone matters, that I should remain in this situation.

Harris's offer of resignation was refused. Portal backed down. And so Harris was able, for all but the final few weeks of the war, to exercise his judgment about targets. This meant all too often (from the point of view of his critics then and now) that he chose area bombing of cities and towns in preference to aiming for specific industrial and oil instal-lations.

Harris's stubbornness on this issue was not necessarily the whim that it might appear. He had been subjected to constant pressure from his superiors, government advisers, and "ideas" ministries such as the Ministry of Economic Warfare (an organization he clearly loathed), to bomb this or that target, concentrate on this or that perceived enemy weakness, throughout the war. He had not liked the idea of the "dam buster" raids on West German dams, and had gone along with it reluc-tantly. He had not thought the concentrated attacks on ball-bearings factories or air-frame plants or other similar "crucial" enemy facilities would end or necessarily shorten the war, and he may well have been right.

In the case of oil, as it turned out, Harris was wrong. Germany in the winter of 1944–45 was afflicted by a real and grave crisis when it came to fuel supplies. German tanks during the Battle of the Bulge had been forced to capture their own fuel from the enemy to keep advancing. The Reich's young fighter pilots, now being trained by the Luftwaffe to replace the catastrophic losses among the German Air Defense's veterans, were forced to learn mostly through simulators; there was scarcely enough fuel for them to fly aircraft. The result, as far as performance was concerned, was predictable. Even the Luftwaffe's operational aircraft had to be towed into takeoff position by horses or requisitioned oxen in a desperate attempt to save fuel. For all the ingenuity of the Wehrmacht's transport experts, the German war effort— especially in the air, where there was no substitute for gasoline—constantly faced shortages that would literally bring it to a halt. Insiders at British intelligence knew this as a certain fact.

One crucial problem might have been Harris's ultimate ignorance of the source of the information about synthetic oil plants. Harris received some ULTRA information. Surprisingly, he had not been initiated into the ENIGMA secret that lay behind it. Harris was therefore not aware of how directly this information, gleaned from German radio communications by means of code-breaking, came to the Allied General Staff from inside the sinews of the German war machine, how reliable it was, and why. Stubborn as Harris was, had he been fully privy to ENIGMA, he might have taken Portal's pleas more seriously, and the dispute between the two commanders could have led to a different policy outcome.

"The matter is critical," as even one of Harris's fiercest detractors admitted regarding the ENIGMA question, "for an assessment of the grave differences of opinion between them about the advisability or otherwise of a concerted attack on Germany's oil industry during the second half of 1944."

For the previous three years Bomber Command and the Eighth Army Air Force had been the main offensive tool available to the western Allies against Germany. The vast expense involved in maintaining and expanding these forces, producing new aircraft, and training their crews had been acceptable because there was really no alternative. Harris wrote that "the education of a bomber crew was the most expensive in the world; it cost some £10,000 for each man, enough to send ten men to Oxford or Cambridge for three years."

By 1944 the combined Anglo-American air forces were massive. In December of that year Bomber Command had at its disposal 1,513 bombers (it would reach 1,609 by April 1945). The Eighth U.S. Army Air Force now had 1,826 bombers, with hundreds of new aircraft being produced *every month*. There were also now ground forces operating successfully on the continent—partly due to the overwhelming tactical use of Allied air power. During the course of 1944 the balance of power in the air had changed even more dramatically than that on the ground. The sheer quantity of new aircraft and trained aircrew—particularly American—would alone have guaranteed such a shift.

However, just as vital was the development of a long-distance fighter escort aircraft, the Lockheed P-51 Mustang. With its British Merlin engine and disposable, wing-mounted drop tanks, this swift fighter could accompany Allied bombers deep into Germany, to Berlin and beyond, and outperform the German fighters that had hitherto inflicted such a terrible toll on the Allied bomber formations. In the first months of 1944, just after Harris had given up his costly assault on Berlin, the Mustang appeared in substantial numbers. In a short time the P-51s all but wiped the German defensive fighters from the sky. German flak defenses remained, but as the ground forces closed in on the Reich from east and west these guns were to find themselves terribly stretched, and many were to be withdrawn from the air war altogether.

For the first time the Allied bomber fleets found themselves, in terms of both numbers and invincibility, truly comparable to the unopposed instruments of destructive power that Douhet, Trenchard, and the other theorists of unstoppable mass annihilation had envisaged back in the 1920s.

So what was now the main use of the mighty, heavy bomber forces that the Allies had brought into being? How were they to be used to maximum effect in this new and, despite the recent setbacks, evidently final stage of the war?

In this respect, so it happened, the mandarins of British intelligence had an idea.

15
Thunderclap and Yalta

THE JOINT INTELLIGENCE COMMITTEE was established just before the outbreak of war, to coordinate the various streams of information that
came in from different branches of British intelligence and to advise
the chiefs of staff accordingly. The JIC existed, technically, as a sub-
committee of the Chiefs of Staff Committee, of which the prime minister was the chairman. To the JIC's meetings came high-level representatives from MI6, MI5, Naval Intelligence, the Air Ministry, and the
Ministry of Economic Warfare.

The chairman of the JIC from 1939 onward was a Foreign Office
man called Victor William ("Bill") Cavendish-Bentinck, nephew of the
seventh duke of Portland. In the course of the war, after starting out as a
rather marginal body, the JIC, and therefore its chairman, became much
more powerful and influential. The JIC's weekly reports summarized a
view of the current situation and prospects for the war, and became a
major source of information and advice for the chiefs of staff, as did the
expert papers it produced on specific subjects.

By New Year 1945 the Germans' Ardennes offensive had ground
to a bloody halt. On January 12 the big Soviet offensive in the east had
begun, thus reducing the pressure on the hard-pressed western Allies.
In order that this relief should continue, it was advantageous to the
Anglo-American interest to ensure that the Soviets, despite their long
lines of communications, should make good progress against the
German eastern defenses (judging from the stubborn German resistance in the West, the Soviets might need help in this matter).

On January 16 a proposal was made to the JIC, through the
medium of the deputy chief of Air Staff (intelligence), that "a report

should be prepared by the Sub-Committee, assessing the effect on the Germans of heavy air attacks on Berlin in conjunction with the Russian offensive and taking into account the timing of such attacks."

In the early summer of 1944, in support of the Normandy invasion, Bomber Command and the U.S. Eighth Army Air Force had been used as airborne "long-range artillery" to pummel the enemy's communications and troop movements. Now, in order to help Russia (whose own heavy bomber force was relatively meager), the air forces were to be called upon to play a similar role, though at even longer range, on the eastern front. It was a novel suggestion.

The weekly JIC report on "German Strategy and Capacity to Resist" (January 21, 1945) showed a little more clearly what this concern with troop movements inside the shrinking Nazi empire was about. The outcome of the Russian offensive hinged, it said, on "the result of the race between the arrival of German reserves . . . and the loss of the Russian advance owing to logistic difficulties and the distraction of forces on the flanks . . ." Ominously, it predicted that substantial German reinforcements could be involved by the beginning of February, and that by mid-March 1945 such reinforcements might reach a total of forty-two divisions—almost half a million men.

On January 22 the JIC's secretary, King-Salter, made a note for guidance on the work still in progress, asking the joint intelligence staff to draft a report specifically assessing the effect on the Germans of "heavy air attacks on Berlin in conjunction with the Russian offensive." In particular, he wanted (on the JIC's behalf) information about how much of the German administrative machine remained in Berlin, whether it had been dispersed around the suburbs as an antibombing measure, to what extent German industry would suffer from a "devastating" succession of attacks on Berlin, and, last but not least, "what morale effect in Germany as a whole it is thought that a catastrophic 'flattening' of Berlin would have." The use of language indicates that what was being considered at this time was a "knockout"-type attack, of a kind discussed in the past, now revived under the very different circumstances of early 1945. Pointedly attached to the secretary's note is an extract from an Air Staff paper of July 22, 1944, when the idea for a massive attack on Berlin, code-named Thunderclap had been under intensive discussion.

The Thunderclap idea floated in the summer of 1944 had envisaged "220,000 casualties. 50 percent of these (or 110,000) may expect to be killed. It is suggested that such an attack resulting in so many deaths, the great majority of which will be key personnel, cannot help but have a shattering effect on political and civilian morale all over Germany . . ."

So the chief of Air Staff, Air Marshal Portal, had written at the time. He also put forward a suggestion that would haunt his reputation for years afterward as historians made the inevitable connections:

> Immense devastation could be produced if the entire attack was concentrated on a single big town other than Berlin and the effect would be especially great if the town was one hitherto undamaged.

However, on August 17, 1944, the joint planners reported that they did not think such an operation "likely to achieve any worthwhile degree of success." Thunderclap was to be retrieved once more for possible use almost five months later, under very different circumstances.

The JIC's general report on bombing and the eastern front was delivered on January 25, 1945, accompanied by the report on the bombing of Berlin that the committee's secretary had asked his staff to prepare. The latter went into considerable detail about how such a Thunderclap-style "knockout blow" against Berlin might be delivered, and where. It did not explore the overall political and war situation, for that had not been its brief:

> The degree of success achieved by the present Russian offensive is likely to have a decisive effect on the length of the war. We consider, therefore, that the assistance, which might be given to the Russians during the next few weeks by the British and American strategic bomber forces, justifies an urgent review of their employment to this end.

Significantly, straight after this basic statement of principle, the report made a proviso that, since existing attacks on oil targets were causing such problems for the enemy, "attacks against oil targets should continue to *take precedence over everything else.*"

Though keeping options open, the general tone of the main paper is markedly different from the "knockout" implications of King-Salter's note of a few days earlier (and of the report which the joint intelligence staff dutifully delivered in response). It was clear to the JIC's mandarins that to inflict such an annihilating blow against Berlin would take both the British and American air forces a number of days of continuous bombing. Something around twenty thousand to twenty-five thousand tons of bombs in three to four consecutive days was the general idea (between twice and three times the total dropped on Hamburg over several days in July 1943 with such catastrophic results for that city). It is difficult to see how a massive, sustained Thunderclap attack (or rather, series of attacks) on anything like such a scale was compatible with idea that raids on oil targets should continue to *"take precedence over everything else,"* as the JIC main paper so emphatically put it, or that attacking tank and aircraft production facilities should remain of the highest priority.

The main JIC paper to the chiefs of staff actually says it is time for major strikes behind the Russian front, the air forces should move from their habitual strategic campaign against industry, infrastructure, and housing toward using their overwhelming strength to create chaos in areas behind the battlefield. In other words, the air forces should assume a quasi-tactical role. This alone represents a substantial change in policy, Thunderclap or no. On the other hand, the JIC says, we cannot commit ourselves to the extent of abandoning the oil program, and we do not expect to destroy German morale by our use of air power (no longed-for "knockout blow"). We just expect powerfully to help the Russians on the eastern front and by this to further discourage German resistance.

This is tough, hard-headed advice, and by no means kindly meant from the point of view of German civilians—ruthless use of the Allied air forces' altogether awesome destructive power has become accepted by now as a key war-winning tool, and enemy refugees are seen as just another element in the equation.

The paper laid out a timetable of movements of German reinforcements suspected to be for the Russian front, the areas from which they might be drawn, and the times within which (subject to operational difficulties) these reinforcements might reach that front. The list of German troop movements that the Allied air attacks would disrupt—based on ENIGMA intercepts of German signals—ran as follows:

Germany	By rail	(This movement is probably completed)
Norway	By sea and rail	(1 division every 14 days)
Latvia	By sea	(1 division every week)
Italy	By rail	(11 divisions, 3 every fortnight)
Hungary	By rail	(6 Panzer divisions by 15 February)
The West	By rail	(7 divisions of which 6 Panzer by 15 February)

The crunch point fell around mid-February. Time was short.

IT WAS ONLY A MATTER OF HOURS later that Bottomley, Portal's deputy, discussed the JIC's analysis and suggestions with Harris. The AOC Bomber Command suggested that Chemnitz, Leipzig, and Dresden be added to the list. Harris had been pushing these targets forward for some time, and it must have seemed a logical occasion on which to bring them up again. Even the JIC had admitted that an attack, however massive, on Berlin alone would not decide the war. The Saxon cities were closer to the front, and in order to really damage the Germans' ability to move large amounts of men and matériel between the fronts, their rail networks (as well as Berlin's) would have to be attacked.

Bottomley's conversation with Harris was not the only immediate consequence of the JIC's paper on bombing the eastern front. On the evening of January 25, before going off for a drink with President Roosevelt's envoy, Harry Hopkins, Winston Churchill, who read such documents as a matter of course, spoke on the telephone with his Liberal secretary for air, Sir Archibald Sinclair. With his habitual colorful turn of phrase, the prime minister demanded to know what plans Bomber Command might have for "basting the Germans in their retreat from Breslau." This apparently tasteless remark may have been misunderstood. The most common use of "baste" is to "moisten during cooking with hot fat and the juices produced" but its second meaning (less common, but then Churchill relished unusual words) is "to beat thoroughly, thrash."

The largest city in Silesia, just a hundred or so miles east of Dresden, Breslau was now under direct threat from the Russians—

though it was to be another three weeks before it was cut off. It had been declared a fortress the previous autumn, but until a few days previously its citizens had seemed unaware of imminent danger. Its fanatical Gauleiter, Karl Hanke (a former close associate of Goebbels at the Propaganda Ministry) was reluctant to spread "defeatism." Whatever the reasons, things changed quickly on January 20, when loudspeakers throughout the city blared out the message: *"Achtung! Achtung!* Citizens of Breslau. The Reich Defense Commissar and Gauleiter announces that Breslau is to be evacuated. There is no reason for alarm . . ."* Universal alarm duly ensued.

Tens of thousands of civilians, mainly women and children, immediately fled Breslau. Outside temperatures were twenty below zero, and the snowbound roads were already choked with refugees from eastern Silesia. Soon bodies—especially those of babies—were being returned to Breslau for burial. Eighteen thousand refugees died of privation on the march to Kanth, twenty miles southwest of Breslau, where they had been told there would be transport.

The *London Times* quoted a radio broadcast by the German journalist "Hans Schwarz van Berg" (actually van Berk, a close associate of Goebbels and longtime political editor of the propaganda minister's weekly newspaper, *Das Reich*) describing women and children crowding the couplings between railway coaches and wagons, despite the bitter cold. Other papers were filled with similar refugee stories. Van Berk, a prominent German propagandist, was hardly likely to describe troop movements, especially retreats, with quite the same care or pathos, but his message was clear. The objects of the "basting" would be mostly refugees.

Whether—since the "fortress" was not yet surrounded and would not be until the middle of the next month—the prime minister could have safely assumed that the entire human tide pouring westward either from or via the Breslau area excluded the military, it is hard to say. The German Seventeenth Army, for instance, was in the process of being pushed out of Lower Silesia (southeast of Breslau), and within forty-eight hours of Churchill's remarks was in headlong retreat to the west. One thing is sure: Churchill, in typical aggressive fashion, was not happy with the air minister's guarded reply to his demand.

Sinclair said that German forces retreating from Breslau might

more appropriately become a target for tactical forces rather than
"heavy" bombers operating from altitude. He felt that the best use of
the "heavies" could be in continuing attacks on German oil plants, but
conceded that if weather conditions prevented these, area bombing of
"Berlin and other large cities in eastern Germany such as Leipzig,
Dresden and Chemnitz" might be considered. These were after all
"not only administrative centres controlling the military and civilian
movements but . . . also main communications centres through which
the bulk of the traffic moves."

Sinclair's proposals were sensible enough, but unexciting—at least
for Churchill's current mood. The prime minister wanted to know that
decisive action was in hand—perhaps because he would soon be away
from London for some weeks, preparing for and then attending the "Big
Three" conference at Yalta. Churchill responded:

> I did not ask you last night about plans for harrying the German
> retreat from Breslau. On the contrary, I asked whether Berlin, and
> no doubt other large cities in East Germany, should not now be
> considered especially attractive targets. I am glad that this is "under
> examination." Pray report to me tomorrow what is to be done.

A foray into eastern Germany to cause serious mayhem and
thereby support the Russians was looking like a certainty, especially
now that the prime minister was taking an interest in the matter.

With a minute about all this from Portal on his desk, and
Churchill on the warpath, Bottomley went ahead and issued orders to
Bomber Command. His letter to Harris, dated January 27, 1945,
attached a copy of the JIC paper of January 25 (pointing out that it
had not yet been considered by the chiefs of staff):

> The opinion of the Chief of the Air Staff, however, is that it would
> not be right to attempt attacks on Berlin on the "Thunderclap"
> scale in the near future. He considers that it is very doubtful that an
> attack even if done on the heaviest scale with consequent heavy
> losses would be decisive. He agrees, however, that subject to the
> overriding claims of oil and the other approved target systems
> within the current directive, we should use available effort in one
> big attack on Berlin and related attacks on Dresden, Leipzig,

Chemnitz or any other cities where a severe blitz will not only cause confusion in the evacuation from the East but will also hamper the movement of troops from the West.

I am therefore to request that subject to the qualifications stated above, and as soon as moon and weather conditions allow, you will undertake such attacks with the particular object of exploiting the confused conditions which are likely to exist in the above mentioned cities during the successful Russian advance.

Sinclair sent a minute to Churchill, conceding that "available effort should be directed against Berlin, Dresden, Chemnitz and Leipzig or against other cities where severe bombing would not only destroy communications vital to the evacuation from the East but would also hamper the movement of troops from the West." The secretary for air added: "The use of the night bomber forces offers the best prospects of destroying these industrial cities without detracting from our offensive oil targets . . ." Churchill acknowledged the communication without comment on January 28. That same day Portal and Bottomley talked the plans over with Spaatz, who was on a brief visit to England from SHAEF HQ. Spaatz and Bottomley, it was decided, would consult with Air Chief Marshal Tedder, Eisenhower's British deputy at SHAEF.

Churchill's departure from London happened a little earlier than planned. Forecasters warned of a storm, approaching from the Atlantic. To keep ahead of the bad weather, the prime minister set off at around 9 A.M. the next day, January 29, from Northolt Aerodrome, and flew to Malta. There he and his chiefs of staff would be spending six days with President Roosevelt and his senior military men, preparing for the conference with Stalin at Yalta in the Crimea. The chiefs of staff would be out of the country until February 11. Churchill himself would not return to London for three weeks. From now on, communication with superiors would be by signal. And decisions would have to be discussed and confirmed at a distance.

On January 30, at a meeting in Whitehall of the Chiefs of Staff Committee, Sir Douglas Evill, vice chief of the Air Staff, confirmed that the Air Staff had "studied the feasibility of carrying out an attack of the scale suggested." The minute of Evill's statement continued:

At this time of year it was most unlikely that the weather would be such as to permit concentrated bombing on four consecutive days and nights. In view, therefore, of the priorities recommended for attacks on oil and tank factories, he believed that a "Thunderclap" attack would not be feasible at present. On the other hand, the Air Staff agreed with the Joint Intelligence Sub-Committee that an attack, even on a lesser scale, against Berlin, would play a considerable part in assisting the military campaign on the Eastern Front.

In other words, Thunderclap was to be replaced by a number of very powerful—but not, in numbers of aircraft dispatched, freakishly large—air raids on eastern German cities, including Dresden. The statement was copied to Portal in Malta for his approval.

In a further note to the Chiefs of Staff Committee on February 1, Sir Douglas Evill spelled out the priorities in rather more detail and with added frankness, beginning with oil targets and tank factories before making the nature of the big-city attacks in eastern Germany quite clear. The note was headed "Evacuation Areas":

> Evacuees from German and German-Occupied Provinces to the East of Berlin are streaming westward through Berlin itself and through Leipzig, Dresden and other cities in the East of Germany. The administrative problems involved in receiving the refugees and re-distributing them are likely to be immense. The strain on the administration and upon the communications must be considerably increased by the need for handling military reinforcements on their way to the Eastern Front. A series of heavy attacks by day and night upon those administrative and control centres is likely to create considerable delays in the deployment of troops at the Front, and may well result in establishing a state of chaos in some or all of these centres.

Evill makes clear that disrupting mass movements of civilians is one element—perhaps the key one—in the calculations of damage to be inflicted during the next weeks. The chilling implication is that, if the plan is to cause maximum disruption to movements of any kind within and through a given city, the place to strike that city will be in its heart. In the center are to be found not just the main railway junctions, but also the communications systems, the administrative headquarters, the utilities, the key junctions of all those networks of streets and wires and pipes and cables. Destroy and disrupt these, and you have chaos not just in the city but also in the entire surrounding area, even perhaps the broader region that depends on these services.

This was one of the key lessons that British planners had learned long ago from the German bombing of Coventry. The damage inflicted on the city's infrastructure had lasted far longer and caused more long-term difficulties for war production than the actual bombing of the industrial plants.

The first objective would be the hearts and brains of the eastern German cities, then their viscera—the transport links—and finally any industrial manufacturing.

WEATHER CONDITIONS in the first and second weeks of February were, as so often in the winter of 1944–45, poor. Despite this, a massive attack was indeed mounted against Berlin by the U.S. Eighth Army Air Force, just as General Spaatz had promised. On February 3, almost a thousand B-17 Flying Fortresses attacked Berlin in daylight. Marshaling yards and railway stations throughout the vast urban area were the official targets, with the importance of the raid underlined by the belief that the Sixth Panzer Army was moving through the German capital on its way to the Russian front.

The first wave of more than four hundred B-17s of the First Bomber Division, escorted by P-51 Mustang fighters, arrived over Central Berlin at one minute past 11 A.M. and bombed the entire area beneath them. The attack definitely caused massive damage and started major fires. However, by the time the second wave—this time almost five hundred aircraft of the Third Division—swept in, a little less than half an hour after the first bombs had fallen, a strong southwest wind had blown clouds over Berlin. These, combined with the smoke and fug from the first attack, inhibited the kind of accurate bombing that their comrades had achieved just a short while earlier. Instead of adding to the blazing inferno in the city center, as had been planned, the bomb aimers of the second wave scattered their loads over wider areas of Berlin, including the working-class residential districts farther east.

The Berliners' relative good fortune once again provides vivid illustration of the role of chance in deciding whether a population exposed to bombing lives or dies. Over two thousand tons of air ordnance were dropped in less than an hour, including more than six thousand high-explosive bombs, a thousand air mines, and about the same quantity of liquid incendiary canisters. A witches' brew fit to make a firestorm. But the failure of the second wave to capitalize on the concentrated bombing achieved by the first saved central Berlin from the fate of Hamburg and Kassel.

Later claims by the American authorities put the death toll at a cataclysmic twenty-five thousand. German estimates are lower and almost certainly more accurate: just fewer than three thousand dead, with two thousand injured. This figure nevertheless represents the largest number of Berliners killed in a single air raid—during a war in which the city was bombed 363 times by the British and the Americans over a period of almost five years, losing a total of more than fifty thousand of its citizens. Most German cities died a death of a thousand cuts; only a handful suffered swift execution.

One thing was true: The February 3 raid on the heart of the Reich capital was awesomely destructive. Many familiar streets and quarters, churches and landmarks known to Berliners for decades, even centuries, were destroyed. The royal Hohenzollern castle was burned to a shell, which was eagerly dynamited by the German Communists after the war. Some claimed that this was "area bombing" by the Americans in all but name, and in this they had some justification.

The next day, February 4, with the ruins of Berlin's Regierungsviertel still smoldering, the "Big Three"—Churchill, Roosevelt, and Stalin—began their first formal meeting, in the Grand Ballroom of the Livadia Palace near Yalta, summer residence of the last czar of Russia. If the notoriously poker-faced Stalin was impressed by what the Eighth Air Force had done to Berlin, he did not see fit to mention the fact. But he did have some wishes, as the Allies had already guessed.

THE QUESTIONS dominating discussions between the Allied leaders meeting in the threadbare splendor of the Crimea had much more to do with the postwar settlement than with immediate wartime exigencies. The fate of Poland, where Stalin was busy installing a government of Russian stooges, in a coup that was to dictate a pattern in postwar eastern Europe. The final boundaries of the occupation zones in Germany, the feeding of the German population, and the treatment of Nazi war criminals. Soviet entry into the war against Japan. The problem of Greece, where since the German withdrawal a developing civil war between Communist guerrillas and British-backed royalists threatened to create the first fratricidal skirmish of the cold war. But there were military discussions and briefings, both in the main sessions and in subsidiary talks between the chiefs of staff who had accompanied their leaders to this balmy prerevolutionary playground.

As far as bombing was concerned, there were two main issues, both connected with the recent rapid Russian advances. The first had to do with the establishment of a "bomb line," attacks to the east of which would require consultation with the Soviets. The purpose of this was to avoid accidental bombing of Russian positions. The second issue was, how could the western Allies' air forces best support the Red Army's advance into Germany? General Antonov, the Red Army's deputy chief of general staff, first broached the issue of support for the Russian advance in a lengthy presentation to the plenary conference on the afternoon of the first day. After giving his view of the military situation, he asked that the western Allies "by air action hinder the enemy from carrying out the shifting of his troops to the East from the Western front, from Norway, and from Italy. In particular, to paralyse the junctions of Berlin and Leipzig." Antonovs last request was in line with decisions already authorized by SHAEF even before the British delegation left London.

And Dresden? There is no formal mention of the city in the official transcript of the session, except when referring to the bomb line, which the Russians requested should run through Berlin, Dresden, and Vienna to Zagreb.

But Hugh Lunghi, the young army officer who was one of the interpreters accompanying the British delegation (and who translated from Russian into English for both the prime minister and the chiefs of staff), firmly maintained that the idea of bombing Dresden was brought up by the Russians and discussed on two separate occasions:

I was very much involved in the talks about the bombing of Dresden, which the Russians had asked for, both at the plenary ses-

sion, the opening plenary session, where General Antonov . . . laid
out the military position and mentioned this; because Dresden was
an important junction, they didn't want reinforcements coming over
from the Western front and from Norway, from Italy and so on; and
similarly on the following day, when there was a meeting of chiefs of
staff in Stalin's quarters in the Kareis Palace, where Antonov very
clearly said, "Well, we want Dresden . . . the Dresden railway junction bombed because we are afraid the Germans are putting up a
resistance, a last-stand, as it were." And we agreed to this, we agreed
to pretty well everything . . .

It might be argued that the question of whether the Russians
specifically requested a major air raid on Dresden is merely a technical
one, especially since the attack had already been ordered in
Bottomley's letter to Harris of January 27, 1945. But what we do
know is that the combination of the bomb line demand, and the
accompanying discussions about targets in eastern Germany close to
and behind the front, caused a flurry of communications between
Portal and Bottomley. Portal was certainly unhappy about the bomb
line, fearing that it was too far west and would inhibit Anglo-American
attacks against important targets. On February 5 (before the second
reported mention of Dresden at the chiefs of staff meeting) he signaled
Bottomley, asking him to supply some good objectives east of the
bomb line that it would be desirable to keep bombing until the ground
situation dictated otherwise. Bottomley replied the same day with a
list that, along with a selection of oil plants and tank and aircraft factories, specifically included the cities of Berlin and Dresden.

Portal replied on February 6, agreeing that a new directive should
be issued to enshrine the new priorities. So why issue a new directive,
when Harris already had his orders to attack Dresden when the conditions were right?

First, because (if Major Lunghi's memory serves him well) the
bombing of Dresden, as a specific target, was now a direct matter of
high policy, decided among all three great power representatives,
political and military.

The idea of bombing eastern Germany to help the Russians had
originally come from London. A new, joint Allied directive would
broaden responsibility and give the planned attacks on Dresden and

the other cities behind the Russian front a coalition, rather than a predominantly British, stamp.

"The appearance of Dresden as a specific target for attack came as a surprise not only to Harris and Saundby but also to the Command's Intelligence Staff," said the historian David Irving. According to Irving's information, Harris was so displeased that he drove to London to query the order. Yet Harris had specifically mentioned Dresden as a possible, in fact desirable, target in his correspondence with Portal as far back as November 1, 1944, and had further requested its inclusion on the target list when told of the JIC's proposals to bomb Berlin and the eastern front. If the information is accurate, then his behavior was highly uncharacteristic.

In the end, no new directive was handed down, but a new target list was definitely adopted at the Targets Committee meeting on the afternoon of February 8 at the Air Ministry in Whitehall. The next day, Thursday, February 8, copies were formally sent to SHAEF, Bomber Command, and the U.S. Strategic Air Force Command. It declared: "The following targets have been selected for their importance in relation to the movements of Evacuees from, and of military forces to, the Eastern Front."

Berlin was in first place, Dresden second, with Chemnitz third.

From now on, the attack on Dresden was simply a matter of weather and timing.

16

Intimations of Mortality

DESPITE THE ATTACK on Freital in August, in which a few stray bombs fell on the city's southern suburbs, most Germans (including the civilian population of Dresden) still saw Dresden as a "virgin" city that had somehow remained safe from harm, and would continue to do so.

Not so the professionals in charge of Dresden's air raid measures. On September 21, 1944, the city's police chief presided over a day of meetings and technical rehearsals aimed at sharpening the preparedness of the emergency services. The men at the meeting were clear about two things: first, that sooner or later Dresden would be subjected to a serious air raid; and second, that given the population density in the center of the city, the inadequate air defenses, and the continuing (because of wartime shortages, now irreparable) lack of properly constructed air raid shelters, the consequences would be grave.

The police chief grimly summarized the conclusions. They would need to designate more than two cemeteries for the dead, "since we shall have to reckon on huge losses."

Since the citizens of Dresden could not be adequately protected, the authorities could only hope to minimize the destruction. They organized skeleton service staffs to be housed in safe shelters, prepared alternative offices for key staff members (mostly in the thinly settled outer suburbs) in case of major destruction of inner city areas, provided for emergency communications via a network of preselected messengers, and made plans to restore basic utilities and services as quickly as possible after any attack. These provisions were based on observation of the measures taken in the previously "air raid–endangered areas" of western Germany. From the middle of 1942, vital documents from various city departments were filmed and/or transported to safe sites outside the city.

It was easier and cheaper to generate paper and assign jobs to civil servants than to undertake the vast planning and construction task involved in actually providing the necessary air raid shelters for the people of Dresden, on a scale that might have given them the same degree of protection as the inhabitants of Berlin and the Ruhr.

ON SATURDAY, OCTOBER 7, 1944, in pursuance of the ongoing assault on Germany's supplies of synthetic fuels, the Eighth Air Force planned a mighty raid, using its entire available bomber fleet against the still-productive oil processing plants in central and eastern Germany. Of 1,422 four-engined aircraft, accompanied by more than seven hundred escort fighters, 1,311 reached their target.

The First Bomber Division was given the longest-range missions. It was divided into two independently operating groups for its attacks against oil plants in far eastern Germany. In the event, it turned out to be a messy and somewhat costly day for the Eighth Air Force. More than fifty bombers and fifteen fighters were lost, half of those to ground flak emplacements, which were numerous and accurate around the crucial oil plants. Many aircraft could not find their targets, which were relatively small and in some cases geographically isolated, or considered conditions too difficult for meaningful attack. Under such circumstances they tried, where possible, to attack the secondary targets allocated in their field orders. In the case of the First Division, this meant easier-to-hit urban targets, either Zwickau or Dresden, depending on weather conditions.

Lieutenant Colonel Walter K. Shayler, commander of 303rd Group—nicknamed "Hell's Angels"—decided on Dresden. Shayler led his thirty B-17s toward the city, where they were to attack "military installations"—the Friedrichstadt marshaling yards and nearby facilities. Unlike most of the other units that day, they had no trouble finding the city. Prepared to undertake "radar bombing," the three squadrons of Flying Fortresses found surprisingly light cloud over the city and very little industrial haze.

At the Ufa cinema on the Postplatz, a few hundred yards to the east of the Friedrichstadt rail complex, the second showing of that weekend's main feature had just ended when the sirens were heard. This was the 111th alarm of the war, and *none* so far had seen bombs fall on the city proper. Now, for the first time, the trek down into the shelter was not to be in vain.

On Shayler's orders, the B-17s flew northeast across the city and dropped almost three hundred five-hundred-pound general purpose bombs, a task in which "moderate to intensive and accurate flak" greeted them. One of the B-17s dropped its bombs short of the target. When the aircraft got back to base at Molesworth in Norfolk, ten had suffered substantial damage from antiaircraft fire. Hell's Angels was the first unit to bomb Dresden accurately and with intent.

It was all over in a few minutes, the bombs being dropped between 12:35 and 12:40 P.M. The surviving U.S. Air Force photographs show target markers still descending and bombs exploding on the southern perimeter of the Friedrichstadt marshaling yards. Some of the bombs intended for the yards strayed northward onto the neighboring Seidel and Naumann factory, where once typewriters, sewing machines, and bicycles had been made, but where production was now devoted to armaments.

Bombs also hit the Wettinschule district school, near the Wettiner Bahnhof station, demolishing part of the building, tearing doors and windows from their fixings, and one even exploding in the emergency toilets next to the school's air raid shelter. Fortunately, only a scratch air raid crew of teachers and older pupils was on duty there—it being a weekend—plus some civilians from the neighborhood, who preferred the custom-built shelter to the inadequate basements of their buildings. No one was killed, though there were some serious injuries. Less lucky was the Municipal Business School, a few hundred yards to the northwest, where a bomb careened down a light shaft and exploded in the cellar, killing a similar air raid team and some residents.

There were numerous detonations in the heavily built-up blocks between the two school buildings. A total of fifty buildings were totally destroyed, twenty badly damaged, twenty-five subjected to medium damage, and several hundred to light damage. A municipal office block not far from the historic Zwinger pleasure gardens was destroyed, and the former freemasons' lodge (now a craft and trade museum) in the Ostra-Allee directly opposite the Zwinger was seriously damaged. A few bombs a little farther along the street, and one of Dresden's most-loved historic landmarks might also have suffered destruction.

Victor Klemperer reported how he and the other Jews in the community house (Jew house) at No. 1 Zeughausstrasse went down into the cellar when the warning siren sounded at noon.

> Then there was anti-aircraft fire, then we heard clear loud explosions, evidently bombs, then the light went out, then there was a swelling rumbling and rushing in the air (bombs falling a short distance away). I could not suppress violent palpitations, but retained my composure . . .

The impression of "bombs falling a short distance away" was deceptive. It seems the nearest exploded around half a mile west of the Jew house—a foretaste of how violent a raid can feel, even from a distance. The Americans' goal had been to concentrate the bombs in the triangular complex of tracks and platforms bordered by the Friedrichstadt goods station, the Hauptbahnhof, and the Wettiner station, but this was only partially achieved. As so often, military and industrial targets had been hit, but a lot of "spillage" (in the official phrase of the time, roughly equivalent to "collateral damage") had also inevitably occurred.

Dresden had been through its baptism of fire. Around 270 victims were claimed by the wandering trail of destruction that the 303rd laid across the face of the city that day.

From the authorities' point of view, the "crisis management" experience of the raid was a success. Most of Dresden's preorganized systems had been shown to work in satisfactory fashion, though not quite as perfectly as the reports to Gauleiter Mutschmann implied. It helped that most of the six hundred coffins stored for the eventuality of air raid deaths were kept in the basement of the company Guhr & Stein, No. 8 Kleine Zwingerstrasse, close to the main areas affected by the raid.

The strange thing was this: The political authorities and Nazi Party behaved as if the raid had never happened. In the October days after the bombs had fallen, not a single word about it was to be found in the two remaining Dresden daily newspapers, the *Dresdner Zeitung* and *Der Freiheitskampf.*

Officially, no bombs had fallen on Dresden.

Every day between then and the beginning of November, a small number of the civilian dead from October 7, 1944, were permitted to have their death notices printed, until all had been dealt with in a way that would not cause unnecessary alarm among the public. The words "air raid" or "terror raid," the latter almost obligatory in official propaganda, were not used in the notices. No cause of death was given. The victims died of "tragic fate" or "a hard blow of fate." On October 12 the *Dresdner Zeitung* announced that a "communal burial ceremony" could take place, tickets obtainable from the Nazi Party office by the Zwinger. Still no open acknowledgment of the cause of death, only general platitudes about "our dear fallen" and "sacrifices for Germany."

But in the speech by Gauleiter Mutschmann that followed, there was a hint, which no one with a grip on reality could ignore. He proclaimed, "No one should live in the illusion that the place where he lives, his town, will not be attacked . . . There are no islands of peace in Germany."

On serious consideration, the experience of October should have caused real concern. Fewer than thirty American bombers, diverted from other targets, had in an improvised attack killed 270 people in Dresden during those few minutes on Saturday, October 7, 1944.

Dresden's vulnerability was beyond dispute. In London in 1940, at the height of the Blitz, where the entire might of the Luftwaffe was thrown against the British capital, the daily death toll averaged 250. Yet Dresden—and Germany—wilfully ignored this telling prelude.

FOR DAYS AFTER the October 7 raid, sightseers came from Dresden and the surrounding country to examine the damage. There was a certain novelty to it. Pastor Hoch was one of the sightseers. "Dresden had never seen a ruin until then," he recalled. "It said in the paper would Dresdeners please not go running over there to take a look. Of course, we could not help ourselves."

Most Dresdeners still thought the raid an anomaly, and considered themselves safe. Not so the few remaining Jews in the city. No one wanted to die under the bombs, but for Jews the attacks—however terrifying—heralded and encouraged the breakdown of the system that would one day certainly murder them and those they loved.

Klemperer's neighbors in the Jew house confided in him a few days later:

> "Every day we wait for the aircraft as we used to wait for Clemens and Weser" (the Gestapo bloodhounds). I: "Then I prefer the bombers."—Which is also true. But the present state of affairs also gets terribly on one's nerves. One hears awful details about the mutilations and deaths on Saturday, but the most divergent figures as to the numbers of dead . . .

The bomber war had crept eastward. The Ruhr and the western German cities were still important targets, but the Allies knew that German industry was being dispersed into Saxony and Thuringia, and knew also that the fighting in the east might decide the war. With the expanded numbers of aircraft now at the Allies' disposal (almost four thousand bombers by the winter of 1944–45), they could now divide up their fleets and still send hundreds over any one of several selected targets. The Americans especially were hitting Berlin, and the central and eastern German armaments-producing areas, and the synthetic oil plants in Thuringia and Saxony-Anhalt (above all the Leuna Works). More ominously, both the British and the Americans were now attacking the oil plants in the Sudetenland and eastern Saxony, which regularly brought them over Dresden. Such long flights— almost unheard of before the end of 1943—were now becoming relatively routine.

All the newly important targets were within around thirty minutes' flying time from Dresden, which was turning into an obvious transport and communications center behind the front. Klemperer, reflecting the talk he heard in the streets and shops, noted on October 17, 1944:

> It worries me greatly that our personal situation has become so very much altered by Hungary's elimination. Now Dresden may become a transport junction behind the front which is most threatened, and that in a very short time. Then we shall get heavy air attacks . . . there will be an evacuation and at the same time the mixed marriages will be separated and the Jewish parties gassed . . .

But there were no more air raids on Dresden that autumn. The winter came, and with it especially bitter weather, which kept the Allied aircraft away from eastern Germany on many nights when they might otherwise have ventured there.

As Christmas approached, Gauleiter Mutschmann, addressing his underlings at a meeting of Ortsgruppenleiter, could declare, "This Christmas will be beautified for us by the fact that we can see our people back on the offensive."

Mutschmann was referring to the Battle of the Ardennes at a time when the German advance was clearly on the ebb. Nevertheless, police reports confirmed that the temporary successes had caused "the bad public mood to disappear at a stroke. People are greeting each other again with a free and open gaze." Most Dresdeners were by now employed on war work or in armaments factories, where a sixty-hour week was standard. Many of the city's schools, including the Vitzthum-Gymnasium in the city center, had been turned over to military hospitals, and the boys and girls likewise put into war work as flak helpers or other kinds of welfare work. Nevertheless, the city seemed eager to enjoy the sixth Christmas of the war as much as it was able. Families were encouraged, as in all belligerent countries, to take young soldiers and walking wounded into their homes at Christmas. And they made the best of it.

The famous boys' choir of the Kreuzschule, the city's oldest educational institution, sang and played the story of the Christ Child under the high dome of the Frauenkirche in utterly traditional fashion, as it had done for many, many years before Hitler or Mutschmann had troubled the world, but elsewhere there was strong control of content. At the opera house, closed for months now to regular performances, the Hitler Youth staged their "Christmas concert." The curtains parted to reveal a stage dominated by a large iron cross, in memory of the war dead.

"Reich German children" (and those children only) were allowed 125 grams of sweets. Children, teenagers, and pregnant or nursing mothers also received half a kilogram of apples. And there was an extra ration of 250 grams of meat. All had to be ordered in advance through special coupons attached to the seasonal issue of ration cards.

Victor Klemperer and his wife had no such prospect:

We can give no presents, neither to ourselves nor to others. This year the only Christmas trees are given by the Party to large families. The special rations—the first for several months and naturally only for Aryans—consists of half a pound of meat and two eggs.

Writing some months later, after all their worst fears had been realized, a mother wrote to her married daughter about that final Christmas in old Dresden:

Do you remember, Christmas? The film, "Philharmonic." I had seen it already on the Thursday, as a kind of introduction to the Christmas festival. But when you came the next day, you wanted to see it too. We were together in the lovely "Capital" movie theatre, seated on the left in the last but one row. And I was so glad that you were interested in that film above all others and that we could be together. Then Christmas came . . . the little tree from Chemnitz, the candles and a whole table of gifts. And how Lütte managed to scrounge those gingerbreads from Herta's, and the gold-framed photograph on her table. No one could imagine that it would be our last Christmas, or at least the last at the Struverstrasse—four weeks later we were all homeless, our home city gone, torn apart . . .

BY THE MIDDLE OF JANUARY 1945 the Christmas and New Year celebrations were indeed just a memory, as were any lingering hopes for the success of Hitler's counterattack in the Ardennes. In the east, a new Russian offensive had begun on January 12. The Red Army quickly achieved massive advances along a front that stretched hundreds of miles, from Hungary in the south to East Prussia in the north.

In Dresden itself, scarcely a day had gone by in which its people had not been subjected to an air raid warning. Then, suddenly, came the real thing.

January 16 was a Tuesday. A cold snap had brought temperatures of minus 7 degrees Celsius and thick snow. At 11:20 A.M. a general warning was sounded—indicating enemy bombers heading for central Germany—but at 11:50 the steadier howl of the full alert—*Fliegeralarm!*—rose over the wintry city. Those who could scrambled into cellars and air raid shelters—mostly unheated—and shivered and crouched and waited. Then came the roar of engines, the shudder of bombs, and the explosions.

Like the October 7 raid, it lasted only a few minutes and was concentrated on the west-central part of Dresden. When the explosions stopped, the drone of the B-24 Liberators began to recede into the distance, and the all-clear was broadcast, Dresdeners emerged into the midday chastened.

There was no escaping the conclusion: One raid might be an accident, but two started to look like planning.

Once more Dresden had been designated as a secondary target—to be attacked if the more urgent targets were too difficult to bomb accurately. The lead group of the USAAF's Second Bomber Division was the Forty-fourth ("The Flying Eightballs"), under Colonel Snavely. The Forty-fourth had been directed against the Ruhland oil plant as well as the fighter airfield and repair hangars at Alt-Lönnewitz, near Leipzig. Visibility was poor over the fighter base, and the flak—perhaps radar-guided—fierce and surprisingly accurate. There were similar problems at Ruhland. Codeword for the alternative attack: "tough times." Operations code for Dresden: GH 584.

Snavely led his group, and the others raggedly followed, some homing in from Ruhland to the north-northeast and some from Alt-Lönnewitz to the northwest: the 491st Bomber Group, the 392nd, the 93rd and the 446th, the 448th, the 466th, and the 467th (the last named the "Rackheath Aggies" after their base, which was on the land of a Norfolk gentleman farmer). These aircraft represented around a fifth of the bomber force that had taken off from England to attack targets in central and eastern Germany. Almost half had the luck to be assigned Dessau: They just bombed the aircraft factories and went home.

Altogether, 127 Liberators came at Dresden from various directions, this last factor dictated by the poor weather, which had made it hard for them to keep together. From between twenty-two thousand and twenty-six thousand feet, they dropped a mix of high-explosive and incendiary bombs.

The January raid was a much heavier attack than the one three months earlier, though less concentrated, in fact a little messy. This was reflected in the distribution of the damage. The actual dropping area from north to south was spread over more than almost four miles;

east to west, two and a half miles. The bombing reached a level that could be called "concentrated" around the actual aiming point—once more the Friedrichstadt marshaling yards—and the "Hecht Quarter," north of the Dresden-Neustadt railway station and just south of one of the main concentration of military barracks in the city. Stray clusters and individual bombs landed at the Altstadt goods station and just by the Hauptbahnhof too. Some of the heaviest damage was to the Inner Neustadt Cemetery.

There was no mention of the Second Division's raid in the Wehrmacht High Command's report the next day, but the Reich Ministry for Armaments and War Production briefly mentioned damage to rail facilities.

The Americans' own photographic evaluation of the raid led them to the conclusion that the following military targets had been hit: railway viaducts, central section of the Friedrichstadt marshaling yards, rolling stock, industrial facilities, densely built-up area, a fuel storage container, railway junctions. This was pretty accurate.

The official toll for that day reached 376, making almost 650 dead in the two air raids on Dresden so far—secondary raids that were not even thought important enough to be mentioned in the German High Command's daily report. This time there was no attempt to hide the fact of the raid. Death notices in the local papers were permitted to describe the deaths of the victims as due to "air raid" or "terror bombing." The mourning ceremony for the victims was fully reported—the city orchestra playing Beethoven, the Kreuzschule choir in full song. Dresden's *Kreisleiter* (district leader) told the mourners that "Life is granted to us only so that we may give it to Germany."

And to soften the blow for the newly embattled population of Dresden, the paper carried a (literally) fantastic article about the horrendous destruction being wrought on London by the new German "wonder weapons." Reporting an alleged speech by the "Mayor of Manchester" about the effects of the V-1s and V-2s, the *Dresdner Zeitung* told its readers that the "revenge weapons continue unceasingly to hammer" the British capital:

> Many of London's inhabitants are living in half-collapsed houses and have to spend the nights in underground shelters, just to get a few hours' sleep. Often no fewer than 185 bombs have fallen in the

space of 24 hours. The population has declined to a half of its previous level, and a third of all material assets have been destroyed in just this short period of time.

Again, the damaged buildings in Dresden—this time spread much more widely through the western and west-central areas of the city— attracted sightseers. "You think, something has happened in town, and you want to see it," said Nora Lang. "It was something that hadn't happened before, you want to see it." Anita Kurz thought each shattered building looked "like a doll's house ... you could look into every room."

There was another difference between the Eighth Air Force's January raid and the October attack. In October there had been resistance from antiaircraft guns stationed in Dresden. By the middle of January, when the Americans struck again, most of those guns had been removed and sent to areas where they were more urgently needed. When the Liberators offloaded their bombs onto the Friedrichstadt marshaling yards, guns and equipment from the last flak battery in Dresden were packed into wagons there, waiting to be transported away from the city.

At the end of April 1944 Dresden's flak had fired its first salvos at an enemy aircraft over the city—a damaged B-17 from an American raid on Berlin. The antiaircraft defenses consisted of seven 88mm heavy antiaircraft batteries (each with seven guns) and five units of so-called Russian flak (five to six guns per battery), made up of captured Soviet guns rebored from 85mm to 88mm to fit German specifications. The summer of 1944 saw Dresden's antiaircraft defenses at their strongest, with an extra battery of German 88mm guns and some more light batteries. The big oil raids on central and eastern German targets were now well under way, and Dresden's guns saw use against American and British formations heading for Ruhland, Brüx, and the other synthetic fuel plants in the vicinity.

By this time most of the work on these antiaircraft batteries was done by boys. In the words of the Führer-Order of September 20, 1942 it was "a flak militia made up of juveniles." The aim had been to release 120,000 adult Luftwaffe personnel for the eastern front.

Fifteen- and sixteen-year-old schoolboys were now conscripted to "man" the flak. Except for the battery commander and, for the heavy guns, the loaders, most of the other functions were taken over by these teenagers. A skeleton course of lessons was kept up, with teachers in many cases coming to the emplacements to do their job.

Much of the air battle in the last two years of the war consisted of fifteen- and sixteen-year-olds on the ground, their ranks stiffened by a scattering of experienced veterans, aiming and firing guns that were trying to bring down the eighteen- and nineteen-year-olds of the Eighth Air Force and Bomber Command. Götz Bergander, born in 1927, was one of those so-called Luftwaffe helpers (*Luftwaffenhelfer*). So was his boyhood friend Steffen Cüppers. Bergander was to leave the flak before it left Dresden, but Cüppers stayed with his own flak battery until the bitter end.

Cüppers was fifteen when he and others from his class at the prestigious Vitzthum High School in central Dresden were conscripted. Young Steffen was a "K-3" gunner, responsible for loading the twenty-shell magazines into the machine gun as it fired, and every hundred shots changing the hot barrel, which came on and off in a rapid bayonet action. He was transferred to an 88mm battery just in time to be involved opposing the October 7 American raid, the "first small attack on Dresden," though they were still in training and did not do any actual shooting.

By this time the Dresden flak was already depleted. Between then and the next American raid in January 1945, the evacuation process accelerated. Steffen was hospitalized shortly after with scarlet fever, so he knew little of the transfer order and the organizational maelstrom that resulted, though he rejoined his battery in the west when he recovered.

Much of the artillery, with its teenage crews, was turned over to ground use—in Silesia, in Vienna, eventually in the battle for Berlin. Many Luftwaffe helpers from Dresden, scarcely old enough to serve at the front, died far from home, overrun by American and Soviet infantry or vaporized in ground-guided air attacks. Many of those guns that kept their antiaircraft functions were transferred to oil and artificial fuel plants, which retained a high priority: Brüx, Zeitz, and others.

At the same time that Dresden was being stripped of its last antiaircraft defenses, in January 1945, the big hydrogenation plant at Brüx was still guarded by 166 guns—down from a peak of 260, but a formidable "flak zone" for the Allied fliers to reckon with. Others ended up in the Ruhr and central Germany, trying to ward off the final, devastating assaults.

And so it was that the American crews on January 16 reported "non-existent to weak and extremely inaccurate" flak. Given the chaotic circumstances of the raids, the height at which the planes operated, and the speed with which an aircraft could pass into another flak zone, it is common to find recollections of flak over specific targets where there was verifiably none.

Dresden is often spoken of as "defenseless," which was largely true by February 1945, but the Allied air forces planning to attack the city could not know this. Nor, indeed, could they know for certain that the Luftwaffe would not send up fighters to harass the Allied bombers by day or night.

It is also said that the withdrawal of flak indicated the low status of Dresden as a military or industrial center. Not necessarily so. No one would deny the industrial status of the city of Chemnitz, to the southwest of Dresden, a place just over half the Saxon capital's size but known as an industrial center, especially for tank production. Yet Chemnitz too lost its antiaircraft batteries in November-December 1944 and became equally "defenseless." The Third Reich, bereft of labor and fighting men, had become desperate. Towns that should have continued to be defended were not. Dresden and Chemnitz ended up among those places. Their flak units were redistributed according to harsh new priorities, along with manpower and other vital, rapidly diminishing resources.

THERE WAS ANOTHER general warning on the evening of January 16, 1945, which once again sent Dresdeners scurrying wearily for their basements. The fear was that the British might be coming by night, to track Dresden by the still-glowing fires from the American raid and bomb it again. It was a technique that had become familiar over the past few months.

That night the British in fact carried out an area raid on the center of Magdeburg, 120 miles to the northwest. A total of 371 aircraft dropped a lethal mix of high explosives and incendiaries that lit numerous fires. These in turn rapidly combined to form one huge conflagration. The resulting firestorm was claimed to have destroyed 44 percent of Magdeburg's historic heart. Some 4,000 people died, more than 11,000 were injured, and 190,000 were made homeless.

Unlike Dresden, Magdeburg had an extensive and efficient air raid shelter system. The relatively high fatality rate might have been due to the fact that the alarm was raised only a few minutes before the British bombers appeared over the city, so that many citizens did not have time to reach the safety of the public bunkers before the destruction began—and the sealed doors were closed.

THE NEXT DAY, January 17, after more than five years of unimaginably cruel occupation, the last German forces withdrew from Warsaw. For most of the eighteenth century this had been Dresden's Polish sister city and alternative residence of the Wettin monarchs—indeed their refuge during the years of Prussian occupation and exploitation of Saxony. After bloodily suppressing the Warsaw uprising the previous autumn— forty thousand civilians were executed, a quarter of a million died in the subsequent fighting—on Hitler's orders German forces proceeded to systematically raze the Polish capital to the ground. Between October and January they destroyed 80 percent of its center street by street, building by building "as if it were a latter-day Carthage or Persepolis."

The Poles of Warsaw, with the Red Army at their gates the previous August, had hoped to liberate their own capital before the Russians arrived. But all those months, as the unequal fighting raged inside the city, Stalin's forces had waited on the eastern side of the river Vistula, close enough to hear the gunfire and see the smoke rising from the burning buildings. On January 17, as the Germans left, the Russians and their puppet Communist government took charge of a city that, thanks to Himmler's SS killers, had been transformed into one vast, wretched, unresisting ghost town.

Meanwhile in London, responding to the request by the deputy chief of the Air Staff, the Joint Intelligence Committee had begun to draw up a paper on the possibility of using Bomber Command to help the Russians on the eastern front—in the first place asking about an intensive bombing of Berlin but also stimulating other suggestions that boded ill for the people of Dresden.

17

Time and Chance

WHILE THE STATESMEN and their entourages pondered and haggled in the mild, almost balmy conditions of Yalta, much of Europe still lay swathed in cloud and snow.

From a meteorological point of view, for the Anglo-American air forces this was the most difficult winter of the war. February's operations started a little better than January's, but briefings followed by cancellations remained depressingly frequent. Three days into the Crimea conference, on February 7, the weather over Germany lifted, allowing substantial attacks. The first were in support of British land forces, directed against the fortified west German towns of Goch and Kleve. Major RAF operations on Thursday, February 8, included an attack by almost five hundred bombers on the Pölitz synthetic oil plant close to the Baltic coast, which put the works out of action for the remainder of the war. Then 228 aircraft, mostly Halifaxes, attacked the fuel storage facilities at Wanne-Eickel near Gelsenkirchen in the Ruhr, while 151 Lancasters bombed railway yards near Krefeld.

The original Dresden plan had been for an attack similar to that on Magdeburg the previous month—a midday visit by the U.S. Eighth Air Force, to soften up the city and get some fires burning, followed by a nighttime "double punch" from the RAF, guided by the still-blazing daytime targets. This was supposed to be precision followed by area bombing, but of course these definitions always did depend to a great extent on the weather. And the weather turned bad for both forces.

February 9 was the day after the Target Committee's revised target list, with Dresden prominently displayed, had been issued to the Anglo-American air forces. The RAF was pretty much grounded by bad weather all over northwest Europe. A Halifax of 100 Group (support) flew a routine RCM ("radio counter-measures") jamming operation. Night operations connected with the resistance in occupied Europe were undertaken by seven Stirlings of Third Group, of which one was lost. More RCM operations. On February 10, during the day, RCM. On the night of February 10–11 eighty-two Mosquitoes went to Hanover and eleven to Essen. There were patrols and more of the relentless RCM. On the night of Monday, February 12, seventy-two Mosquitoes went to Stuttgart—another large "nuisance" raid—while small numbers of the high-flying fighter-bombers visited other cities, presumably to make sure the inhabitants had to trek down to their shelters for at least part of the night. This was not the impressive crescendo that many had expected to coincide with the conference of the "Big Three."

The Eighth Air Force was in scarcely better shape than Bomber Command. After the big Berlin raid on February 3, there was a three-day lull. On February 6 the Americans, unable to go for precision oil targets because of the weather, attacked "marshaling yards" at Chemnitz and Magdeburg. The official history reported coyly that they destroyed "structures of cultural and historic importance as well." On February 7 and 8, U.S. bomber forces left for Germany but were recalled because of extremely poor conditions. On February 9 a window of opportunity resulted in some fairly successful daylight attacks on the oil plant at Lützkendorf between Leipzig and Halle, an oil storage depot at Dülmen, and an ordnance plant at Weimar, the town where the poet Goethe found refuge and usefulness for most of his life. Here the feared Me 262 German jet fighters put in an appearance, dancing in among the bomber stream, which was shepherded by P-51 long-distance fighters in the now-standard fashion, but dancing a little too fancily to shoot down more than one of the American intruders. The jets were a reminder that the once-mighty German fighter defenses had not been eliminated from the equation, but for the next three days the chief enemy was once more the weather.

As February 13 dawned, it was clear that the situation remained hopeless, at least as far as daylight sorties were concerned.

The Yalta conference had been over for two days. For a little more than a week, Churchill and his aides had slept in their Soviet-assigned residence, the Gothic/Moorish Vorontsov Villa. Prince Vorontsov, the czar's ambassador to Great Britain at the time of Queen Victoria's coronation in 1837, had been much impressed by the variety of building styles in England. He had built his curious summer residence in the fond belief that he was creating a piece of architecture *à l'Anglais*. From there the slightly bemused British delegation had been shuttled every day to the conference itself.

As soon as the talks were concluded on the afternoon of February 11, Churchill left the Vorontsov Villa and transferred to the British naval vessel *Franconia*, two hours' drive away in the harbor of Sebastopol. The former Cunard luxury liner, used as a troopship throughout the war, had recently been adapted as HQ ship for the conference and as alternative VIP accommodation. There Churchill stayed for almost three days, relaxing and enjoying lavish meals prepared by navy cooks with unrationed ingredients. It must have been a hard place for the prime minister to exchange for an English winter—and indeed he was to decide on visits to Athens and Cairo before finally returning to London. If such a prospect was a trial for him, how much more so for the shivering air force personnel at home, for whom the weather had been grim reality since the previous November.

At that time of year the eastern counties are the least hospitable part of England, open to penetratingly cold winds that seem to sweep right in off the Russian steppes. It was bad enough for the idle aircrew, often itching to get their "ops" done and finish their tours of duty, but as twenty-seven-year-old Flight Lieutenant Lesley Hay of 49th Squadron, 5 Group, observed:

> All the work the ground crew were doing, the necessary work, all had to be undone when the bombers couldn't take off. We went to bed or somewhere whenever we could, but the ground crew couldn't do that—and the winter was no time to be working outside on aircraft . . .

Added to this was the fact that most aircrews had just had their tours extended. A bottleneck in the training schedule had apparently led to personnel shortages. Lesley Hay, for instance, completed the final (thirtieth) operation of his tour just before Dresden. This achievement, since crews were usually given a six-month respite before starting a twenty-op second tour, would have seen him and the

other crew members of his aircraft, "U for Uncle," out of active danger
for the rest of the war—and would have meant no trip to Dresden—but
it was not to be. The same went for another crew whose bomb aimer,
Miles ("Mike") Tripp, later became a well-known English writer. His
crew had their tour extended twice. He and the other noncommis-
sioned members of the crew had slept in following a night raid over
western Germany in early February. Shortly after noon their
Australian skipper, nicknamed "Dig," came into their hut and woke
them with the unwelcome news:

> He sat on the side of a bed, lighted a cigarette, and began, "You're
> not going to like this."
> No one spoke.
> "Another directive came through from Group today. That order
> about extending a tour to thirty-five ops has been amended. The
> order is now 'forty sorties over enemy or enemy-occupied territory.'"
> George broke the dreadful silence. "But that'll leave us with four-
> teen to do! We're back where we were two months ago!"
> "That's right, mate."
> "We shan't make it," said George.

They were, as it turned out, about to be grounded for several more
days and nights. Before the tour extension order, the heavy bomber
crews affected had been eager to get on and finish their tours. Now
they were back in the long-term routine of survival, of getting the job
done and getting back—and blanking out every other consideration.

WHILE THE AIRCREWS cooled their heels and the bomber squadrons'
maintenance mechanics chilled their bones dealing with the stop/start
whims of an air campaign in bad winter weather, the high-ups and the
planners were perfecting the details of the plan for the big attacks on
eastern Germany. Above all, the meteorological officers were scouring
their reports for a possible break that would enable those attacks to
take place with a chance of real success.

The weather was the most important factor conditioning the destinations and the styles of attack that the planners would approve.
Nevertheless, there were other influences. It was Air Marshal Harris's

job, supported by his command staff, to make those operational decisions, within the framework supplied by the Air Staff in Whitehall, and ultimately the Royal Air Force's political masters, the secretary for air, Sir Archibald Sinclair, and the prime minister. Churchill gave no further impetus to the Dresden project, at least on the record, once Sinclair and the chiefs of staff had agreed and Bottomley had issued the order to Harris. Whether the prime minister was consulted or involved in any way while at Yalta was not known, though it seems both unlikely and unnecessary.

In practical terms, the absence of Churchill and Chief of Air Staff Portal meant that Deputy CAS Bottomley and Air Minister Sinclair were minding things on a day-to-day basis. Harris and his planners, as long as the order remained in force, stayed in charge of the details of the raid.

Apart from the weather, Harris had other practical problems, which helped dictate his actions. With large expanses of western Germany, Berlin, and the central German cities now in ruins, incendiary bombs were not the ideal tool for continuing destruction. Rubble burns poorly. If Harris was to continue attacks on the already much-devastated urban areas of the Ruhr and the Rhineland—which he would expect to do with the Allied ground forces now so close—with any hope of effect, he had to drop a preponderance of high explosives. The same applied to the key oil plants, which were truly vulnerable only to the penetrative power of high-explosive bombs.

Of these last items there was now, owing to the RAF's Herculean bombing activities since the previous autumn, a shortage. During the last six months of the war Harris, as he admits in his memoirs, had to beg supplies of high-explosive bombs from the Americans. This made "virgin" or near-virgin targets such as Dresden and other, smaller cities (Pforzheim and Würzburg were two of the most notorious cases) correspondingly more attractive. Such targets burned easily and well. Raids against them ate up mostly supplies of incendiaries, which were relatively plentiful.

As for aircraft numbers to be allocated, it was already apparent that the coming raids on eastern Germany, including Dresden, were going to be big ones. Not freakishly huge, though, by the standards of the time. Raids involving six, seven, eight hundred aircraft were now fairly commonplace. For the attack on Dresden, 796 Lancasters would take to the air (making over eight hundred aircraft altogether if we include the nine Mosquitoes of the Pathfinder force and the master bomber who coordinated the attack). Bomber Command's schedules between mid-September 1944 and mid-March 1945 evidenced fifteen attacks involving more than seven hundred aircraft, and nine calling on more than eight hundred. The most powerful forces dispatched during this period, against the Ruhr cities of Essen and Dortmund on March 11 and 12, consisted of 1,079 and 1,108 bombers respectively.

With the numbers of aircraft now at his disposal, Harris could afford to think big on more than one target per night. He could use largely for decoy purposes substantial forces that two years earlier would have represented Bomber Command's entire operational strength.

Just how routine the sending of these kinds of numbers against distant enemy targets had become was indicated that same February. As part of an ongoing discussion about using the Soviet air facilities at Poltava in the western Ukraine for "shuttle bombing" against eastern Germany, a note to the Air Ministry, signed on Harris's behalf by his deputy, stated, "It will seldom be worth bombing any target in Eastern Germany with less than two hundred heavies. Usually I shall want to employ at least four or five hundred."

The Americans' now-delayed role was still dependent on the weather. They had nevertheless, on February 12, let the Soviet General Staff know, through their military mission in Moscow, of their plans to bomb "the marshaling yards" in Dresden "on the following day." This was strictly in accordance with the bombing lines agreement demanded by the Soviets at Yalta. The British were criticized for arrogance in not also formally advising the Soviets of their own linked raid. If the Russians, as Churchill's interpreter, Hugh Lunghi, asserts, specifically requested on two separate occasions at Yalta that Dresden be bombed, then there would be even less reason for the British to feel compelled, less than ten days later, to dispatch a formal notification of such an intention. The object of the agreement was, in any case, supposed to avoid accidental bombing of Russian forces, and the front was still at least sixty miles distant.

When Harris and his staff assembled for the early conference—known to the regulars as "morning prayers"—at Bomber Command headquarters on February 13, it was clear that the original Magdeburg-style collaboration with the Americans for the bombing of Dresden was not going to be possible. Should the raid against Dresden be confirmed, the British would be going in that night and going in first. Also up for discussion were the supplementary raids, mostly intended to draw the remaining German fighter protection away from the main Dresden-bound force.

As it happened, the senior meteorological officer's weather forecast for the night to come was now reasonably optimistic. Cloud cover was expected to be 10/10ths (total) over most of the route, tops lowering to 6,000 feet beyond 05-07 degrees East, with a chance of breaks to 5/10ths (medium) in the Dresden and Leipzig areas. Just as importantly, conditions over the home airfields were predicted to be good enough, toward the end of the night of February 13–14, to allow homecoming bombers to land safely on their return from their long, fuel-hungry flights.

The meteorological officer's prediction of weather over Dresden doomed the city. A little before 9 A.M. Bomber Command's liaison man at SHAEF headquarters confirmed the Supreme Command's approval for the night raid. Harris would later state that the attack on Dresden—for which many in the following years have held him totally, indeed personally, responsible—was "at the time considered a military necessity by much more important people than myself." This was certainly literally true. Orders had now officially come from the very highest level. Further execution of them was now down to the planners and aircrew over whom Harris exercised his appointed command.

In the final phase, this is their story—and the people of Dresden's.

ACCOUNTS OF THE AIRCREWS' preparations for the Dresden raid vary in detail, but the basic sequence is clear, and it wasn't much different from the usual.

Preflight meal was in the midafternoon, and briefings—depending on whether the crews were going in with the first or second waves—took place from the late afternoon into the early evening. Many would already have found out that their aircraft were being filled with a maximum load of fuel, indicating a "deep penetration" trip, but the actual destination was supposed to be a secret until all the aircrew had entered the briefing room.

Lesley Hay describes the Dresden briefing as he recalls it today, with the careful precision of the civil servant he had been before the war and afterward became again:

> First of all when you get into the briefing rooms the huge wall chart is covered by a curtain and everybody goes in and the doors are closed and the police are put on the outside of the door and the squadron commander draws back the curtain. And he said, It's going to Dresden. Right at the back of Germany.
>   And my heart sank and I thought, Crumbs, that's a long way, and even although a great deal of France had been liberated we hadn't even crossed the Rhine—we hadn't even got to the Rhine.

All the aircraft going in the first wave were from Bomber Command's 5 Group. This force's reputation stood very high. Sir Arthur himself had commanded the group for fourteen months in 1939–40, before being posted to Washington—from where he returned to take overall charge of Bomber Command. Based from the end of 1943 at Swinderby in Lincolnshire, the group had become renowned for bold, innovative, and very skillful operations. These included the Bremen raid of October 1941 (the largest single raid so far in the war), the attack on the Schneider armaments factories in central France (October 1942), and the famous "dam buster" operation, in which the Möhne and Eder dams in western Germany were breached using a specially designed "bouncing bomb."

In 1944 a system of low-level target marking was developed, largely by Wing Commander Leonard Cheshire, who had now taken over what became 5 Group's Pathfinder squadron (627th). This routine, peculiar to 5 Group, distinguished its operations from those in which the general Pathfinder force (8 Group) prepared the targets. The system enabled it to bomb enemy communications targets in occupied France during the period before D-Day with greatly enhanced accuracy. In June 1944 aircraft from 5 Group dropped the first twelve-thousand-pound "Tallboy" bombs, blocking the Saumur tunnel and thereby cutting off the area of the Normandy landing from southern France. In August they caused heavy damage to the all-but-impregnable U-boat pens on the French Atlantic coast. In November bombers from 5 Group also sank the Tirpitz in Tromsö fjord in

Norway, one of the great precision attacks of the Second World War. For this they flew a round trip of more than two thousand miles from an advanced base in Scotland.

Inasmuch as the RAF had an elite bomber force, 5 Group was it. Its fourteen squadrons, theoretically totalling around 250 aircraft (244 took off for Dresden), represented a formidable attacking force. Operating as a unit against Darmstadt in September 1944, a total of 226 Lancasters and 14 Mosquitoes of 5 Group had unleashed a savage firestorm, which had devastated the center of this city of 120,000 inhabitants and caused the deaths of around 12,000 people. Their lethal effectiveness was put down to the group's sophisticated marking methods, which "produced an outstanding and concentrated raid on this almost intact city." Ominously, this degree of mayhem had been caused by just one wave of bombing. In the case of Dresden, also an "almost intact city," there was a second wave planned for three hours later, involving more than twice as many aircraft.

But for now the men of 5 Group were concerned with finding out the whys and hows of this new deep-penetration excursion into Germany. Like Lesley Hay, bomb aimer John Aldridge, a quietly spoken Norfolk man, was also at 49 squadron's briefing. The intelligence officer concentrated on the supply and reinforcement aspect of the coming attack on Dresden:

> Well, they said the reason for the raid was chiefly the supply to the Russian front, blocking the supply to the Russian front. It was quite feasible to us that it could be a main supply center for the eastern front and we were out to knock it out so far as we were concerned. But I don't remember anything about industry. Anyway, that's all I remember about the target. Well, I also remember about the refugees coming the other way, and that wasn't a very pleasant thing, but there we are.

He was not to be the only man listening to the briefings who felt a little uncomfortable at the mention of refugees—had not the Germans been pilloried earlier in the war for their ruthless strafing and harrying of refugee columns? But most, perhaps, accepted as John Aldridge did that in a situation such as this, the lines of supply and of retreat were the same. Incidental destruction was inevitable.

Lesley Hay also registered the talk of German reinforcements and supplies, but also a little more detail of the intelligence officer's talk and his references to a large map showing the city's position as a center for both north-south and east-west movements:

> He says, not a great deal is known about Dresden. But says of course as you see and you can appreciate, it is a vital supply route. We know from experience that bombing a place like Dresden must cause great disruption right the way up to the front and would certainly hinder stuff going up there. Intelligence also said that the Germans consider it a safe city, and it immediately came to my mind, if they thought it a safe city, what were they making there? Because we'd had the V-1s coming and the V-2s coming over lately, and what were they preparing after them? If they thought it a safe city, it could be a very good place to make these things, so I thought it was a good target . . . They didn't know what the armaments were surrounding the place. Normally they could tell you, they have four hundred guns and so many searchlights, but he said we have no information.

The fullest intelligence from that time is preserved in the Dresden target information files that can be consulted in the London Public Records Office. It is undated (probably 1942—an ancillary document, dealing with the Friedrichstadt marshaling yards, originates from February 27 of that year) but includes a large "zone map" of the city specifying (accurately) the density of buildings in various parts of Dresden, fanning out from the very high-density Altstadt and also marking areas of industrial development (again also matching the actuality), plus the garrison areas and the main utilities.

Such zone maps were produced in 1942–43 for most major German towns in accordance with the new policy of "area bombing," where population and building density became key factors in target choices. Attached to the zone map is an "information sheet" (three pages foolscap). This contains general "Bomber's Baedeker" information about Dresden, its area and population, and "Lay-out."

"Dresden," it tells us accurately, "is the historical center of Saxony, its present administrative center, and an industrial center of considerable importance. It is one of the finest residential cities in Germany."

The city is divided on the map into five color-coded areas or zones:

1. Central city area (very densely populated).

2. Compact residential areas (divided into fully built-up areas and partly built-up areas).

3. Suburban areas ("they consist mainly of single family areas and are very open and scattered").

4. Industrial areas.

5. Railways and ports.

Public buildings are also listed and located, though not color-coded. Hospitals, which had been marked on earlier maps—supposedly as places to be avoided—no longer appear, in recognition of harsher military realities.

What the maps cannot show is what was actually being produced in those industrial areas and factory buildings. It seems likely that the best British intelligence could do in this respect was to extrapolate from prewar patterns of production. This, as has been seen from the quite radical nature of the switch from the manufacture of consumer goods to armaments and war-related production in Dresden since 1939, would have given a less than adequate picture. There is no mention in the accompanying information sheet of war industries as such.

Two years later a separate target information sheet dated September 29, 1944, correctly surmises—and marks on an excellent July 1944 aerial photograph—the position of two war-related factories in the shape of the Universelle J. C. Müller Factory (Target A) and the neighboring aircraft components production facilities in the former Feldschlösschen brewery (Target B), a few hundred yards southwest of the Hauptbahnhof. It is noted that "in Target A the buildings have been darkened to tone in with the surrounding area. No camouflage is visible on Target B."

The box factory in the area between Friedrichstadt and the Alberthafen river port gets it own target file (dated December 15, 1944), which also describes the oil storage tanks in the neighborhood, their approximate capacity (38,165 tons), their lack of camouflage, and the fact that, although a protecting wall was being erected around one of the forty-foot tanks at the Rhenania Ossag depot, "the other tanks remain unprotected against blast."

The railway stations and yards are also fairly well covered, but there is almost no detail about the other major factory complexes in Dresden.

It seems that toward the end of 1944 Bomber Command started to take a little more interest in Dresden—perhaps due to a growing awareness of how much of German industry was being transferred to "less endangered" areas—though intelligence on Dresden was sparser than on most other cities in eastern Germany. Leipzig and Chemnitz were of course well known for their involvement in the large-scale manufacture of aircraft and tanks, so would naturally have attracted more attention than Dresden's light industries and smaller, more "high-tech" factories.

So briefing officers on the afternoon of February 13 kept it general. What they didn't say was that the aircraft being dispatched to Dresden on the night of February 13–14 had the task of simply destroying as much of the vital center of the city as possible. That didn't mean that Dresden was not a significant industrial and military center, and thereby a "legitimate" target. But this attack was about creating overwhelming disruption, as near to a perfect state of chaos as could be inflicted.

It is true that the Pathfinders had to be issued with a fairly crude photomontage put together in November 1943 instead of the more usual sophisticated multicolor target map showing full details of defenses and individual targets. This first comprehensive photo map was the result of a Mosquito photographic mission in September 1943, probably connected with the coming attacks on Berlin (other major east German cities such as Leipzig and Stettin were systematically photographed at the same time). There was actually a lot of photographic material about Dresden in British and American hands at the time, although it had in many cases not yet been processed sufficiently to provide the basis for the more usual kind of target map. Aerial photography had continued throughout 1943 and 1944, and then there was the material from the American raids on Dresden in October and January. Such files often took some time to be processed, evaluated, and used at the central photographic evaluation office at Medmenham, near High Wycombe, where pressure of work ensured ruthless prioritization, but by February 13, 1945, the Allies knew enough about Dresden to attack it successfully, as events would prove.

After the intelligence briefing came the meteorological officer. He summarized the official Bomber Command forecast that Harris and his staff had been given that morning when pondering their decisions. At first, there was little to cheer the crews, as Lesley Hay remembers:

> The met officer comes on . . . and he says I'm afraid I've not got good news for you. It's that the cloud base is down to about three or four hundred feet and you've got to take off in that. And you can't climb above it, you will be in cloud the whole way, and you will all be together in cloud. You've got your work cut out . . . that's the command order. And when you get there, they think there could be a break around ten o'clock in Germany. But they're not too happy about it and we may need to bomb on H2S.

Then the meteorological officer added a personal touch. Hay, who trusted him and often chatted with him in his office between operations, listened keenly:

> However, my forecast is that about ten to fifteen miles from the target it will break. And you'll get for a short period a clearing. You might be lucky then and you might not be, but that's my forecast.

And that was that. There was some more technical stuff, but soon it was time to leave the briefing room and get the crews together.

At this point, the aircrew about to go on an operation would all empty out their pockets and put the contents in a bag. This they would then hand in, so they carried nothing that would give information to the enemy if the crew was brought down over Germany. For the Dresden raid, since it was so close to the Russian lines, they had each also been issued one special piece of equipment: a Union Jack to put across their chests, printed with the words—in Russian—"I am an Englishman."

These items were supposed to ensure their safety if they landed in Soviet-held territory. As John Aldridge commented dryly, aware of the Russians' trigger-happy reputation: "We thought they were not much of an asset—rather they would present a better target!"

Finally they collected their flying gear, received their flasks of hot tea and their sandwiches from the mess, and headed out to their aircraft.

The long night had begun.

• • •

On the way out there's usually some comedians trying to put a brave face on it, but it wasn't a very happy-looking trip. All the same, we were pretty experienced at flying groups, we were a specialized flying group and we knew our job.

When they got out to their aircraft—about 6 P.M.—the pilot started the engines and left them to warm up. Then, as commander of the aircraft, he had formally to sign for the 90,000 pounds' worth of taxpayer-funded machinery (1945 prices—multiply at least by twenty for a modern equivalent) and thereby take full responsibility for it. The procedure, on the chilly, overcast February evening in question, then went like this:

We shut down and wait for the green Verey light to go up from the control tower for takeoff, and while we're there the station commander and the padre usually come down—always thought it was the last rites, myself—and the wing commander comes and says, all right, Uncle? Everything okay? And I say, yes, sir, all right. Padre exchanges a few words and that's it and away they go.

Then the green light goes up and it's all aboard. You start up, wave chocks away, and taxi and then slot yourself in between any who are coming round, so the whole squadron moves away in line going down to the far end of the long runway, which faces five to six degrees—something like that—straight for Lincoln Cathedral five miles distant.

At the end of the runway you queue up and wait; if there's no hold-up you're okay but if there's a hold-up you've got to turn into the wind so that the wind comes into your engines because otherwise you'll overheat. But fortunately we go up quite nicely and take off on time, and up we go. The Lancaster goes up and at the end of the runway, with brakes full on, you start opening your throttles full. Then you suddenly release and . . . ssshhh . . . she goes forward and the right throttle goes because the torque takes you off to the right-hand side . . . up you go down the runway until you feel she begins to lift.

Lesley Hay took off at 6:18 P.M. According to squadron records, that put him and the rest of the crew of "U for Uncle" almost exactly in the middle of the squadron (the first left the ground at 6:10 P.M., the last at 6:25). And when they got to five hundred feet, then turned to port, they met thick cloud. It made the trip down to Reading, though less than an hour, a little nerve-wracking.

The large town of Reading, situated on the Thames thirty miles west of London, was a regular assembly point (or "meet-up" as the Americans called it) for RAF bombers bound for Germany. The assembly was due to be completed at 7:13 P.M., by Hay's records. Despite the conditions, he and rest of his squadron made it on time. They maneuvered into position at their allotted heights and speeds. Then they took off two degrees east, on a course that would take 242 remaining Lancasters across Beachy Head and over the English Channel.

A few minutes before 8 P.M. they breasted the French coast, halfway between Boulogne and Le Havre, and the real game began: the deadly game with the German air defenses and their controllers—one that became more lethal, and more complicated, the further the British were due to penetrate German airspace. For five nerve-wracking hours they would be over enemy territory and so, even at this late stage in the war, exposed to the German flak and the enemy's night-fighters.

The complex system of feints, bluffs, and "spoofs" that the Allied air forces had developed over the years would be fully engaged tonight. In all, more than fourteen hundred aircraft would be in operation over Germany, almost half of them chiefly for the purpose of confusing the enemy. To provide convincing cover for the Dresden attack, Bomber Command was putting on another big raid, one that the German defenders could not possibly ignore.

A force of 368 aircraft would hit Böhlen, the hydrogenation plant north of Leipzig, which turned brown coal into engine fuel. The force, taken from parts of 4, 6 (Royal Canadian Air Force), and 8 Groups (the rest of which would go in the second wave to Dresden) was scheduled to be over the Böhlen at 10 P.M., fifteen minutes before 5 Group was due to start bombing Dresden. This would give the enemy air defense controllers a lot of different things happening at once in much the same area of the country, and therefore a great deal to think about. To complete the picture, seventy-one Mosquitoes would bomb Magdeburg in two waves, sixteen would visit Bonn, and smaller groups of Mosquitoes would also attack Misburg, Nuremberg, and Dortmund. Plus there would be sixty-five RCM sorties and fifty-nine Mosquito patrols. Tonight there would be more British aircraft of all kinds in operation over Germany than since the night of October

14–15, 1944, when devastating raids had been launched against the Ruhr town of Duisburg and the city of Braunschweig.

The actual route plan for 5 Group, heading ultimately to Dresden, was a carefully constructed minor masterpiece of deception. Once the aircraft reached the continent, at a point southeast of Boulogne, they would fly due east to Belgium, and from there, northeast to an area just short of the Ruhr. Then suddenly they would fly up, heading north-northeast for a couple of hundred miles on a trajectory just north of Kassel. From here—this was the point—you could still be heading for any number of places: Berlin, Leipzig, Chemnitz, or one of the big, well defended oil plants in Saxony or the Sudetenland.

The whole scheme was about keeping the German defenses guessing until the last moment. That was why the bombers needed so much fuel. They would hardly ever be flying direct. They would always be involved in feints and decoy schemes, which stretched a twelve-hundred-mile trip into one of seventeen hundred miles.

> They could pick us up from around here. Their Freya would get us . . . We would have a Mandrel screen going down, and we would break his screen, and the German controller would say to himself hello, there's something coming now, what is this? Is this Five Group coming on one of their advanced raids, or a few Mosquitoes windowing just to make them look like a bomber stream?
>
> He sits tight and waits until we get to two degrees here and we turn east for Frankfurt . . . He's probably well experienced, these night controllers, they have a huge team around them watching everything, reports coming in from everywhere, what information they've got . . . So we fly along due east and if he's guessed right we turn north and he knows we're going for the Ruhr, and he knows it's not the Dortmund-Ems canal because we've already done it and it would be another two weeks before we went there again. He knows it isn't that. But it could be any target up around here, could be Münster . . . and then we suddenly turn east, a little bit northeast of east, could be Cologne. Will he raise his fighters—can he get his fighters off the ground? We don't know.

Even the craftiest German air defense controller could not be sure of 5 Group's destination. Such practiced deception, which the British learned when the German night-fighter force was in its prime, remained routine. Only occasionally during this winter of 1944–45 did the Germans manage to get their defenders up into the air—but when they did, appearing among a bomber stream when least expected, they could still ensure that a lot of young men never made it back to Lincolnshire for their bomber's breakfasts.

Tonight, as 5 Group passed near heavily defended areas like the Ruhr, or Frankfurt, or Hanover, they were to a great extent protected from the equally lethal flak by the very cloud that made flying to the target in formation, next to their invisible comrades, such a matter of skill and anxiety. German antiaircraft gunners at this stage in the war did not like to waste their ammunition. They preferred not to shoot at things they could not see. So every circumstance has its good side and its bad. In the end no German fighters appeared at any time on the way out, to the relief of the British aircrew:

> It is 8:51 P.M. My navigator says I'll be turning in about sixty seconds, Okay? I know the course, but he'll give it me according to what he's worked out in the wind. He finds a new wind every six minutes. He's working hard all the time . . .
>
> We're heading for Magdeburg, so the enemy thinks, are they going to Magdeburg, or is it Berlin? Then he thinks we're going for oil, probably Leipzig. Or Böhlen. Or Rositz. So he could bring his fighters in here . . . whatever he's put up earlier will have to refuel. But he might bring some more in here. On the other hand, he might be thinking about getting us on the way back. Anyway, we keep going. It's 22:00 . . . And here we're heading straight in for Dresden . . .

They are still navigating by instruments.

We are still in cloud here. But as we get down toward Dresden, all of a sudden we run out of cloud. It isn't completely cloudless, but ahead of us is a big hole and I think, by golly, he's right! Our met man is right!

Meanwhile, at Böhlen, just a few minutes' flying time to the northwest, the slower, more vulnerable Halifaxes are making their bombing run on the well-defended oil plant. For them there are no fortuitous holes in the cover. The target is, according to the operational reports, covered by 10/10 stratocumulus cloud, allowing only a few of the initial green TI (target indicator) markers dropped by the Pathfinders to be seen.

The master bomber of the Böhlen force decided it was pointless to drop any more markers, given the very poor view of the ground. He ordered his remaining marking aircraft to save their flares. He instructed the attacking force to "bomb near the edge of the green TIs ... then towards the center."

Rashes of dummy flares were dimly discernible on the ground, lit by the defenders in an unsuccessful attempt to confuse the intruders. "One big dull orange explosion" lit up the target for several minutes, and fires were started, but crews considered the attack to be "scattered."

But then the Böhlen raid had been designed almost entirely as a decoy move. Attacking a relatively small, precision industrial target like this in such weather would never normally be considered.

Böhlen was lucky. And Dresden, forty miles or so to the south, correspondingly unlucky.

18

Shrove Tuesday

ON JANUARY 1, 1945, unknown to the people of Dresden, their city had been secretly classified as a military strongpoint, a "defensive area" (*Verteidigungsbereich*).

The order for this had been issued by no less a person than Colonel General Heinz Guderian, chief of the Army General Staff. The difference between a "fortress" and a "defensive area," as defined over the next few weeks by a trickle of secret communications from the OKH (Army General Staff) and the OKW (Wehrmacht General Staff), was a question of practicality rather than principle. Fundamentally, a fortress was a town or city with permanent fortifications, while a "defensive area" was a town provided with defenses that, however formidable, were of a temporary nature. Both had senior officers specifically assigned to them, whose function was that of a fortress commander and whose orders took precedence over those of the ordinary civil authorities. Both were intended, in late-Hitlerite Götterdämmerung fashion, to be defended to the bitter end.

Dresdeners continued in the illusion that their city still enjoyed "special" status because of its cultural distinction. They did not know that Berlin had appointed a certain General Adolf Strauss as "commander in chief of the eastern fortifications" and that he had been ordered to create an "Elbe line," running from Prague (via its tributary, the Vltava) through Dresden and on to the mighty river's mouth at Hamburg. Chief among Strauss's tasks was to strengthen the "defensive areas of Magdeburg, Dresden and Prague," where the blood of countless Red Army soldiers would supposedly be spilled in costly house-to-house fighting amid warrens of streets—just as it had been in the already established "fortresses" of Breslau, Königsberg, and Posen.

So from the first day of 1945, "Florence on the Elbe" was designated a candidate for the fate that Breslau and Königsberg were already suffering—besieged, starved, shelled, and bombed to ruins—should the Soviets reach the Elbe line. It was just that the authorities had decided not to tell Dresden's inhabitants. For their own good.

The existing city commandant of Dresden, General Karl Mienert, was in his sixties and had seen no frontline action since the First World War. Mienert was temporarily left in charge of the Dresden defensive area, with power over the city authorities and police, while Berlin sought a more appropriate replacement. He still had not arrived on February 13. Building of antitank ditches started, and a special staff was set up in the basement of the extravagantly rococo Taschenberg Palace, next to the old royal *Schloss*.

Meanwhile the authorities tried, without giving too much away, to spread martial feeling. They were going to have to inject a touch of steel into the daily life of Dresdeners if the de facto defensive area was to be "sold" to a population still floating happily under the illusion that their city was too beautiful and too famous to suffer as other population centers of the Reich had suffered.

Accordingly, the women's section of *Der Freiheitskampf* carried an article that took the form of a dialogue between a war-experienced woman from Cologne and her cosseted Dresden counterpart:

> What would the lady from Cologne probably say to her friend in Dresden? "Don't get upset, dear Dresdener! Concrete bunkers, tank barriers and detonation slits in all our bridges—we got used to those even in peacetime. Such security measures have always been a source of reassurance to us on the Rhine, and have enabled us to live safe from the enemy. Therefore, dear girlfriend from Dresden, accustom yourself to all those things your men are doing for you. Love of the Fatherland is born in a strong heart. Don't let yourself be weakened by the chattering of illegal radio listeners, faint-hearts, and rumor-mongers."

The date on this edition of the party newspaper was February 14, 1945. It had been printed on the evening of February 13, 1945, and was awaiting distribution when the first British aircraft appeared over the city.

• • •

MUCH MORE IMPORTANT to the vast majority of Dresdeners, who knew nothing of the enhanced military status of their city, was the visible, tangible evidence that the Soviets were nearing Saxony.

This proof was amply provided by masses of refugees from the east. In late January the Russians had invaded the neighboring province of Silesia. Almost the entire German population of the province had begun to flee westward, bringing what few possessions they could carry. When the Silesians, and others expelled from the recently "Germanized" parts of occupied Poland, arrived in Dresden, the beautiful, all but undamaged city must indeed have seemed an "island of peace."

A great *Völkerwanderung* (movement of peoples) was under way, marked by expropriation, rape, and murder—what the late twentieth century would call "ethnic cleansing." Millions of Germans, whose ancestors had lived in these areas since the Middle Ages, were fleeing the Russian advance—and the revenge of their other Slavic neighbors. Two hundred thousand civilians remained encircled in Breslau when the Soviet trap closed in the first week of February. Those who had escaped had no choice but to head west, which in practice meant either Berlin or Saxony.

For the Nazi Party, the acceptance of "racial comrades" from the east was an essential duty. Of course, for propaganda reasons it had to be implied that the presence of such folk was a purely temporary phenomenon, until the inexplicably mislaid provinces were recovered. The official Nazi organ for Saxony, *Der Freiheitskampf*, urged citizens to offer temporary accommodation.

> There is still room everywhere. No family should remain without guests! Whether or not your habits of life are compatible, whether the coziness of your domestic situation is disturbed, none of these things should matter! At our doors stand people who for the moment have no home—not even to mention the loss of their possessions . . .

Dresden had been accepting refugees from the devastated cities of the Ruhr, and from Hamburg and Berlin, ever since Bomber Command's bombing campaign began in earnest. By late 1943 the city was in fact already overstretched and finding it hard to absorb more outsiders. Now, in the sixth winter of the war, the pressure was coming from the opposite direction—the east—and the human beings affected were numbered no longer in thousands but in millions.

Many offices and organizations existed to aid and find quarters for the refugees from the east. The NSV (National Socialist People's Welfare) and the women's and youth organizations were especially involved in caring for new arrivals. These organizations worked tirelessly and well to provide temporary accommodation, soup kitchens, and basic medical treatment.

There remained, however, limits on the city's capacity to take in outsiders. Dresden had always suffered a shortage of housing, even in normal times, and by the end of 1944 the situation had reached saturation point. The air raids in October and January had further diminished the stock of accommodation. The Saxon capital naturally acted as a magnet for refugees, but the local authorities—patriotic appeals in the party-controlled press apart—didn't actually want them. At least for more than a day or two. Soon notices to that effect went up all around Dresden.

On February 8 the Red Army crossed the river Oder, on the tenth capturing the ancient west Silesian town of Liegnitz. Here in the thirteenth century the flower of the Polish aristocracy had fallen in the attempt to keep the Mongols from Europe. Before the Russian offensive began, Liegnitz had been declared one of the safe assembly points for westbound refugees. It had been assumed that, if worse came to worst, the Soviets would halt at the Oder, but the unstoppable pace of the Russian advance—from Warsaw to the edge of Saxony in just over three weeks—proved overwhelming. Thousands of refugees ended up trapped in Liegnitz at the mercy of the Soviets. Nowhere seemed safe. And increasingly, no one believed the authorities' assurances.

The human tide of misery could only swell, fed further by reports of brutal orgies of murder and rape in the German territories now at the mercy of the Soviets. The stream of dispossessed German civilians from the east had become a flood. The numbers of people on the move probably exceeded even the most extreme projections that the planners on the Joint Intelligence Committee and the Bomber Command staff had conjured up the previous month when they first discussed possible attacks against eastern Germany.

• • •

IT WAS NOT EASY, late in the war, to become a Dresden resident. In December the city had been declared *Zuzugssperrgebiet* (zone forbidden to new residents). The authorities enforced these restrictions by means of the ration card system. While for existing residents the cards were distributed, as usual, through the local Nazi Party organs, new arrivals had to seek out ration card offices. They would receive cards valid for Dresden only if granted permission by the local housing office or the municipal police to live in the city. This in turn depended strictly on their showing a specific reason for moving to Dresden (say, a job) and proof of a place to live.

Fleeing the Russians was not a valid justification for seeking residence. In the latter case, the regulations told the authorities to "direct the persons concerned immediately to the NSV railway station service, who will arrange their passage to other appropriate places or districts . . ." This prompt expulsion to the west was the fate of the vast majority of refugees. On February 6, the regulations were tightened even further. Individuals who had no permit to reside in Dresden must be given only sufficient food to sustain them until transport out of the city was provided, and under no circumstances should ration cards be distributed to such individuals.

It was, in general, the aim of policy to have refugees on their way to the west within twenty-four hours. In the meantime they might be accommodated with local families who had answered the patriotic call, in schools and other suitable public buildings, or—especially if they had arrived by train—overnight on station platforms. In the case of the Hauptbahnhof, the storage cellars had also been crudely converted into a warren of by no means state-of-the-art air raid shelters.

Dresden historian Matthias Neutzner calculated the number of refugees and displaced persons in Dresden during the turn of the year 1944–45 at "maximally in the order of several tens of thousands of persons." While the numbers must have increased as a result of the increased flow of refugees during the weeks that followed, some of these, at least, would have found accommodation in private homes. Most of the rest would probably have been quickly on their way. Tellingly, there is no indication of compulsory billeting of refugees, or of the provision of camps for their long-term use.

In the city center, only the schools, which had been closed the previous summer, would have provided extra short-term accommodation. An extra few tens of thousands of refugees would have been present in these transit camps and on the stations on a nightly basis . . . After, in some cases, weeks of trekking on foot or days-long train journeys, they would reach the city, to find that they would be cared for, provided with accommodation, and fed—and then sent on their way again as soon as possible.

One of the actual carers at the Hauptbahnhof, who during this time spent long days and nights working with the refugees there and at the Neustadt station, was Götz Bergander. He had turned eighteen on February 11. Released from service with the flak auxiliary in October 1944, and awaiting induction into the Wehrmacht, Bergander spent time on fire-watching duties and on refugee relief work. He described the situation in and around the city's main station at 9 P.M. on February 13, 1945, just as he was finishing his shift:

On the station, with its dim blackout lights, a confused mass of human beings flowed slowly this way and that, and the waiting rooms were, as ever, full to bursting. Outside on the Wiener Platz, there were groups of people standing around. However, neither on that night, nor in the days and nights previous to this, were there tens of thousands camping in the open air on the streets and squares, on the meadows by the Elbe or in the Grosser Garten. Horses and carts naturally were resting in the open, but never in such a concentration that the streets were blocked. For there to have been a half a million or more refugees in the city, there would have to have been hundreds of thousands of them swarming around outside.

The horse-drawn vehicles were there because the station had water, which the refugees could use to drink and to bathe their children. And they could use the station lavatories. The notion that the streets and squares and parks were teeming with thousands upon thousands of anonymous peasant families and their horses and carts Bergander dismissed as "complete fantasy."

Another eyewitness told a similar story:

The city was not overflowing with people, or at least not to such an extent that all the streets, squares, and green spaces were packed with refugees. At that time we used to go to school some way beyond the Hauptbahnhof, and have no memories of such masses. The station, however, was full of refugees. That is true.

Bergander estimates that around a hundred thousand might have been provided for in this way—mostly arriving on trains, but also including thousands of "trekkers" arriving by horse-drawn means or on foot with handcarts. But they were not all, by any means, within the area later affected by the air raids. Accommodations were available not only in central Dresden. On the contrary, they were spread around the city in districts that had not been bombed and never would be. Typical was the spacious outer suburb of Loschwitz, where Pastor Hoch's prosperous family had several refugees staying with them in their large villa at the time of the air raid. Another high school student gave an account of escorting a Silesian farmer, his family, and their cart some miles out to a distant waterside suburb.

Bergander considers the total number of households in greater Dresden, and ends up with the estimate that for anything like the supposed five hundred thousand to one million refugees to be accommodated within the city limits, *each* existing family would have had to take in *several* refugees on a semipermanent basis. This would have required a massive government program of compulsory billeting, which was never even contemplated. It also ignores impossible difficulties with the ration card system.

"We were expecting friends from Silesia," says Gertraud Freundel, then eighteen years old. Her family lived in a pleasant roomy flat just south of the Hauptbahnhof.

> Refugees, yes. We had arranged extra beds. My father came from Oberlausitz, that's close to the Silesian border. We had left some of our belongings with relatives. Mother had gone to fetch them so we would have something to put on the beds. She saw the fire that night from a distance ...

So in this case, on the night of the raid, two outsiders had been due to be added to the population of Dresden, while, as chance would have it, one who *should* have been there was not. Multiply such haphazard circumstances by tens of thousands, and accurate calculation is difficult. Bergander's best estimate is that during the previous days and weeks a further hundred thousand refugees might have been accepted into the homes of existing Dresden residents, because they were friends or relatives, or friends of friends, or out of common human kindness. This makes a total of some two hundred thousand nonresidents in the city on the night of February 13–14. Many of these would have found, or been assigned, quarters away from the vulnerable center of Dresden.

None of these estimates by an acknowledged expert plays down the number of refugees who passed *through* Dresden and its surrounding area during those weeks. The numbers probably ran over the million mark. But they were not all there at the same time; and only a fraction is likely to have been unlucky enough to be lodged in the affected districts of Dresden on the fatal night of Shrove Tuesday.

ON THE AFTERNOON of February 13, 1945, Victor Klemperer was also engaged on government business.

The previous day he and other elders of the now tiny Jewish community in Dresden had been charged with delivering to every remaining Jew a deportation order. The order, signed by Dr. Ernst "Israel" Neumark of the Reich Association of German Jews, told them to report on Friday, February 16, at 6:45 A.M. to No. 1 Zeughausstrasse (one of the "Jew houses" opposite the now nonexistent synagogue) for "a work detail outside Dresden." They should bring hand luggage and provisions for two to three days' march. They were to bring work clothes, blankets, sheets, and shoes, but no money, foreign currency, bankbooks, matches, or candles.

> I must emphasize that this order is to be obeyed *unconditionally* without regard to all existing conditions of employment. Otherwise, state police measures are to be anticipated.
>
> I ask you to confirm the receipt of this communication on the attached slip.

Such was the letter that Fräulein Henny "Sara" Wolf received at the family's flat in the Glasshüter Strasse, where they were now living in severely reduced circumstances. An identical instruction was delivered to her mother. Both of them, and Henny's "Aryan" father, knew exactly what this meant. Every Jew or part-Jew knew. Whoever's signature lay underneath it, this order came from the Gestapo. It promised at best transportation to the Theresienstadt ghetto, at worst a death march of the kind that had already consigned tens of thousands of Jews to a bitter and brutal fate just as the new Allied advances seemed to bring deliverance so tantalizingly close.

Early in January 1945 Henny Wolf, now grown into a pretty young woman of twenty, had been ordered to an interview with the Gestapo for the first time alone. Her mother stayed home, helpless with anxiety. More than one acquaintance had committed suicide rather than answer such a summons. Henny's father accompanied her there, but had to remain outside. The doorman bellowed at the young woman: "*Sara* Wolf, upstairs that way!" She obeyed, sick with fear.

> There were four or five men in the room, smoking cigars, seated in club armchairs. They asked me about things they already knew, whether my parents' marriage was a mixed marriage or not, why I wore the yellow star, and similar questions. All just pure, unadulterated harassment. I shall never forget the moment when I emerged from the building and saw my father standing outside. In the hour I had been with the Gestapo, he had aged years, convinced that they intended to hold me there.

For the first and only time, Herr Wolf exercised his right as an Aryan to take the tram home. He had to leave his daughter, for whom all public transport was forbidden, to walk back to the apartment alone. They agreed that this was in order, so that the agony of Henny's waiting mother might be ended as soon as humanly possible.

Also at the beginning of January the night shift at Bauer's box factory had suddenly been abandoned, leaving the Jews without employment (except for bomb-site clearing and the like). No one knew why. Or not until the deportation order came just a few weeks later.

The Wolf family had less than three days to think about their response. They decided immediately that rather than allow the family unit to be broken up, and Henny and her mother to be shipped off to a concentration camp, the women would abandon their yellow stars and they would try to "go underground." There was little prospect of doing this successfully, because without Aryan papers, how would they obtain sufficient rations, or a place to live? And what if there were identity checks? But it was better to perish like that, as a family, they told themselves, than separated and lonely amid the obscene, impersonal violence of the concentration camp system.

So came the evening of February 13. Henny's father, still in shock from the news of the deportation order, lay stretched out fully clothed on his bed. This was the first time his daughter had ever seen her highly fastidious father do such a thing.

The air raid siren sounded. At first they thought it was yet another false alarm. The Wolfs had no radio, because in a house containing Jews this was not permitted, so they received no warning of the approaching British bombers. Henny and her mother were not permitted to use public air raid shelters.

A few minutes later, however, their doorbell rang. It was the air raid warden, a decent older man who had known the family for years. He told Henny's father to bring the family down into the air raid shelter. Herr Wolf said this was forbidden, but the man gently insisted. They filed down behind him into the cellar beneath their apartment building.

YOUNG GÜNTER JÄCKEL, seven years old and just starting school when Hitler came to power, was now eighteen and a half and in uniform. The fortunes of war had taken him far away from his hometown of Dresden, and then back again in the space of a few months.

In the summer of 1944, along with most of his schoolmates, Jäckel was conscripted into the armed forces. He managed to join the Luftwaffe. Jäckel had left school at sixteen, taking a job as a trainee clerk at the local government office in Pirna, just outside Dresden. Still in the Hitler Youth, he also undertook gliding lessons. This, he believed, was why he was sent to the Luftwaffe when he reached military age rather than as cannon fodder to the eastern front.

Not that Jäckel became a flier. By mid-1944, with the air war now horribly one-sided and ground troops desperately required, he and his comrades were sent to what was left of German-occupied France to fight as infantry. German resistance had stiffened, and fierce fighting was taking place in the eastern part of the country. Jäckel sardonically recalls his bemusement, as a boy who until now had scarcely left the safety of Dresden, when he realized how hated he and his young fellow country-men really were in the places their elder brothers had conquered. He remembered his unit's withdrawal, as night fell and the danger of Allied aircraft diminished, from a small town northeast of Dijon:

> The last of the fighter-bombers had trailed away into the twilight; the route eastward across the plain lay open. But the street, which led down from the higher part of the town, was lined with women and children, a scattering of men. They stood there silently and just stared at us as we marched by. Expressionless. I have never again been looked at like that; and I don't think I have ever felt such shame—without even knowing why.

October 1944 found his unit entrenched in an apple orchard near Belfort, facing strafing attacks from American Thunderbolts and Mustangs, and—as the front line got closer—systematic air-guided shelling. It was during this fighting that Jäckel and his comrades came under fire from American tanks. His arm was filleted by shell splinters. The boy next to him was almost certainly killed. Only the fact that Jäckel had ducked down at that particular moment saved his own life.

Jäckel was evacuated out of the combat zone and taken to a military hospital run by nuns in Colmar. There, things turned bad. He got a high fever, and his arm swelled up. The Germans had no penicillin. The wound had become infected. He was shipped back into the Reich, and operated on in a Bavarian military hospital. It was there, starting to recover, that he heard his mother had died suddenly. Scarlet fever. A chance civilian death, from natural causes, in the midst of war.

Two days later Günter Jäckel arrived back in Dresden for her funeral. And also for a last reunion with his father, who had been con-scripted earlier in the war. After the ceremony, the middle-aged former tram driver went back to his unit on the eastern front. His son never saw him again.

The evening of February 13 found Jäckel, ambulant but still some-what feverish, in a convalescent hospital in the outer reaches of the Südvorstadt. The dirty work there—dealing with the laundry and changing the urine bottles and bedpans—was done by Italian prison-ers of war. They were a friendly bunch, men who had refused to serve the fascists after Mussolini's fall and so had ended up in what seems to have been a relatively relaxed, if not exactly fragrant, form of captivity. Then came the sirens.

> Just near me lay a Rhinelander, an Unteroffizier with decorations and so on, and when the first alarm went, he immediately packed a few things together. We Saxons laughed and smiled and said, "Oh, we're always having warnings!" But he was from the Rhineland, he knew . . . the Italians carried the badly wounded . . . and we followed reluctantly down into the cellar.

HANNELORE KUHN was born the middle child and only daughter of a prosperous middle-class household in the suburban area surrounding the Münchner Platz. Her father had studied law, but after service in the First World War went into real estate consultancy and management. Her mother, a teacher until she had her first child, was now a housewife. They owned a substantial villa in the Bamberger Strasse.

As 1945 began, Hannelore Kuhn had already lost her elder brother in the war. Her younger brother had been conscripted into the Wehrmacht. She was a diligent student. After gaining her high school certificate (*Abitur*) in 1942, instead of going to college she had first to complete a stint at a work camp in the mountains. Then, as part of the compulsory war aid service (*Kriegshilfsdienst*), she was sent to work for some months as a tram conductress in Breslau before being allowed to return home.

February 13, Hannelore recalls, was one of those mild prespring days.

> I had sinus trouble and I had been signed off by the doctor, so I was at home on 13 February . . . we thought, let's celebrate Fasching, go for a walk. We went up onto the heights . . . There was a road to the top and from there you could look right down over Dresden. We had a really good view. The sun was glinting on all the windowpanes in the city. And the snowdrops were already in bloom. Yes . . . and back down at the Münchner Platz, just nearby here, we said our farewells. One of us was a chemistry student who had been invalided out of the army after being almost burned alive in his tank. He lived down here in the bar-

racks, but studied . . . and he said, think of me tonight, because I'll be
in the guardhouse. I have gone past the curfew hour so it will be the
guardhouse for me. And we said good-bye. There were those three or
four friends present as we parted. I never saw any of them again. So . . .
Then I came home to here, just in the neighborhood, and there were
still a few kids around in the street, celebrating Fasching with silly
games. Then they went inside.

At ten that evening, when the air raid warning sounded, she was
resting in a chair, holding heated pads over her cheeks to relieve her
aching sinuses, just as the doctor had told her to.

THERE WERE FOUR CHILDREN from the Dresden inner suburb of
Johannstadt—two girls and two boys. They grew up together within
the same few hundred square yards north of the Grosser Garten
between 1930 and 1945, went to school and played together.

Nora Lang and Anita Kurz had been friends since before they
were old enough to remember. Nora had two brothers, one almost the
same age but the other just six years old. Anita was an only child.
Nora's father worked as a welder at the Gussstahl plant in distant
Freital, traveling there by bicycle for every shift, night or day. His was a
skilled job, but the family counted itself as working class. Anita's parents were more upwardly mobile, her father working at a bank before
the war, as a salesman for their security and safe-deposit services.

Johannstadt was a socially mixed neighborhood, tall castellated
apartment houses in an elegant late-nineteenth-century style grouped
around leafy squares. There were large and small apartments, but all
were reckoned light and airy. Shops and workshops had been incor-
porated into the blocks, so there were plenty of crafts people, even
small factories in the inner yard areas. It was a nice part of town to live.

On February 13, 1945, there was no school, because there was no
coal to heat the classrooms. Such shortages were now a frequent problem. Even when the coal could be mined, it could not necessarily be
transported. As so often in the past few months, they had collected
assignments from school. These they would deliver when circumstances
allowed. Things had gotten so bad that Anita's grandfather, who worked
in the railway yards in Friedrichstadt, had to help out with the family

rations by getting biscuits from the British prisoners of war he worked with, who continued to receive quite lavish parcels from the International Red Cross. Her grandmother, in exchange, did the prisoners' laundry. That night the children in the block had been playing Fasching with improvised masks, for a while after dark within the apartment house. Then they all went back to their homes, including Nora:

> My father had no night shift. He was home . . . we listened to the news before we went to bed. I didn't go to bed because the radio had said that enemy formations were approaching via Hanover and Braunschweig. And that was a direction where they could have been heading for Saxony. Could also have been Berlin . . . but we waited. First my little brother was put to bed. Then we listened to the reports but somehow we didn't realize . . . this was just before the air raid warning came. A terrible surprise. We all rushed to get our coats on, and then mother had to get the little one out of bed. And everyone had their piece of luggage.

The place where they hurried to take refuge was actually just a cellar.

> I was always frightened of the basement. There was no electricity. Not even in peacetime. You had to go down there with a candle or a lantern. There were about fifteen or sixteen of us in the basement. There was a table in the middle and we sat around it. In the middle a candle flickered. So, fifteen or sixteen people, all of those who lived in the building. There were a few children, two older men, and women. The youngest child was about six weeks old. Her mother was a young woman of twenty, twenty-one.

Almost exactly the same situation occurred in the basement of the nearby building where the family of Anita Kurz—Nora's best friend—lived, though it had a street entrance for the public and was signed as an air raid shelter. The one advantage was that it had electricity. Anita remembered:

> There was an old bathtub filled with water, and buckets filled with sand. You can't compare it with modern basements. It was more like a vaulted cellar. Like in a church . . . it was sandstone. Partly plas-

tered. And then there was a light bulb . . . Apart from me, two other children from the block. There were very few men. But my father was there. He worked as a civilian clerk with the Wehrmacht and was also usually on night duty, but that night he was there. So was the shopkeeper from downstairs. Otherwise mostly women and children . . . Once we were all down there, two of the men went up and looked out through the basement doors that led out into the yard, then immediately returned to say that they had seen "Christmas trees." Then it all started, the explosions. And it happened very, very quickly.

For the two Johannstadt boys, things were different, partly due to more pronounced class divisions.

Christoph Adam's family were middle-class. A thoughtful, serious boy of fourteen, he lived on the Dürerplatz, one of the largest squares in Dresden. Its solid apartment houses, at that time about fifty years old, were grouped around a green central square, running a hundred yards along the south side and two hundred yards from south to north, almost a village within the suburb. Academically clever, Christoph was a student at the famous Kreuzschule, a high school attended by the sons of Dresden's middle and upper classes and internationally famous for its choir:

So far, Dresden had been preserved from harm, and as children we had no conception . . . a few bombs had fallen in the city, but really we knew about bombing only from the weekly newsreel. The only concrete thing in those days was when parents were killed in action . . . but in our immediate circle that had not occurred up to that time. I must say at this point that our parents did not enlighten us. There was the radio . . . and they listened to foreign broadcasts, though they didn't tell us at the time. You see it was a time when people didn't trust each other. You couldn't confide in each other.

As far as he recalls, life went on with something approaching normality. Even with the Russian advance units just a couple of hours' drive away, there was talk of whether, under the tenancy agreement, the landlord was due to repaint the building. Dr. Adam—as he now is—adds that "things were rationed, but it's not as if Germans went hungry. I didn't. I was well clothed and well shod." And it was

Fasching, and there were festivities—decorations in the stairwell, children in improvised costumes.

Günther Kannegiesser's father was a fitter. By the latter part of the war he had been conscripted into the Wehrmacht. To help make ends meet, his mother worked at a camera shop. Günther was a handsome boy with a roguish smile—a tough, resourceful city kid. He was also fourteen, but his parents didn't have the money to finance any further education. Soon he would be looking for a job.

Meanwhile, young Günther was his mother's mainstay. He used to take his four-year-old little brother to kindergarten in the mornings so that his mother could go to work. His eight-year-old sister went to the same school as he did. After lessons, he then cleaned, fetched coal, and often did the shopping, before he could play with his friends.

Since the previous autumn, Günther had also been a war worker, acting along with some of the other neighborhood boys as a messenger and emergency helper for the local police headquarters, Police District Four, in Dinglinger Strasse. Every other evening the boys would report to the police station. There, though sleeping in bunk beds in the basement, they were on call in case of emergency. In exchange they received 1.50 Reich marks and a small extra ration token. The night of the RAF raid followed a regular sort of day:

> On the evening of 13 February my friend Fritz and I were still hanging out at the bowling alley in the Quellmalz pub on our street . . . So I wasn't in bed when the alarm sounded. I had already told my mother about the large numbers of enemy aircraft coming into the area. But that happened almost every day. After helping Mother get my younger sister and brother down into the air raid shelter (it was beneath the Suchy slaughterhouse), I headed for Schumannstrasse, where I met my other friend, Siegfried, and we then set off for Police District Four.

As they approached the police station, the British aircraft were already overhead. The boys could see them and hear them. They gazed up in adolescent wonderment at the display. Then they realized they could also hear explosions.

They hurried on, toward the deceptive security promised by the headquarters of Police District Four.

19

"Tally-Ho!"

BY THE TIME THE 244 heavily laden Lancasters reached the end of their five-hour flight and started to curve down the Elbe toward the night's target, 5 Group's master bomber and his team of eight marker aircraft had been waiting and making their preparations above Dresden for about ten minutes.

At 7:57 P.M., the master bomber had taken off from Coningsby in Lincolnshire. He flew in Mosquito KB 401-E, accompanied by navigator Pilot Officer Leslie Page. They quickly climbed to thirty thousand feet in their fast, wood-framed aircraft, and set course for Germany. At exactly the same time, Flight Lieutenant William Topper, in Mosquito DZ 631-W, and his navigator, Flight Lieutenant Davies, left Woodhall Spa. Topper, as lead marker among eight Mosquitoes of 627 Squadron, would have the job of marking the aiming point for the first wave of the main force. He and his fellows would carry out the low-level marking operation in a fashion peculiar to 5 Group.

To further confuse the German air defenses, the master bomber and the marker aircraft flew in close parallel with part of the diversionary Mosquito force heading for Magdeburg before slipping away over south-central Germany to make for Chemnitz and thence Dresden. Despite leaving almost two hours later than the main Lancaster force, 5 Group's master bomber and his marking team, riding the brisk northwesterly winds, got there before them, all according to plan.

Wing-Commander Maurice Smith, the master bomber, had already gained experience controlling attacks on sizable German cities, including Karlsruhe and Heilbronn. He was considered a specialist in "sector bombing," a refined form of area bombing practiced by 5 Group. Smith's job was to remain in direct contact both with HQ in England and with the aircraft taking part in the raid (in the latter case through the newly developed VHF short-range radios that the bombers now carried on board). He was therefore not just High Wycombe's liaison man on the spot but, in effect, director of the first act of the drama due to take place over Dresden shortly after 10 P.M. that night.

Smith's Mosquito, and the marker aircraft, also carried the new navigation aid LoRaN (*Long Range Navigation*), an American-developed device, similar to the Gee target-finding system but operating over more than twice the distance. Gee's use was limited by the curvature of the earth. Loran operated on a much longer wavelength and therefore was not restricted in this way. The disadvantage was reduced accuracy (its best use as a navigational aid was at sea), but loran was precise enough to enable lead and pathfinder aircraft—which on distant trips like Dresden would otherwise have been out of range of Allied transmitters—at least to be sure they were over the right city before beginning their marking operations.

Specifically, it was Smith's task to ensure that the marking was carried out accurately and then, observing the waves of Lancasters embarking on their carefully choreographed bombing runs over the target, to correct any mistakes, or bomb-sighting errors that might threaten to diminish the effectiveness of the bombing. The master bomber's duties were not only supremely responsible but also highly dangerous. Whatever perils arose from the enemy's flak or fighter defenses, he was duty-bound to remain in the target area for the entire duration of the attack, often flying at low altitude to observe the outcome of the bombers' efforts.

From its bunker deep beneath the Albertinum, in the heart of Dresden, the Local Air Raid Leadership (Örtliche Luftschutzleitung = ÖL) had been tracking the Mosquitoes for the past ten minutes or so, since they had overflown Chemnnitz.

For an hour it had been clear that a *dicker Hund* ("fat dog"), as the German air defense controllers called a big enemy bomber formation, was on its way to central or eastern Germany. On the basis of information from the Reich Air Defense Leadership, Dresden's Air Raid Police had already been placed on alert at 9:15 P.M. At 9:39 P.M. a general *Fliegeralarm* (enemy aircraft warning) was issued for the city, though it was still far from clear where exactly the bomber force was headed. Leipzig still seemed the most likely target. It was probable that this warning would count, for the Dresden population, as just another among scores of false alarms. Then, at 9:59, the ÖL reported: "Enemy combat units in the area of Dresden-Pirna, circling."

The main Lancaster force had also been spotted twenty minutes previously, but the dropping of window and other deception/evasion measures were still proving successful in confusing the enemy. Specific final-stage air raid alarms had been issued in Leipzig, but still not in Dresden.

Meanwhile, acting on orders from the Luftwaffe's First Fighter Division HQ at Döberitz near Berlin, the small German night fighter force at Klotzsche airfield had already been alerted. Its "A" group of ten Messerschmitt BF 110s had been "scrambled" in anticipation and was soon in the air, but it would take the night fighters up to half an hour to climb and get into an attack position. Even as these few pro-tectors started to gain height, however, the marking of the target was under way, and the main bomber stream drawing close. It was already unlikely that the tiny gang of night fighters would trouble the execu-tion of the attack, and so it proved.

One of the Messerschmitts was downed by unknown fire as it climbed. One account has it that the light flak still remaining at Klotzsche panicked and mistook it for an enemy intruder as it circled at some height. Another suggests the fighter might have had the mis-fortune to fly through a dense hail of just-released British incendiaries. One British Lancaster was also lost in this way over the city.

At 10:03 P.M. the first marker group started to get to work. These were Lancasters from 83 Squadron, the advance guard of the main force. Crisscrossing the city, they dropped green marker flares to delineate the city area, and the first of almost one thousand white magnesium para-chute flares cascaded to illuminate the ground. These latter were the notorious "Christmas trees" that so many Dresdeners who survived would later recall seeing from half-open cellar doors or glimpsing as they rushed to public shelters. Almost simultaneously, the ÖL received con-firmation that "the attack is intended for Dresden."

Three minutes later, at 10:06 P.M., the ÖL finally broadcast over the *Drahtfunk*, the wire-borne radio found in shelters and other pub-lic places, its final-stage, definitive warning:

Achtung! Achtung! Achtung! The lead aircraft of the major enemy bomber forces have changed course and are now approaching the city area. The dropping of bombs is to be anticipated. The population is required to make immediate use of air raid protection facilities.

Dresden had heard this twice before, on October 7 and January 16. It was not until this moment that they knew their city was going to be bombed. But by now the Mosquitoes of 627 Squadron had already begun to swoop down on the agreed aiming point, just to the west of the city center. Here, led by Flight Lieutenant Topper, they began drop the red flares that would guide more than 240 bomb aimers during the attack itself.

THE OSTRAGEHEGE STADIUM of the Dresden Sport Club (DSC), located by the main railway bridge, the Marienbrücke, was one of several large sports grounds in the city, and the home ground of the city's most popular soccer team, winner in both 1943 and 1944 of Germany's national soccer championship. In 1945 the DSC ground was selected as the aiming point for the first wave of the bombing of Dresden.

The wind that night was in the northwest, the same direction from which the bombers were approaching. The technique for the first wave of attack was sector bombing, first used in 5 Group's awesomely efficient raid on Braunschweig in October 1944. This involved marking a fixed aiming point and then assigning to the individual aircraft not just different headings for their approach—two degrees' variation at a time—but also progressively differing timed overshoots of the aiming point. The aim was to ensure an even and devastating density of bombing over a fan-shaped sector. If achieved, and with the right mix of high-explosive and incendiary bombs, this might well result in a firestorm. It had almost happened at Braunschweig—only the swift intervention of the experienced fire brigade had prevented it.

As soon as the first red-flare Mosquito went down to two or three thousand feet to mark the aiming point at the DSC stadium, what had been suspected for a while now became obvious. Dresden was undefended by antiaircraft artillery.

One of the three "Link" aircraft, all Lancasters, attached to the Pathfinder force, had a wire tape recorder aboard. From this we have a record of the exchanges between the master bomber and his markers. It was Topper, marker leader, who took this first aircraft down. At three thousand feet the master bomber asked him, "Do you see the green yet?" "Okay, I see it," confirmed Topper. And then the job was on.

As Topper dove down to two thousand feet and then less, he called into his VHF microphone: "Marker leader: Tally-ho!" This was the agreed signal to deter other markers who might think it was time to make their run. It was also a call closely identified with British blood sports.

At less than eight hundred feet, Topper opened his bomb doors and released a thousand-pound target-indicator canister, which was set to burst at seven hundred feet, scattering a cascade of red plumes of light. His Mosquito swooped across Friedrichstadt—over the Friedrichstadt hospital complex, over the stadium, then over the hospital's rail siding, where a train was currently unloading. As it did so, a special camera fitted to the bomb bay, using a flash cartridge system developed to automatically take a photograph every second over the aiming area, was doing its work. There were three flashes in quick succession on the south side of the Elbe, followed by a fourth as he passed over the river into the Neustadt, catching an image of a locomotive puffing along near the Japanese Palace.

Topper turned away from the city, climbing swiftly as he did so. Immediately the second marker followed him, ready to check the leader's indicator for any overshoot. A second target-indicator canister was loosed. The leader's marker had hit the ground just a hundred yards east of the exact aiming point. Then came the next and the next, each repeating the litany: "Tally-ho!"

The process would take two to three minutes. The area of the DSC stadium was turning into a flickering forest of red markers.

Down below in the Albertinum, the ÖL had announced—one minute after the final-stage air alert for the city—the "first bombs dropped on the city." The big marker canisters must have been mistaken for bombs.

On the first night of the series of Allied raids on Hamburg in July 1943—it was the second night that produced the infamous firestorm—the Pathfinder markers had fallen anything from half a mile to *seven miles* wide of the official aiming point. The intervening year and a half had brought huge improvements in technique and equipment. Compared with the early years of the war, this was sensationally precise, even pinpoint marking. No one could ever say that what happened at Dresden was an accident.

Now, as the last markers went in, it was a question of checking visibility, to ensure it was sufficient for the waiting Lancasters of the main force to see the markers and therefore to bomb as accurately as the indicators had been placed. A Lancaster of 97 Squadron, waiting at eighteen thousand feet, had been equipped and positioned to this end. Its name for the night was Lancaster Check 3.

The wire recording captured the moment Check 3 joined the process:

> Controller to Check 3: Can you see the glow? Over.
>
> Check 3 to Controller: I can see three TIs through cloud. Over.
>
> Controller to Check 3: Good work, can you see the Reds yet?
>
> Check 3 to Controller: Can just see Reds. Over . . .

Communication continued between the marker aircraft and the controller, simultaneous to that with Check 3. After a brief discussion, the marking was considered complete. The master bomber told the markers to drop any remaining flares and get out before returning to his dialogue with Check 3:

> Controller to Check 3: Can you see the Red TIs? Over.
>
> Check 3 to Controller: Can see green and Red TIs. Over.
>
> Controller to Check 3: Thank you.

This confirmation of sufficient visibility was the signal for the attack proper to begin. The 244 fully laden Lancasters of 5 Group, codenamed "Plate Rack Force," could now start bombing. The master bomber issued the order that sealed Dresden's fate:

> Controller to Plate Rack Force: Come in and bomb glow of Red TI
> as planned. Bomb the glow of Red TIs as planned.

•   •   •

"U FOR UNCLE," as part of 49 Squadron, was one of the early attackers. The bombing run was almost always the most tense part of a trip, though with tonight's weather better than expected and no sign of antiaircraft fire or enemy night fighters, the pilot could concentrate just that little bit more completely on the task in hand. The precision sector bombing system developed by 5 Group was especially demanding. Each squadron of sixteen aircraft was given an arc of thirty-two degrees to cover. They operated in exactly this close formation throughout the attack, ideally maneuvering as one.

Lesley Hay takes up his story once more:

> I'm watching my height and speed and ready for the turning point. This has to be done very exactly, because each squadron has been given an arc of thirty-two degrees. Anyway, we have to bomb on this football pitch. So one aircraft bombs on that heading. And the next had to bomb on that heading plus two degrees, the next on heading plus four, so we're all going out on two degrees different so each squadron fans out in spokes of a wheel from that point on the football pitch out onto the perimeter of Dresden.

This was where cooperation between pilot and navigator mattered to the fraction of a second. On either side of "U for Uncle," just yards distant, was another Lancaster at the same altitude but two degrees off, preparing to go in. The pilot's job was to avoid collisions and be ready to turn onto the "heading" from which the bombing run would be directed. The navigator's job was to do his pinpoint calculations of wind and height and speed and position, and tell the pilot, his "skipper," when that precise moment for the turn had come.

"Thirty seconds," said the navigator. "I'll be turning you on thirty seconds."

Over the radio, the calm, authoritative voice of the controller was encouraging the aircraft to go in, but the navigator took his time, held out for his calculated course. After months of training and an entire tour of duty survived together, the relationship of trust between pilot and navigator was intense, almost spiritual, and at such moments they paid little attention to outsiders, no matter how senior—and no matter how much, deep down, they themselves longed to get this job over.

"Turn now," the navigator stated simply.

And Hay turned the aircraft smoothly onto course for the approach, briefly tipping into the curve. He straightened out and saw his two companion Lancasters still two degrees either side of him after performing the same maneuver with the same skill—"not wingtip to wingtip, but that was how it felt!" as Hay recalls. This was the precision 5 Group was known for, and which had cost so many thousands of German city dwellers their lives.

By now the bomb aimer had slipped down to the bomb bay. Hay, the pilot, concerned himself only with acting on the data he was receiving about the aircraft's height, course, and speed. "Getting everything just on the right heading." He went lower in his seat, and paid close attention to his instruments.

All this was happening two minutes before the scheduled bombing time of 10:15. As the urgings over the radio had indicated, the marking had been so quick, accurate, and trouble-free that the master bomber had decided to send the Lancasters in early. Why leave them waiting over the target, especially when it was impossible to tell if the break in the weather would last?

So in they went, crossing the city from northwest to southeast, the prevailing wind directly behind them. Their crews' specific instructions would see the Lancasters traverse the city center area at slightly but precisely differing angles. The strict, exact attack sequence of aircraft and squadrons was intended not only to guarantee the intensive bombing of the given area but also to avoid the ever-present danger of collision, or of aircraft bombing one another as they approached and released their loads from different directions and at different heights.

The bombing area, shaped like a wedge, extended just over one and a quarter miles along each edge and a little under one and three-quarter miles at its widest point. It took in the railway lines immediately in front of the DSC stadium, then its northern boundary traveled across the Elbe to clip a corner off the ministerial buildings on the Neustadt side of the river, violently embracing the Augustus and Carolabrücke road bridges (though it did not include the Marienbrücke railway bridge). Between this boundary and the southern edge, it then spread itself over the entire Altstadt, almost as far south as the Hauptbahnhof. The railway station was, however, like the Marienbrücke, not quite included in the designated bombing area.

"U for Uncle" carried one thousand-pound high-explosive bomb;

the rest of her load were four-pound incendiaries. All of 49 Squadron went in at between 12,000 and 13,500 feet—relatively low, because of the absence of flak. One aircraft every 7.5 seconds. All bombed between 10:14 and 10:16 P.M. except for one aircraft which, according to the record, went in at 10:22.

During the approach, pilot Lesley Hay was no longer strictly in control of the aircraft. Once the bomb aimer felt able to assure his skipper that he had the heading and had set his bombsights accordingly, he was in charge. For however long it took for the bombs to leave the aircraft, he would direct the movements of "U for Uncle" and hold the lives of his fellow crew members in his hands.

On most trips, this was the hardest time. The voice of the bomb aimer as he gave running instructions over the intercom: "Steady, steady, steady, port, steady. Steady, port steady, steady, steady—and starboard steady, steady, steady, steady . . ."

The whole crew crouched tensed in their seats. They just wanted to get out of it. And they knew that even after the drop they would have to hold it for a minute for the photo flash. The bomb aimer's instructions were drawn out, in this case, because he had a problem. Visibility had seemed good, but just as "U for Uncle" started its run-up, a thin veil of low cloud suddenly drifted over the target. Ground visibility was reduced to the dull glow of spot fires through this low haze—no view of the city itself. It was taking the bomb aimer a little time to work out from the dim pattern below where the aiming point must be. Then, as they approached, the marker flare force, seeing the difficulty for the incoming aircraft, put down some lights to illuminate the area a little better. A few moments later, the aimer opened the bomb doors, there was the loud *bm-bm-bm* of the load being released into the night.

Bombs gone.

The sudden lightening of the aircraft's load, as several thousand pounds of bombs and incendiary sticks spilled from its belly and began their tumbling journey to the earth, caused "U for Uncle" to shoot up very quickly. Hay, an experienced pilot, knew what was coming and steadied his aircraft. Soon he started to curve away again, leaving the bombing area. Hay headed southeast, following the stream of other Lancasters that had finished their job for the night.

Behind "U for Uncle," more aircraft were making their runs under the watchful eye of the master bomber, who hovered over Dresden at a mere three thousand feet. His aircraft, like all the small group of Pathfinder Mosquitoes, was unarmed, relying for safety on its exceptional speed, maneuverability, and ability to fly above the flak and the night fighters.

Way above the master bomber, the Lancasters fanned out over the bombing area. Down in the city, the lights of a thousand new fires burned. High-explosive bombs had punched holes in roofs and blown doors and windows out of buildings to provide the necessary draft. While the civilians crouched in their cellars registered mostly the terrifying force of the explosions, the four-pound Thermite-based incendiaries (wrongly referred to by German civilians as "phosphor bombs") clattered down by the tens of thousands, lodging in roofs and attics and upper rooms and catching on whatever furniture and beams and household items made suitable fuel.

Already the bombers of "Plate Rack Force" were finding it hard to discern the individual red markers dropped by 627 Squadron. And already, thousands of feet below, tiny fires were starting to cluster and melt into ever-brighter conflagrations, breeding like glowing bacteria under a microscope.

"Good work, Plate Rack Force," said the marker leader as he prepared to go home. "That's nice bombing."

He was, by his own lights, correct. In Bomber Command's attack on Dresden, everything had so far gone terrifyingly right. On this night of February 13–14, 1945, the planners and the markers, the master bomber and his aircrews—not forgetting the ground crew back at base and the "lassies" who filled and packed those incendiaries in their little factories tucked away among the Scottish hills—had together created the rarest of things.

The perfect firestorm.

Aerial view of the Dresden Altstadt, 1943.
(Sächsische Landesbibliothek—Staats- und Universitätsbibliothek Dresden; Deutsche Fotothek/Walter Hahn)

Prewar Dresden by night.
(Sächsische Landesbibliothek—Staats- und Universitätsbibliothek Dresden; Deutsche Fotothek/Richard Peter)

Dresden schoolboys
in Nazi regalia, 1937.
*(Fotoarchiv Dresden
Stadtmuseum)*

The Schloss Strasse
bedecked with swastika
flags for Hitler's visit to
Dresden, June 1934.
(Stadtmuseum Dresden Bibliothek)

The Dresden synagogue in 1870. It burned down in 1938.

Market stand
in Dresden,
September 1940.

*(Fotoarchiv Dresden
Stadtmuseum /Kurt
Schaarschuch)*

Dresden high school students train as antiaircraft auxiliaries—
on the left is the future historian Götz Bergander.

(Christian Köster/Götz Bergander)

Children evacuated from Dresden to the countryside, 1943. By 1945
most would have returned to face the firestorm with their parents.

(Fotoarchiv Dresden Stadtmuseum/Kurt Schaarschuch)

Production line work on radar instruments at the Sachsenwerke factory, 1944.

(Presse-Foto Koch/Dresden Stadtmuseum)

The bombproof Zeiss-Ikon Goehle-Werk factory, where Henny Brenner and the other young people of the "Kindergarten" did forced labor.

(Sächsische Landesbibliothek—Staats- und Universitätsbibliothek Dresden, Deutsche Fotothek)

Gauleiter Mutschmann and his entourage inspect products for the military at the Radio-Mende factory.

(Sächsisches Hauptstaatsarchiv Dresden)

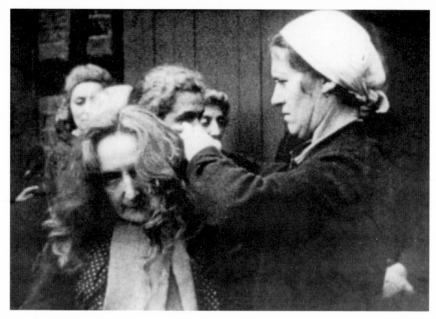

The Hellerberg Jews are subjected to "delousing" at the City
Disinfection Institute after being ordered from their homes.
(Stiftung Sächsischer Gedenkstätten)

The head of the Dresden Gestapo's Jewish Department (*Judenreferat*),
Untersturmführer Henry Schmidt (*second from the left*) jokes with Dr.
Hasdenteufel (*second from the right, with cigar*) of the Zeiss-Ikon manage-
ment as they watch the "delousing" process with two Gestapo agents.
(Stiftung Sächsischer Gedenkstätten)

Jews arrive at the Hellerberg camp. They hoped to be safe there, but the next spring they were transported to extermination camps. Few survived the war.

(Stiftung Sächsischer Gedenkstätten)

Alfred Goldschmidt (*left*) was a Jewish childhood friend of Günther Kannegiesser in Johannstadt. They played football and hung around the local streets together. One day Alfred disappeared. This was where he "disappeared" to. Alfred died in Auschwitz-Birkenau.

(Stiftung Sächsischer Gedenkstätten)

Mass declaration of loyalty to the Führer in Dresden
the day after the attempt on Hitler's life, July 21, 1944.

(Presse-Foto Koch/Dresden Stadtmuseum)

Günther Kannegiesser (*left*) and family on a
wartime swimming excursion. His elder sister
(*center*) would be absent from the Dresden when
it was bombed, but his mother, little brother, and
younger sister would all die during the second
RAF attack. *(Günther Kannegiesser)*

Nora Lang and family. Her brother (*right*) was called up shortly before the
bombing. The family was separated during the firestorm. Nora (*center*) and
her little brother (*in mother's arms*) wandered the blazing streets all night.

(Nora Lang)

Anita Kurz and her parents, winter 1943–44, in the Dürerplatz. Her father worked as an army clerk.

(Anita John)

Henny Wolf at seventeen, in the courtyard of the Zwinger Palace, just before being forced into labor at the Zeiss-Ikon factory.

(Henny Brenner)

Seventeen-year-old Günter Jäckel in uniform as a Luftwaffe infantryman, Toul, eastern France, summer 1944. The photographs were taken, he believes, so that if they died these boy soldiers' parents "would at least have something." Two months later Jäckel was wounded and invalided back to his hometown of Dresden, where he witnessed the firestorm.

(Günter Jäckel)

District Target Map
No. G. 82

DRESDEN
(GERMANY)

Illustration No.
D.T.M. G. 82/1

Illustration No.
D.T.M. G. 82/1

0   500   1000   1500   2000 YARDS
0   1 MILE
(1 : 24,800)   approx.

Issued November 1943

The Dresden Target Map.

(Public Record Office, London)

Air Marshal Sir Arthur Travers Harris, Air Officer Commanding, Bomber Command—the man many would see as Dresden's nemesis.

(Hulton Getty)

Miles Tripp (*left*) and the crew of the Lancaster bomber "A-Able" after the final operation of their extended tour, March 1945. They had bombed both Dresden and Chemnitz three weeks earlier. *(News Chronicle)*

"A-Able" over the Ruhr.

Lieutenant Alden "Al" P. Rigby of the 8th Fighter Air Force at the time of his escort flight to Dresden. The names on his P-51 Mustang are those of his wife and baby daughter.

(Alden P. Rigby)

Just after 10 P.M. February 13, 1945. The RAF marker flares known as "Christmas trees" begin to illuminate the target. This was the viewpoint of Dresdeners who watched these in horrified fascination from their cellars and shelter entrances. Within moments the first RAF bombs will fall on Dresden.

(Archiv der Interessengemeinschaft 13. Februar 1945 e.V./Hans Überschaer)

The firestorm at its height between midnight and the small hours of Ash Wednesday. The thousand-degree heat could be felt by RAF aircrew at more than ten thousand feet. *(Family of Peter Firkins)*

Firefighters at the old Technical University in the Bismarckplatz, in the neighborhood of the Hauptbahnhof. The time is around midnight. The second wave of the RAF attack is still to come, and it will almost certainly cost these men their lives. *(Hans-Joachim Dietze)*

Historic Dresden still burning. A photograph taken from a German reconnaissance aircraft forty-eight hours after the firestorm. *(Ingeborg Grossholz)*

Ruins of Augustus the Strong's pleasure palace, the Zwinger.

(Heinz Kröbel/Dresden Stadtmuseum)

A 1920s picture postcard shows the long sweep of Striesener Strasse, facing west through Johannstadt, with the spire of the Johanniskirche on the right and the center of Dresden in the far distance (tiny and faint on the left is the dome of the Frauenkirche). In and around this central artery, Günther Kannegiesser and his friends lived—and died.

An almost identical view of Striesener Strasse after 1945. The ruins of the block where Günther Kannegiesser lived are to the left of the distant parked truck.

(Günther Kannegiesser)

Old Dresden: The Münzgasse, once known for its restaurants, photographed from the ruins of the Elbe Terrace, with the remains of the Frauenkirche's dome in the background at the left.

(Heinz Kröbel/Dresden Stadtmuseum)

Industrial Dresden: The Universelle cigarette machine factory, long since converted to making armaments, after the RAF raid February 13–14, 1945.

(Sächsische Landesbibliothek—Staats- und Universitätsbibliothek Dresden, Deutsche Fotothek/Richard Peter)

In the Altmarkt soldiers and rescue workers gather up and burn the dead to minimize the risk of epidemics. SS units with experience from the death camp at Treblinka are said to have supplied expert help.

(Sächsische Landesbibliothek—Staats- und Universitätsbibliothek Dresden, Deutsche Fotothek/Walter Hahn)

Nearby, many died trying to take refuge in the water of the fountain in the middle of the square. The water boiled dry.

(Sächsische Landesbibliothek— Staats- und Universitätsbibliothek Dresden, Deutsche Fotothek/Walter Hahn)

The Angel of Dresden.

(Sächsische Landesbibliothek—Staats- und Universitätsbibliothek Dresden, Deutsche Fotothek/Richard Peter)

20

"Air Raid Shelter the Best Protection"

THE EXPERIENCE of the first wave of the raid on Dresden was terrible, but did not yet seem cataclysmic. The beginnings seemed often to resemble something out of a beautiful but frightening fairy tale. Stories were told of sneaking outside to watch the glittering fall of the silver-trailed "Christmas trees" illuminating the wintry city beneath them— before the whistle of the first bombs sent even these bold observers back into their basements, hastily closing the doors behind them.

For most civilians in Dresden the raid was undergone in a modestly converted basement or cellar beneath an apartment block or a private house. There would be the sacks of sand, the buckets of water, the basic tools, but no firewalls, no air filters, no custom-built sealed doors. In the big public shelters of most major German cities, great portals would close inexorably when the raid was on. Filters and seals would keep breathable air circulating and poisonous gases out. Layers of reinforced concrete could withstand all but the most powerful, direct hit. Not so in Dresden. You took shelter in your immediate neighborhood. Occasionally someone would be caught away from home and take refuge with strangers, but mostly the people seeking safety beneath the city were people who knew each other well from years of everyday friendship and contact, who lived together. Tonight, in many cases, they would also die together.

Though it was just outside the bombing sector, plenty of bombs fell in Johannstadt during the first attack. Nora Lang recalls:

> They were mostly incendiary bombs where we were. High-explosive bombs exploded only occasionally. You know incendiaries fell in

great masses. They would penetrate the roof. It felt as it someone directly above me was shaking out coals or potatoes onto the roof. *Boom-boom-boom.* Then sometimes would come this hiss and an explosion. You would think it had hit the house but actually it had hit somewhere else. And I shook so much. My whole body was shaking. Mummy was sitting there with my baby brother, and the owner of the building said, "Take care of your daughter." But my mother couldn't.

Her friend Anita Kurz, a couple of blocks away, was in their basement-cum-public-shelter:

My mother had first-aid training. And at some point someone had taught her that when this happened you should all throw yourselves on the ground. We practiced that. So we all lay down, and she lay on top of me. And Father on top of her. I can't recall that there were individual explosions. It was just constant. And I was there at the bottom, with my mother above me, and I had this feeling that everything was shaking . . . right down to the foundations. The light went out, so it was dark. And then actually we lay there like that until quiet returned. Then my mother started screaming. Until finally someone said—I don't know who— "Frau Kurz, calm yourself. When Anita is back at her dancing lessons, all this will be over!"

People prayed and wept, bit their lips in silence, winced at the shudders of nearby bomb impacts, worried about the fates of their homes, their possessions, and their pets.

Günther Kannegiesser and his friend Siegfried never made it as far as Police District Four. They were compelled by the growing force of the bombing to take refuge in the air raid shelter in the Schumannstrasse. The lights went out. They felt repeated explosions. Then things seemed to go quiet. Eager to see what was happening, they sneaked out of the shelter exit and up to street level to examine the damage for themselves. They arrived in the lobby of the building just in time for more bombers to come overhead. Within seconds, tile-splinters crashed through the windows. Günther still has headaches from the tiny fragments that embedded themselves in his temples. The boys retreated hastily back down into the shelter and did not reemerge until the first raid was over.

A mile and a half to the west of them, in Friedrichstadt, Götz Bergander had returned home from his refugee work at the Hauptbahnhof only minutes before the first air raid warning sounded. He and his family tramped dutifully down to the custom-built shelter beneath the Bramsch distillery, where Dr. Bergander, a chemist, worked as technical director.

It was a very good air raid shelter. Gas-proof, with steel blinds, and rubber buffers, and with real bolts so that you could really seal the place off. Steel girders. I mean, really the bunker couldn't be bettered. It was constructed around the time Mutschmann built his personal bunker. There must have been some kind of appeal at the time for Dresden to prepare better air raid defenses. And you know, people mocked this shelter at our place. They would come visiting and be shown around, and they would say "Oh God, what's this for? We don't need anything like this!" And yes, you could say this cellar saved our lives, because we in the house had three near-hits, one at each corner . . . if it hadn't been for the steel blinds, then the impact of the explosion would have burst into the cellar and done for us. We would not have survived.

The Bergander family lived in an apartment on the premises, as did the managing director. A number of other civilians from the area also came to take advantage of the shelter whose sturdiness they had once mocked. These were middle-class people—there was considerable industry in Friedrichstadt, but also the hospital complex and some attractive apartment blocks dating from the early nineteenth century. The composer Wagner had lived just along the street while royal master of music in the 1840s. The shelter served them all well—now and over the next hours it almost certainly saved their lives. The factory was only a few hundred yards from the aiming point for 5 Group's bombers.

One Dresdener who, by coincidence, found himself in the center of the city that night was the anti-Nazi artist Otto Griebel, who had witnessed the burning of the Dresden synagogue almost seven years before. Now in his forties, Griebel had avoided the clutches of the Gestapo, but not of the Wehrmacht. He had carried out his army duties, as a technical draftsman attached to a company of engineers in

Poland, with appropriate ill grace and almost exultant inefficiency. The big Russian offensive enabled him to make his way back to Dresden, where he arrived on the last day of January.

On the evening of Tuesday, February 13, Griebel took a tram into the city from his apartment east of the Grosser Garten and got out at the Neue Gasse, on the edge of the Altstadt. As it happened, this was hard by the Bauer box factory, Henny Wolf's last place of employment. The landlady of a local bar, a friend from prewar bohemian days, had invited Griebel to a small party. The place was full of people Griebel knew and was pleased to see again. They drank and chatted. At around 10 P.M. Griebel picked up his hat and coat and made ready to go home.

As Griebel went to pay his tab, the air raid warning sounded. One of the company, who had left her children at home nearby, turned pale and rushed from the room, but most still refused to believe this was anything more than the usual false alarm. When the full alert came, they piled down into the basement of the building, which was deep and capacious. Soon the first bombs started to drop on the center of the city:

> A series of whistling sounds sliced the air, then the building shook from a quick succession of steadily more powerful explosions, which drove us into the corner of the basement . . . The roaring fall and crash of the bombs now just didn't seem to stop. The air pressure blew open the iron doors of the beer cellar and the flickering electric light suddenly went out . . . the landlady assured us that the basement was sturdy enough to withstand a direct hit. A few of the impacts felt literally like blows to the back of the neck. We just hunkered down ever lower and waited as one shudder succeeded another. At one point it felt as if the whole building was rocking on its foundations . . .

Not all the people caught in Dresden that night had even the faint comfort of the company of friends and neighbors. At the overflowing Hauptbahnhof, the final air raid warning had caught hundreds of travelers and refugees on the concourse and the platforms, in the waiting rooms and restaurants, or crowded into departing trains. The express to Munich, which had been about to depart when the British raid started, was stuck inside the station. Panic-stricken passengers pushed toward the supposed safety of the station's basement complex, in some cases forced to negotiate steep stairs, for the Hauptbahnhof was built on several levels.

Even though the main station was beyond the south edge of the designated bombing sector, inevitably quantities of both high explosives and incendiaries began to fall in and around the buildings. Perhaps this was when the master bomber had chided his aimers: "Try to pick out the red glow. The bombing is getting wild now . . ."

A fire started on the steps down into the cellars. Some passengers were trampled, others crushed to death or suffocated by billowing smoke in the enclosed space. A few tried to hide under the coaches of the trains. Some succeeded in reaching the cellars, only to find them already packed to overflowing with people and luggage; the access corridors were blocked by hordes of sheltering humans and their belongings.

So much for the "shelter" under the Hauptbahnhof. The authorities had been warned, but had done almost nothing to improve the public's chance of survival in Dresden's most vulnerable location.

The chaos had been bad enough during this first, fairly brief attack. There was a great deal worse to come.

THE TENS OF THOUSANDS of civilians crammed into basements and cellars throughout the inner areas of the city could only wait and pray. Torrents of high explosives and incendiaries were falling on the Altstadt, the heart of the RAF's bombing sector. There was some drift to the north as the raid went on. Although only the immediate riverside had been included in the original plan, bombs also hit the residential and official quarters of the Neustadt, several hundred yards from the river. There was also some "overshoot," taking the bombing eastward from the ordained curved edge of the wedge and causing quite extensive damage and fires in the area of the Grosser Garten and Johannstadt, but the bombing was on the whole executed with extraordinary efficiency.

Part of the reason for this near perfection of destruction, in the case of Dresden, was the absence of "creep-back." The phenomenon, common throughout the war, came from a tendency by bomb aimers to drop their loads too soon. It could cause serious cumulative inaccuracy as the raid progressed, and turn a potentially effective night's work into what the RAF's planners judged as a waste of men and aircraft. "Creep-back" was one of the reasons that those planners had learned to set their aiming points squarely in the centers of cities. Were an aiming point to be set in the suburbs—where industrial and similar targets might well, in fact, be situated—accumulated "creep-back" (the first aircraft drops a little soon, the next even sooner, and so on) could quite soon lead to bombs falling not onto urban or even suburban targets but onto open fields.

The problem was not officially admitted at the time, perhaps because it was felt to smack of "cowardice" on the part of aircrew. Air Marshal Harris admitted its existence only in a secret account written for Air Ministry reference, referring to it with unaccustomed delicacy as "undershooting."

> This undershooting was due partly to the relative visibility of the markers, those that were lying short of the target in open country being much more easily seen than the more distant ones of the target itself, which were often partially hidden by buildings or obscured by smoke. An additional factor was the not unnatural desire to bomb as soon as possible.

At Dresden there was not the usual intense pressure to bomb and get out. As everyone soon realized, there were no enemy antiaircraft guns and apparently no night fighters. The weather was also sufficiently benign to allow low-level, accurate marking. The crews themselves, relieved of the necessity of staying toward the limit of flak range, could moreover bomb in a careful and relatively relaxed fashion at a lower altitude than usual, in this case mostly between ten thousand and thirteen thousand feet. Perfect conditions. So no "creep-back." And an extra reason that the bombing that night was so devastatingly accurate and effective.

In fifteen minutes that February night, 5 Group had unleashed a huge, carefully mixed quantity of air bombs on Dresden. This provided the high concentration required if there was to be a chance of a firestorm.

Through the Lancasters' bomb doors tumbled 172 four-thousand-pound air mines ("cookies"), 26 two-thousand-pound air mines,

72 one-thousand-pound high-explosive bombs, and 648 five-hundred-pound high-explosive bombs. In addition, the aircraft dropped, usually in the same loads, 128,550 four-pound stick incendiary bombs (individually released), 8,250 four-pound stick incendiary bombs fitted with explosive charges, and 68,628 four-pound stick incendiary bombs packed into cluster containers. The last were fitted with a barometric pressure trigger. The containers blew open to release dense showers of the sharp-nosed little fire raisers, usually at around a thousand feet from the ground. This ensured concentrated and even distribution. Four-pound incendiaries dropped individually from ten thousand to twenty thousand feet were notorious for drifting, in their long descent to earth, way off target and landing so widely scattered as to be all but useless.

Altogether, 881.1 tons of bombs fell on the central districts of Dresden between 10:13 P.M. and 10:28 P.M. Around 57 percent by weight were high explosive, 43 percent incendiaries.

The big air mines were not just for blowing apart buildings or causing huge craters in the streets, thereby causing access problems for firefighters and other emergency services, though they did both these things. These explosive monsters' function was also to create huge waves of high-pressure air, as they had in the packed tenement blocks of Hamburg in July 1943. Such waves blew out scores, even hundreds of windows and doors, swiftly increasing the through-draft needed for the little fires from tens of thousands of stick incendiaries to spread and combine as quickly as possible.

It was also crucial to the attackers that the density and force of their bombing kept the population pinned down in shelters instead of in their homes, fighting the fires that had started in the roof spaces and attics. Most incendiary bombs were easily extinguished if immersed in sand, which by law had to be kept ready in the familiar sacks distributed around houses and apartment buildings. In some cases, venturesome householders would even pick up the burning four-pound bombs (usually with a shovel or a pair of tongs) and toss them out of the window. There they would splutter their way pointlessly into extinction outside on the street, along with the host of other incendiaries that had failed to land inside a building.

In the case of the RAF's incendiary attack on Leipzig just over a year earlier, the surprisingly low casualty rate had been largely due to the *disobedience* of the city's population. Instead of staying in their shelters until the official all-clear, the Leipzigers quickly emerged and took an active part in extinguishing fires before these could spread and become unmanageable. The Dresden population was more passive and more obedient, perhaps more trusting of the authorities. It would pay dearly for this.

For years Dresdeners had been fed mixed messages. In the press, concerned citizens could read advice from the regional air protection leader, a retired general named Schroeder, about the measures they could take in case of a serious air raid, in which incendiary bombs were used:

> Even in those places where large-scale attacks occur, their effect is not the same everywhere. It is without doubt not easy to leave the security of the air raid shelter while the attack is still raging out-side—the crash of the bombs, the firing of our own flak filling the air—and to go up and check the condition of your house. Not to mention then taking measure to combat fires there, should this become necessary. This requires as much courage as the warrior at the front . . .

But less than two months before the big air raids on Dresden, an article in the party newspaper, chirpily entitled "Air Raid Shelter the Best Protection," had provided its readers with frankly contradictory counsel:

> The air raid shelter is the best protection. The numbers of those fallen [killed] in such shelters is small to the point of nonexistence compared with those whose lives and possessions have been saved by them. Instead of fleeing thoughtlessly into the open, we should rather put all our energy into turning our cellar into a really secure refuge. We should reinforce its structure if necessary, ensure there are emergency exits, and mark the ways to these exits and the exits themselves with luminous paint.

It is not hard to imagine that, given the confusing choice between summoning "as much courage as the warrior at the front" and racing upstairs in the middle of an air raid to check on their homes, or remaining in the shelter that is allegedly "the best protection," the inmates of the average wartime Dresden household (mostly women and children, with a sprinkling of older men) would tend to settle for the latter option. And this is what they did on the night of February 13, 1945.

More than half an hour after the end of the first attack on Dresden, many civilians had still not emerged from the deceptive security of their shelters.

At around 11 P.M., fireman Alfred Birke found himself on the Altstadt's eastern periphery—which was also the edge of 5 Group's bombing sector. He had driven, in a commandeered vehicle, from a fire brigade command center quartered in the refreshment room of the zoo in the Grosser Garten. He was looking to report to the authorities in their deep bunker at the Albertinum, and his news was not good. Much of the equipment and many of the fire-fighting vehicles parked a short way from the command center had been destroyed by the British bombs. He reached the western edge of the park, then searched desperately for a safe way through the burning Altstadt to the river. In his report of that night he describes this eerie journey:

> Flames shoot from the facades of buildings, from the Kreuzschule, from the Waisenhausstrasse. I drive at walking pace into the broad expanse of the Ringstrasse . . . I don't meet a living soul. At the Pirnaischer Platz I encounter three naked bodies, a woman and two children. I take care not to drive over them. At last the smoke thins a little, the flames retreat. The quarter behind the Frauenkirche does not seem to have been affected so badly . . .

The empty streets held an ominous message. The inhabitants of this densely populated area were still underground, perhaps planning to stay there until morning. Which meant that in many—probably most—cases, they were already doomed.

OTTO GRIEBEL tells of his reactions as the bombs stopped falling and the drone of the British bombers faded away and finally disappeared.

> After a while the electric light came back on. Only then did I dare, however hesitantly, to edge my way upstairs. At that moment a sob-

bing woman wearing an air raid warden's helmet burst in through
the cellar doorway. She threw herself into the arms of her husband,
the musician Scheinpflug—who had also been drinking with us—
and cried: "We have lost everything!" Looking out of the pub
through a shattered window, I could now see that the entire Neue
Gasse was on fire and burning bright as day. There were sparks flying in every direction, and through this turmoil were hastening
frightened people, often only half-dressed.

The Bauer box factory, although Griebel probably knew nothing
of it, was among those burning buildings. It was the great good fortune
of Henny Wolf and her fellow Jews that Bauer had shut down the
night shift a few weeks before. They would have been working inside
when the bombs hit. The only employee who was there that night—
his company secretary—died in the blaze.

The pub in which Griebel had been drinking had survived, almost
as miraculously as Herr Bauer's Jews. The relieved landlady opened a
rare bottle of old schnapps to celebrate. Griebel downed a glass to give
himself courage, but he was really concerned only for the wife and
children he had left at home. He bade farewell to his friends—in many
cases, though he could not know it, for the last time—and set off
homewards in the company of a woman friend.

> Everywhere we turned, the buildings were on fire. The spark-filled
> air was suffocating, and stung our unprotected eyes. But we could
> not stay here. Entire chunks of red-hot matter were flying at us. The
> more we moved into the network of streets, the stronger the storm
> became, hurling burning scraps and objects through the air.

They found their way down to the Brühl Terrace and the river.
Everything was ablaze. The confusion and the panic were unspeakable. Griebel saw that strange-colored flames were rising from the
nearby Carolabrücke, and realized that the gas pipe that carried the
mains over the river was on fire. Then he remembered that a couple of
days earlier sappers had been working on the bridge, possibly setting
explosive charges on its pillars. It was time to leave the area. Griebel
took a long, last look along the river in the direction of the Altstadt, all

the buildings he had known all his life burning to destruction, including the Academy of Arts where he had studied, and the galleries around the Albertinum. The familiar skyline was starting to disappear in a monstrous pillar of smoke and flame. Across the river, much of the Neustadt—although not included in the RAF's original sector bombing plan—was also ablaze.

Griebel's eyewitness account was but one among many indicating that the fatal Dresden firestorm had begun to take hold shortly after the first wave of bombing finished. For the Altstadt and its historic core of treasured buildings, annihilation was already a near certainty within a few minutes of the last bomb's being dropped. No sirens informed the public of the end of the air raid. Since they were all electrically operated, they stayed silent for the rest of the night—another reason that so many Dresdeners stayed in their shelters.

The city's communications had cut out in the first minute of the attack. The well-equipped emergency headquarters in the bombproof bunker deep beneath the Albertinum, where many of the city's most senior Nazis had also taken refuge, had all but lost contact with the surface world—except for one phone line that inexplicably still functioned and allowed a dribble of information to and from Berlin. Otherwise, information came in randomly to the Albertinum from those who, like Birke, struggled through to report. How could they gain more than a fragmented picture of what was happening in the city? After parking his vehicle and hurrying down to the control bunker, fireman Birke faced a bad-tempered interrogation about his movements and motives, as if his very appearance there represented some dereliction of duty. They had no idea what it was like out there, he told the high-ups in their secure bunker. They told him to wait in the other room.

To some extent, the ignorance of the crisis staff under the Albertinum was understandable. The raid had, after all, lasted only fifteen minutes. How could so much irretrievable damage be wrought in such a short time? The system had mastered the two previous raids, in October and January. Perhaps the officials directing it now believed that this was a similar situation. Just a bit bigger.

At around the same time, Georg Feydt, director of the city's repair service (*Instandsetzungsdienst*) was also traveling through the inner city, casting an expert eye over the deteriorating situation:

The characteristic thing about a slow-developing area conflagration is that in the early stages the population, intimidated by high-explosive bombs into staying in its shelters, is therefore not available to extinguish the small fires in their infancy. The result is this: the conflagration develops very slowly and only becomes apparent, all at once, when the thousands of small fires have grown sufficiently to burn their way through the ceilings beneath roofs and attics. Then, very suddenly, the third and fourth stories of buildings everywhere start bursting into flames . . .

The near-apocalyptic nature of 5 Group's attack and its continuing consequences were certainly apparent to Otto Griebel, as he paused grimly just outside the building under which the central control bunker lay. He had no choice but to head eastward, through the Grosser Garten, away from the roaring core of the growing inferno.

Despite the danger of more blazing buildings, despite the uprooted trees and the scatterings of maimed corpses—at this stage mostly unlucky people caught in the open by high-explosive bombs—that was the way home. If home still existed.

HENNY WOLF and her parents sat out the first attack in the cellar of their building. Most of the people sheltering with them were women and children. There were small windows, and the full impact of the exploding bombs could be felt.

Once the noise stopped, and it was clear that the raid was over, the Wolfs hastened up the stairs and into the yard of the building. It was on fire. Flames licked through the blasted windows of their apartment.

Herr Wolf did not hesitate. Ignoring the pleas of his wife and daughter, he rushed inside. Minutes passed. The fire burgeoned, until it threatened to consume the building. Just as it seemed to the anxious women out in the street as if the fire had swallowed him, Herr Wolf staggered out of the flames, carrying some money, important correspondence, and documents—including those relating to the cinema and the other property the Nazis had taken from him as punishment for staying married to a Jew. He also brought the deportation order that had been delivered earlier that fatal day.

Everything was secured in a knapsack. Mouths masked against the smoke, air raid helmets firmly in place, the Wolfs prepared to leave the street they had lived in for many years. But first the women tore the hated yellow stars from their coats. The badges joined the rest of the papers in the family rucksack. The Wolfs were orderly people. Only now did they set off through the burning city.

Even before the raid, their hopes had depended on leaving everything they knew and loved behind them. Now, in contrast to all but a few of the threatened human beings in Dresden that night, for the Wolfs safety lay in chaos, their only chance of survival in the destruction of their home city.

The Wolfs first tried to find a way southwest, toward the Hauptbahnhof. They were possessed by a strong desire to see for themselves that the Gestapo headquarters in the Hotel Continental, just behind the station, had been destroyed—along with all its records and files. The fires south of the Grosser Garten were now too intense, but they were encouraged by reports from other air raid victims that the entire area around the Hotel Continental was fully ablaze. Gestapo headquarters and the Nazi Party offices in the Altstadt had both been struck in the RAF's attack.

MANY FAMILIAR SYMBOLS of the regime's power were being razed to the ground.

All the same, it was still quite clear, even at this terrible, bewildering time, who ruled in Dresden. At 11 P.M. on the night of February 13, a squad of firefighters operating in the historic area around the Adolf-Hitler Platz (Theaterplatz), close to the opera house and the Hofkirche, was ordered to another location: the Reich governor's residence.

The governor himself, sixty-five-year-old Gauleiter Martin Mutschmann, was nowhere to be seen. According to postwar accounts, he escaped the effects of the first attack by taking to the controversial, state-of-the-art bunker beneath his palatial villa in the Comeniusstrasse, just north of the Grosser Garten. Wherever Mutschmann might actually have been that night, the account in the Nazi press shows breathtaking insolence in its attempt to paint him—against all the odds—as the hero of the hour:

Despite the raging conflagration, all the men of the local air raid leadership remained at their posts and directed—in as much as this was possible, given the extent of the common murder inflicted on our beautiful native city—the first essential aid measures to be brought in from outside the city.

Then the Gauleiter and his coworkers forced a way through the blazing city into the open air, from where that same night they progressed to the emergency headquarters of the *Gau* leadership. From there they continued their unceasing work to bring in external help.

Such courage! To remain fearlessly in the safest place in the center of Dresden, and from there to rush to an even safer location outside the city—the network of bunkers housing the party leadership's bombproof HQ at Lockwitz, on Dresden's far southern outskirts.

WHILE THE HEART of the city burned, parts had still been scarcely touched by the bombing, the outer suburbs especially. Other districts had been hit patchily, including the area south and southwest of the Altstadt, and the area to the east. Here there was some high-explosive damage, and fires had started, but the situation was comparable to that in an "average" bombed town.

The Johannstadt hospital complex was, after the Friedrichstadt hospital, the largest in Dresden. The Women's Clinic had grown out of a school for midwives toward the end of the eighteenth century, and now found itself in a large building between the Grosser Garten and the Elbe. It became a byword for excellence. The nearby Pediatric Clinic—the first in the world—was founded in the 1890s by a Jewish physician, Dr. Schlossmann, one of a group of doctors who had formed an association for children's health. After existing in temporary locations around Johannstadt for thirty years, in 1930, with government help, a purpose-built clinic came into being.

The clinic was thus only fifteen years old, built in a light, airy Bauhaus style and one of the most modern units of its kind in Europe. The final-stage alarm had led the medical authorities to start evacuating the patients, sick children (many with illnesses such as scarlet fever), and expectant or recently delivered mothers and their babies, but owing to the unusually short interval between alarm and attack, few were brought to safety at this point.

The first wave of British bombers passed over. There was some damage from stray bombs, but the hospital buildings were not destroyed. Even when the attack seemed over, the doctors and nurses continued to evacuate all the mothers and children they could from the hospital. The babies were evacuated, it seems, mostly because of existing damage and the unhealthy environment created, and not because the authorities were frightened of another air attack.

The senior medical officer at the Rudolf Hess Clinic (a naturopathic establishment in the same complex) had offered his villa in the suburb of Blasewitz as a temporary refuge for endangered infants. As in the case of the Royal Warwickshire Hospital in Coventry, some of the mothers could not be moved. Where possible, they were transported down to the shelters beneath the hospital. The important thing was to save the babies, and that was done. Nurses did their best to wrap them warmly and protect them from drifting smoke, then loaded them onto trucks and set off on the journey.

Elsewhere in Dresden, especially in the only partly affected areas, those who emerged from the shelters did so mainly to check on their homes and belongings rather than to flee. Anita Kurz describes the scene in her family's apartment block:

> The curtains were burning. And the windows had shattered. My father wanted to go across the street to find my grandmother, but the fire was spreading. Cars were on fire where just they had been parked out on the street. The fire was spreading. Incendiaries had hit our building. The men had tried to throw them out, but failed . . . My father asked me what I wanted to save from the flat and I said my dolls' pram and my school bag. These were saved. People tried . . . they thought if they could get things into the basement, they would be saved.

Just a block away, Nora Lang and her family emerged to find things worse than at her friend's place.

> My parents decided to leave the building. We were in the back section of the building, so we had to make our way through into the

front. We were sort of hemmed in, but we decided to get through onto the square, the Dürerplatz, and we did. With our luggage. And all around the square everything was on fire . . . there were lots of fires, all at various stages, but I remember this one. And all the people were around the square, milling about in the dark. I don't think a single one thought there would be more bombs.

Also in the same crowd on the Dürerplatz, after the first raid, was young Christoph Adam. His family's apartment was actually on the square. He was lucky too that his father, far from panicking, showed a capacity for improvisation.

I think it can be said that this first attack set around 95 percent of the area on fire. The attack was on such a scale that my father—there were a few older men there—said we should get out of the building. We would suffocate. It was already starting to smell . . . We were fortunate enough to have a large water container for fire-fighting, and my father got hold of some quilts, soaked them, and gave one to each of us. He said we have to get out of here—either in the direction of the Elbe or the Grosser Garten. Then we left, trying to make it through the streets, several of which were impassable. Explosive bombs had fallen and quite a lot of house fronts had collapsed across the street. The rubble was on fire. Some houses, I still remember, were burning in wide swathes—the result of phosphor canisters, as was later explained to me. Of course, at the time I didn't know what everything was . . .

What Christoph Adam saw were almost certainly houses set ablaze by canisters, filled not with phosphor—feared because it was rumored to stick to flesh and be impossible to extinguish—but with gasoline-based stick incendiaries. No phosphor-based bombs were loaded for the Dresden raids, as Allied records prove. Nevertheless, the sight of such sudden and uncontrollable burning unnerved them further. They became if anything more determined to leave their area and seek out a place of greater safety.

In this they followed the instructions of the air raid authorities. The Grosser Garten in the south and the Elbe water meadows to the north were the two places that had been officially designated as assembly areas, in case of people having to abandon their homes.

Rather than head north toward the Elbe, the Adam family decided to make their way south to the Grosser Garten. First, though, they decided to make a slight detour along the broad Dürerstrasse. Wider streets were less likely to be blocked by collapsed buildings. Walking due east, they came to another main street, the Fürstenstrasse (later Fetscherstrasse), which cut sharply south in the direction of the Grosser Garten. They tramped three hundred yards between burning buildings, until they reached the Fürstenplatz, about halfway to the Grosser Garten.

It was here, Christoph remembers, when they met the main Striesener Strasse, the major direct route east from the Altstadt, that "suddenly we were caught up in a whirlwind, a storm." The firestorm was spreading out from its breeding place in the narrow streets and alleyways of the city center, reaching into the inner suburbs. The family, clinging together, struggled on.

How could they predict that things would get even worse?

HAVING SURVIVED with his friend Siegfried in the air raid shelter on the Schumannstrasse, Günther Kannegiesser emerged to see that his own family's building seemed intact. On the other hand, there was smoke billowing from fires further up the Schumannstrasse, where Siegfried lived. Günther accompanied his friend there.

They ran upstairs to the family apartment on the fourth floor. An incendiary bomb had come through the ceiling and was burning inside a cupboard. Using water from the bathtub, which had been left filled in accordance with air raid instructions, the two boys had all but mastered the fire in the empty building when they noticed that smoke was creeping up the stairs. Just in time, they managed to find their way back down to street level. Another apartment in the building was on fire, and the blaze was spreading. The boys separated, Siegfried to look for his mother and Günther to check on his own family.

A few minutes later, Günther reached his home. The block was intact except for the blast-shattered windowpanes. His little sister and brother were still in the basement, but his mother was up in the apartment on the third floor, busily clearing a mess of broken glass. He noticed that it was now exactly twelve midnight. He began excitedly to tell his mother of his adventures, and how much of the city had already been destroyed.

Then the local air raid warden appeared. An officious type, he warned Günther about such loose talk and told him that, instead of spreading rumors, he should be off helping to put out fires. Günther said a half-reluctant, half-eager farewell to his mother and dashed off to the headquarters of Police District Four. There he was told simply to help wherever he could.

> On the Striesener Strasse, there were fire-fighting operations under way wherever houses were burning. A balcony had fallen onto the overhead tramlines. I had come back to look for my friend, but then I saw a group of women with baby carriages taking shelter in the lobby of a building. They wanted to get to the Johannstadt hospital . . . I offered to help. We went by way of the Dürerplatz. All around there were many buildings on fire, some already burned out to the second story. There was a fire engine on the Dürerstrasse, which was pumping water out of a reservoir that prisoners of war had constructed a short while before. I followed the hoses and found the firefighters so that I could ask the best way through. They were dealing with a fire in a house on the Reissiger Strasse. One of them told me they had just arrived from Bad Schandau and had driven into the city via Fürstenstrasse and Dürerstrasse. The truck lay totally wrecked on the Dürerplatz for a long time afterward.

The boy guided his column of women and children east along the Dürerstrasse to the junction with Fürstenstrasse. But instead of turning south, as young Christoph Adam and his family had done an hour or so earlier, Günther led the way north toward the Elbe. The Johannstadt hospital complex lay between here and the river.

BY 11 O'CLOCK the Altstadt was burning so extensively that the blaze was too much for the city force of a thousand firefighters. Auxiliaries and suburban units had quickly reinforced them, and fire trucks were starting to come in from other towns—eventually even from Berlin, almost a hundred miles to the north. It made no difference. As midnight approached, the firestorm had the heart of Dresden in its fierce, pitiless grip, and there was very little that anyone could do about it.

Gerhard Kühnemund, a fifteen-year-old from Leipzig, was staying with an aunt in the Serrestrasse, close to the densely built center of

Dresden, on the night of February 13. Typically for a Dresdener, his aunt did not treat the alarm too seriously. They were just in the process of leaving her apartment when the first bombs struck, rocking the building and blowing out many of the windows. They joined the rest of the neighborhood in a suddenly panic-stricken rush to the shelter. Once down there, the teenager realized that apart from one elderly man, the only other male adult present was a much-decorated young Luftwaffe sergeant. They had not been there long when neighbors pushed their way through from the cellar of the next building, screaming that everyone should leave. The entire Serrestrasse was on fire.

The Luftwaffe sergeant turned to Gerhard, indicated the Hitler Youth uniform he was wearing, and asked him if he was "a man." To which the boy answered, "Of course." They stormed up to the attic, where they tried to put out two incendiary bombs with water and sand. Suddenly one of the bombs blew up—it had been booby-trapped with an explosive charge in its weighted nose. The sergeant clutched his stomach. A splinter had torn out part of his stomach wall. The boy suffered a cut to his shin, which sent blood pouring down into his shoe. The sergeant recovered quickly, however. He shouted to his companion: "Let this rubbish burn." Stumbling toward the stairs, the boy glanced out of the attic window and was appalled.

> The whole of Dresden was an inferno!
> On the street people were wandering about helplessly. I saw my aunt. She had wrapped herself in a damp blanket, and seeing me, called out: "Come to the Elbe terrace!" The sound of the rising firestorm strangled her last words. A house wall collapsed with a roar, burying several people in the debris. A thick cloud of dust arose and, mingling with the smoke, made it impossible to see. Then a fist grabbed me by the neck and pulled me away across the rubble. It was the young pilot, who with his calm composure probably saved my life in this chaos. Time and again we stumbled over corpses . . .

Seventeenth-century laws forbidding the construction of half-timbered (*Fachwerk*) houses in Dresden meant that it was a city of sandstone and brick. But whatever impression their Baroque facades might have given, the frames around which the blocks that crowded the streets and alleys of the city's heart had been built were of centuries-old, dry wood. So were the floors, the panelled walls of their rooms and much of the heavy traditional furniture that filled them.

Once the incendiaries tumbled through into the lofts and garrets of the Dresden Altstadt, and—just as importantly—were allowed to burn unchecked while the inhabitants cowered in their basements and shelters, the scene was set for a conflagration of the same ferocity unleashed in the relatively modern tenements of Hamburg-Billwärder on July 28, 1943, and in the genuinely medieval hearts of Darmstadt and Kassel the previous year.

In the suburbs, it was different. The villa of Hannelore Kuhn's family in the Südvorstadt, for instance, was in one of the more lightly bombed areas, and their preparations had been thorough.

> Someone said, there's a fire on the third story of the house. And it was burning . . . they had dropped incendiary bombs, but onto our loft, and the loft was impregnated. It was huge, but we had cleared it, that was one of the safety measures. And large containers full of water were ready there. And the floor had been treated with some kind of fireproof stuff. But it was smoldering . . . in the basement we took shovels and tossed the incendiaries out of the window. In the loft it was harder; all the windows were broken and there were red-hot fragments flying from other buildings. Anyway . . . somehow we did it. We put out all the fires.

For the moment, they had escaped the fate of the city's center and inner-eastern suburbs. The Südvorstadt would not be so lucky the next day.

THE REICH CHIEF of fire-fighting services, Major General Rumpf, who had by chance been in Leipzig when it had narrowly escaped just such a fate in December 1943, later reported on Dresden as well. He described what happened in the Dresden Altstadt after 5 Group's initial raid:

> The fire-fighting forces, though over a thousand men strong and well equipped, were from the outset completely powerless. The

supporting forces of the regiments and all the neighboring towns, including beleaguered Berlin, struggled through the night on icy roads to the scene. The sights with which they were presented filled these men, hardened on the inside and out by experience of a hundred nights of fire, with horror and dismay. The buildings along the streets, shattered under the hail of bombs and seared by fires from the incendiaries, had collapsed and blocked the exit routes, consigning thousands to death in the inferno.

The rubble from the tall, close-packed buildings both created an insuperable barrier against attempts to escape from the endangered area and also, in many cases, buried the exit routes from the cellars and basements in which the people of the Altstadt had taken refuge when the air raid warnings sounded. Yes, it was sometimes possible, with difficulty, to overcome these barriers, but this took time, and time was too short for the unlucky inner city dwellers.

Rumpf described the rapidly developing firestorm with a seasoned expert's eye:

> Such a natural phenomenon can change the normal qualities of the atmosphere to such a degree that within it organic life is no longer possible and is snuffed out . . . The individual fire centers combine, the heated atmosphere shoots up like a huge chimney, sucking the rushing air up from the ground to create a hurricane, which in turn fans the smaller fires and draws them into itself. The effect of the pillar of hot air produced by such a huge blaze over a burning city would be felt by those in aircraft up to thirteen thousand above ground level . . .

Nora Lang tells of how, with her parents busy in the apartment trying to salvage what they could, she was delegated to look after her little brother. She and the five-year-old positioned themselves in a spot on the burning Dürerplatz, where so many survivors were still to be found after the first raid. Nora waited patiently, clutching the boy's hand and minding a suitcase filled with the family's most precious belongings. Still her parents did not emerge from the building. Eventually a young mother from their block said they must leave and seek the safety of the Elbe. Nora followed reluctantly.

The young woman had her baby in her arms; I was carrying the suit-case with all our family's documents in it, and two blankets. There was so much luggage. I was carrying something on my back too, and as for my little brother, I just dragged him along behind me.

We tried to make progress along the Dürerstrasse. But this was scarcely possible, because there was fire everywhere. We had to walk along the middle of the street, to avoid being hit by flying roof tiles, or burned-out window-frames, or all the stuff that was flying around. It was like a hurricane made of fire. The distance was not far to the Fürstenstrasse—maybe two hundred meters—but we just couldn't get through. The young woman with her baby in her arms urged me on all the time, and tried to counteract my fears. We had to get through, we had to do it—there was no other way to safety. But it was impossible, so we decided to go back into a building and find a basement in which to wait for morning.

Dresden had been hit badly. Still, until the warning sounded for the second raid, most people there thought it was all over. Hundreds—even thousands—had died, but now it was a matter of fighting what fires could be fought, saving the living, tending the injured, and recovering the dead.

After the first wave of bombers had gone, eighteen-year-old Günter Jäckel, recovering from his wounds in the southern suburb of Dresden-Plauen, was soon back on the surface. He suffered from claustrophobia.

Then the warning siren sounded again. Could this really be another attack? Either way, Jäckel decided he couldn't bear being shut up in the hospital cellar again.

There were suddenly people shouting—heavy bomber formations heading this way, make for the cellar . . . I was able to walk, so I put my pajama jacket on . . . and my tunic and my slippers, otherwise nothing . . . it was not cold that night. I went outside, and the sky was on fire. There was a meadow by the school and a few people were heading that way. Then they came, the aircraft . . . there were these flares, the illumination canisters, a soft explosion and then they flared up, and you saw the white banners of this magnesium smoke. They were quite close over the meadow . . .

Günter Jäckel followed his companions into the meadow in Dresden-Plauen. From there he would watch the second, even more devastating air attack on his native city.

> I carried on a bit into the field and then just lay down. Yes, that was it . . . from this grandstand seat a panorama . . . I can't describe it, the marker lights . . . red, blue, yellow, or green. And below us the city blazing away . . . down there the city, you know it was like a nightmare. Like a surreal event . . . you thought: This is impossible. This cannot be.

21
The Perfect Firestorm

BY THE TIME 5 GROUP had left the Dresden Altstadt burning, a second group of more than 550 British bombers was on its way to hit the city again.

The second wave—more than twice the size of the first—started taking off from their airfields in England just as the first wave had begun to turn toward their target. By shortly before 10 P.M., they were all in the air. Like the first wave, they had gathered over Reading, and from there headed down across Beachy Head and over the French coast around Boulogne. It was a vast stream of aircraft, more than 120 miles long.

In contrast to Lesley Hay, the fact that Dresden was a relatively unknown quantity cheered Miles Tripp, a bomb aimer with 3 Group, which formed part of the second wave. It wasn't the Ruhr or Berlin, with their massed antiaircraft defenses and their experienced gunners. It would, however, be the longest trip that the crew of his Lancaster, "A for Able," had flown until this point. They would be carrying one four-thousand-pound "cookie"—a "blockbuster"—and canisters of incendiaries.

Derek Jackson was a nineteen-year-old from Manchester who had planned to become a commercial artist until the war made such jobs, at least for the duration, a thing of the past. He had been an apprentice with a painting company until his eighteenth birthday, meanwhile joining the Air Training Corps, and had applied for the RAF as soon as he was old enough. Like most young men he had wanted to be a pilot, but of these there was an oversupply by this stage in the war. Derek instead trained as a gunner on bombers. In late November 1944 he joined 149 Squadron, 3 Group, in Norfolk.

His first "op" was on December 15, 1944. It was a daylight raid on German troops said to be concentrating at a railhead in the Ruhr area. Of his thirty missions, he reckons half were daylight, usually involving airfields, ammunition dumps, troop concentrations, and so on (especially during the Ardennes offensive). The others, at night, were city operations. The weather was mostly filthy. He recalls trips to Nuremberg, Munich, Cologne, and Frankfurt before Dresden.

> As a youngster, I had heard of Dresden in connection with China pottery, but of course all large German cities had factories working on war production, so we had no qualms about the raid. We certainly didn't think that the war was over at this time.

The main thing that worried him was the length of that night's flight. Gunners harbored a special hatred of long trips. They had nothing to do but watch and wait. Their turret spaces were narrow, confining. Other members of the crew at least got to move about a bit.

At around a quarter past midnight, the main force burst through the Mandrel screen, whose jamming had kept the German defenses unaware of their approach. At this point, in the vicinity of Luxembourg, further confusion was created by a windowing force of sixteen Mosquitoes from 100 Group, which veered off in the direction of Cologne/Koblenz, successfully simulating a significant attack on this area. Further "spoof" attacks would draw the enemy controllers' attention to Dortmund, Bonn, Nuremberg, and Magdeburg. All the same, a force this big was hard to miss. As the great stream of aircraft passed over the Mainz-Mannheim "flak belt," they were protected by the cloud and the continuing lack of moonlight. Their luck would hold good until they were over the target.

The first wave had engaged in quite complex changes of direction to keep the German defenses guessing. This second wave followed the first's course as far as the border of the Reich. But where 5 Group had abruptly moved north-northeast toward the Ruhr, the first of several deceptive maneuvers, three hours later the force kept flying pretty much straight, a little to the south of 50 degrees latitude, until its foremost aircraft were well into Central Germany.

At eight minutes before 1 A.M. on February 14, according to a report found in the Dresden Air Defense Leadership's bunker beneath the Albertinum, the following message was logged:

The spearhead of a new bomber force has reached Bamberg. It has taken a northeasterly course. As well as this, strong bomber forces are reported in the area of Mainz-Aschaffenburg, which have taken an easterly course.

What was not yet clear was its precise destination. Only when its vanguard reached the Thuringian town of Gera, around seventy-five miles short of its destination, and adjusted its course to north-north-east—heading for Dresden—must the target must have seemed sickeningly apparent.

At 1:07 A.M. air raid sirens sounded in the undamaged suburbs of Dresden. A few of those who were on the edge of the afflicted areas heard those often faint, distant wails of warning. Otherwise, with the power supply gone throughout the center of the city, the people knew nothing of the new attack until they saw the new marker lights, heard the monotonous growl of engines and the whistle of the bombs. Until, in many cases, it was too late.

There seems to have been some disappointment felt among the crews of 8 (Pathfinder) Group as they approached the target, this time not down the line of the Elbe but from the southwest. The light from the fires raging through Dresden were visible from fifty or more miles distant. Their rival pathfinders in 5 Group had done their job so well that further marking might be superfluous. In addition, the weather had improved still further since 10 P.M. If the weather for the first wave had been satisfactory, for the second conditions over the target would be almost ideal. The problem now was not cloud but the smoke arising from the burning Altstadt. The strong northwesterly wind was blowing these great, miles-high columns of smoke before it, spreading it over the area to the southeast of 5 Group's original bombing sector and concealing this more effectively than the densest concentration of cumulonimbus. Nevertheless, 8 Group's Lancasters dutifully swooped low to drop their illumination flares over the designated area, though the entire city was revealed in the light of the monstrous fire.

The extraordinary success of 5 Group's marking and bombing had created another unexpected problem. There followed an urgent discussion on VHF radio between the master bomber, the Canadian squadron leader CPC (chief pay clerk) De Wesselow (call sign

"Cheesecake"), and his chief marker, Wing Commander Le Good, as they hovered at around eight thousand feet over the city.

"Dresden. Clear over target, practically the whole town in flames. No flak," reported Le Good, an Australian.

Should they instruct the bombers to drop into the existing area of fires, thus probably uselessly duplicating the destruction, or—since their brief was to maximize chaos—extend the bombing over other areas of the city outside the districts already attacked? The assigned aiming point for the second wave was the Altmarkt—the historic marketplace between the Altstadt and the *Schloss*. But this area, which had lain within 5 Group's bombing sector, was already a sea of flame. To bomb from this point would be to simply drop precious ordnance into the raging inferno.

It was decided, in an ad hoc decision, to move the attack into new territory.

The Blind Illuminator (one of the six categories of supporting crews in each squadron) Lancasters had dropped parachute flares, using H2S, but these had vanished into the maw of the fire. Shortly after 1:30 A.M., De Wesselow ordered green markers to be dropped visually onto the fringe areas. The bombers were to release their loads two seconds after crossing the green markers. This moved the attack away from the Altstadt. The new distribution meant that the very first bombs from the 525 aircraft that had arrived at the target dropped southwest and west of the Altstadt, in Löbtau and Friedrichstadt.

Following the continuing distribution of markers, the next wave hit the Südvorstadt, and with it the Hauptbahnhof, which had been seriously but not catastrophically affected by 5 Group's attack. Now the horror really started for the refugees and all the others trapped there in its underground vaults.

Farther south, the suburbs of Räcknitz, Zschernitz, and Plauen (where Günter Jäckel was crouched in his meadow, watching the "surreal" scene), were subjected to a nasty scattering of bombs.

But then came the really big attack on Johannstadt and on Striesen (just southeast of Johannstadt), where many new storm showers of incendiaries further fed the already awe-inspiring blaze. In Johannstadt were Nora, her little brother, and the woman with the baby who had befriended them, and Anita and her parents, and

Günther Kannegiesser, who had taken makeshift refuge in the entrance to the hospital. They were all caught by the new wave of bombing.

A mixture of high-explosive and incendiary bombs rained down on the Grosser Garten, where Dresdeners, obeying the authorities' instructions, had assembled after escaping their burning homes. Here were Christoph Adam and his family, and the painter Otto Griebel, whose home was close to the park. And thousands more, wrapped in damp blankets and coverlets, handkerchiefs over their mouths, watching the exploding trees, hearing the screams of the injured and dying—and praying that somehow they would be spared, for they had nowhere else to go.

It now seemed that the British were bombing the dispossessed and the homeless. The park, the zoo, the lodges, exhibition buildings, and restaurants were all sacrificed to explosion and flame. This was starting to look like sadism, and that would be the view of many observers after the war. To the aircrew, most of whose knowledge of the cityscape of Dresden and its amenities was at best cursory, the evidence is that it just looked like an area that needed to be bombed.

Squadron Leader De Wesselow's decision to spread the damage was now creating a new, wider arc of concentrated devastation. It would lend the bombing of Dresden, it seems, a dubious distinction: that of the greatest area destroyed in a single night. After the green markers had been bombed, red markers were spread in the burning area and the new Lancasters told to aim for those. At 1:42 A.M., with the bomb aimers no longer able to recognize any markers, De Wesselow gave his final order for them to bomb "the middle of the fires."

The actual bombs dropped on Dresden during the half-hour or so of the second wave tell the story:

8 Pathfinder Group—60 Lancasters (includes master bomber, markers, and illuminators)

Bombed from 1:21 to 1:45 A.M.

Type	Quantity	Tons
4,000-pound mines	27	
1,000-pound HE	94	
500-pound HE	159	
Total explosives	280	125.7

Type	Quantity	Tons
250-pound markers	90	10
Illuminators (Christmas trees)	550	

1 Bomber Group—248 Lancasters
Bombed from 1:23 to 1:52 A.M.

Type	Quantity	Tons
4,000-pound mines	145	
2,000-pound mines	101	
500-pound HE bombs	111	
Total high explosive	357	387.3
4-pound stick incendiaries	219,933	
4-pound incendiaries with explosive charge	28,467	
4-pound incendiaries bundled in containers	70,266	
Total incendiaries	312,666	558.3

3 Bomber Group—151 Lancasters
Bombed from 1:25 to 1:55 A.M.

Type	Quantity	Tons
8,000-pound mines	1	
4,000-pound mines	119	
500-pound HE bombs	84	
Total explosives	204	234.8
4-pound stick incendiaries	23,940	
4-pound incendiaries with explosive charge	1,260	
4-pound incendiaries bundled in containers	105,292	
Total incendiaries	130,492	233.0

6 Bomber Group—65 Lancasters
Bombed from 1:27 to 1:45 A.M.

Type	Quantity	Tons
4,000-pound mines	65	
500-pound HE bombs	374	
250-pound HE bombs	155	
Total high explosive	594	216.8

The infernal glow in the distance was astonishing even to experienced bomber crews as they turned east over the Saxon town of Döbeln, halfway between Leipzig and Dresden, and began their final runs. The fires were totally visible even from that distance. Bomb aimer Miles Tripp prepared to do his job.

> Although we were forty miles from Dresden, fires were reddening the sky ahead. The meteorological forecast had been correct. There was no cloud over the city.
>
> Six miles from the target, other Lancasters were clearly visible; their silhouettes black in the rosy glow. The streets of the city were a fantastic latticework of fire. It was as though one was looking down at the fiery outlines of a crossword puzzle; blazing streets stretched from east to east, from north to south, in a gigantic saturation of flame. I was completely awed by the spectacle.

The second wave bombed from up to twenty thousand feet, much higher than the first. Even the master bomber stayed at around eight thousand feet. According to some witnesses, the gunners from their turrets could see not only the flames of Dresden but, sixty or seventy miles beyond them to the east, the flash of the artillery on the Silesian front, where savage fighting was in progress between Russian and German forces. The glow of Dresden burning was visible far to the east. It is not difficult to imagine the desperate feelings of those hard-pressed German troops as they scanned the landscape to their rear—the homeland they were supposedly defending—and made out the sight of Dresden on fire.

At Klotzsche airfield to the north of the city, morale must have been even worse. The night fighters were once more ready for takeoff. This time, however, not even the privileged "A" Group was given permission to start. Eighteen pilots sat helplessly on the apron in their Messerschmitt Bf 110s, waiting vainly for orders, while on the horizon

Dresden burned and the British bombers fed the flames with utter impunity. The fighters' sense of vulnerability was increased by the fact that, with transport aircraft to and from besieged Breslau regularly using the airfield, their base commander continued to illuminate the runway at intervals throughout the night, almost as if signaling to the British to attack.

A film was shot during the second wave of the attack. Cameras had been mounted on the upper gun turret and crew access door of a specially adapted Lancaster "Y" of 463 Squadron Royal Australian Air Force, which had long been associated with the RAF film unit. The aircraft's cameras had captured the dramatic sinking of the German battleship *Tirpitz* in Tromsöfjord the previous November. Its bombers, which were a normal part of 5 Group, had accompanied the first wave to Dresden on normal bombing duties; "Y" escorted the second wave alone.

The aircraft circled over the doomed city for eight and a half minutes between 1:28 and 1:37 P.M., recording one of the most chilling vistas of the air war. Shot from almost three miles above the inferno through cold, clear skies, the film shows the outlines of streets etched in fire; the "Christmas trees" descending into the great pillars of smoke below; the plainly discernible explosions of the big four-thousand-pound air mines, which rise out of the general conflagration like small mushroom clouds, reminiscent of the atomic bombs dropped a few months later on Hiroshima and Nagasaki. Perfection of destruction. What this felt like on the ground is virtually impossible to imagine.

Not everything went smoothly for individual aircraft. When "A for Able" came in for its bomb run, control of the aircraft passed effectively to Miles Tripp as bomb aimer. The aircraft's Australian pilot, nicknamed "Dig," asked him to give a course. Tripp realized that the master bomber could no longer be heard on the R/T (radio ransmitter). No more calm, precise instructions. Tripp made a decision, and with a clear conscience:

> I told Dig to turn to starboard to the south of the city. He swung the aircraft away from the heart of the inferno and when we were just beyond the fringe of the fires I pressed the bomb release. I hoped the load would fall in open country; I couldn't forget what we had been told at briefing, or the old newsreels of German dive-bombing atrocities.

At the time Tripp erroneously assumed the master bomber had been killed. Comparing his navigator's log with the sequence of events during the raid, he later decided that the radio silence must have occurred while De Wesselow and Le Good were privately discussing whether to change tactics due to the invalidation of the original aiming point. Tripp insisted thereafter that if he had been given instructions, he would have obeyed them unconditionally. But he wasn't, so he didn't.

Doug Hicks, a nineteen-year-old Canadian serving with 550 Squadron, was rear gunner with a "virgin" crew—this was their first-ever operation. Only their aircraft, "V for Victory" was a veteran—it had survived more than ninety raids—and that was considered good luck. Hunched in his tiny rear-gunner's turret for almost five hours, watching the sky for enemy aircraft, Hicks found himself surprised that so little talk went on between crew members; no banter, not even when they saw the huge glow of the target fifty miles ahead. This was unlike anything they had ever heard about:

> Almost daylight conditions prevail. The sky is lit up from the horrendous inferno on the ground that is now the target. In this lighted environment I now see bomber aircraft everywhere. They are to the left and the right, up and down, it seems almost impossible that this zone of air space can accommodate so many bomber aircraft. As one of the bombers passes underneath my turret I can see the crew in the cockpit of the aircraft and notice the red-hot exhaust from the four engines glowing eerily in the sky. I have difficulty in comprehending this vast armada of aircraft converging on the target. As quickly as it started, we have dropped the bomb load and turned to head for home. So this is trial under fire. We did it. We have almost completed our first trip. There is no jubilation from the crew, not even a slight hurrah.

Bomber Command's Intelligence Narrative of Operations No. 1007, sent out the next day by teleprinter to the Air Ministry and all group and station commanders, with copies to First, Second, and Third Divisions USAAF, gives a cool, professional, preliminary overview of the second wave's work:

Crews were able to identify the town and river visually, aided by fires from the earlier attack, but despite illuminating flares, the smoke from these fires rendered it impossible to identify the aiming Point with certainty. After assessing the green TIs from the blind markers, the Master Bomber ordered the Main Force to overshoot them by 2 seconds. Later bombing was directed onto the red TIs and finally the center of the fires. Both red and green markers were clearly seen and the bombing was very well concentrated and ably controlled by the Master Bomber with good and clear instructions.

Many new fires were started and the whole city, particularly the old town south of the river, was reported well ablaze towards the end of the attack with a number of smaller fires in other parts of the town. Several large explosions were reported and one particularly large fire just E. of the Marshaling Yards. Smoke was reported up to 15,000 ft and fires were visible for 100 miles on the homeward route.

The attack is believed to have been highly successful.

22
Catastrophe

DRESDEN—OLD, CLOSELY BUILT Dresden—was on fire well before midnight, and probably most of it could not have been saved, even had the second wave of bombers never arrived over the city. The death toll would have been high, comparable with other similar raids on historic town centers in the air war. Dresden would have joined Darmstadt, Kassel, and the rest of the towns from Germany's golden past as a chilling footnote in the history books. The Saxon capital's special inadequacies in matters of shelter provision would have made rich matter for expert discussion.

It was Squadron Leader De Wesselow and Wing Commander Le Good's joint decision, as the second wave surged toward Dresden, to abandon the designated aiming point, and bomb outside the already burning areas of the city, which turned the raid into a byword for slaughter. Their move doomed not only large areas of the residential suburbs but also the great gathering places of the homeless and the dehoused (as the Air Ministry experts had always called them)—the Hauptbahnhof, the Grosser Garten, and the banks of the Elbe. These would become killing grounds without compare.

Many, many died in the streets as they tried to flee—burned, asphyxiated, dragged into the hot, hungry mouth of the firestorm—but for those who did not even try, death was near certain. The terrible thing about the second wave was that, in the center of the city, it came without warning. Rudolf Eichner was another young soldier from Dresden, wounded at the front and recovering in his home city. He was at the old Vitzthum High School, on the southern edge of the

Altstadt, which had been turned into a reserve military hospital, caring for around five hundred soldiers.

Many of the surrounding buildings in these close-packed residential streets (marked in the RAF zone map as the most densely populated) had been set ablaze by the first wave. The fatal problem, Eichner remembers, was that their inexperienced inhabitants—mostly women, children, and the elderly—instead of fighting the fires, seemed concerned mainly with securing their valuables and prized possessions. All the same, the school-cum-hospital remained more or less undamaged. And since it was full of soldiers, there was at least an element of discipline to the proceedings.

The hospital's trained firefighters had successfully neutralized the incendiaries that had lodged in the attic areas. The wounded who were able helped to move inflammable furniture, beds, and equipment into the corridors. The other inmates stayed in the basement shelter. Had the bombing ceased after the first wave, Rudolf Eichner believes the hospital and those sheltering in its cellars could have been saved. He and his comrades were determined to stop fires spreading to the building, and if necessary prepared to work through the night to this end. Then came the second wave.

> In the middle of clearing activities and other measures to secure the building, the bombs of the second attack began to explode. It was 14 February around 1 A.M. There were no warning sirens. Completely surprised, we rushed back down into the air raid shelter. Soon, since the people of the district could no longer find safety in their burning buildings, it was vastly overcrowded. We stood shoulder to shoulder down there in the cellar, so tight that it would have been impossible to fall over. There came the deafening noise of the bomber formations—much louder than in the first attack—and constant explosions, often simultaneous, which shook the building.
>
> The school received several direct hits. The light went out. Dust from the ceiling, the doors and windows were blown in. Bricks from the safety wall over the windows flew down into the basement. A window frame collapsed onto my back. Dust, chalk, and smoke threatened to suffocate us. People were screaming, lashing out around themselves. A young mother threw herself over her baby carriage in an attempt to protect her child. A direct hit on the sec-

tion for the seriously wounded cost many lives. Then the sound of engines and of bombs exploding seemed gradually to recede. Someone shouted: "The ground floor of the school is on fire!" It was a signal to leave the basement. I stumbled up and forward over dust and rubble, past burning furniture and doors, and managed to get out of the building.

This was just the first stage of Eichner's ordeal. Out on the street, in the Dippoldiswalder Gasse, the firestorm was at full, raging power, the air a maelstrom of burning chunks of wood and metal and paper. Blazing tree limbs flew past the terrified young soldier, all kinds of half-recognizable objects caught up in the searing tornado—including, he realized, helpless human beings. He dropped into a crouch, and then crawled on all fours to the far side of the street. In the front garden of a nearby property he spotted a space that was not only partly sheltered from the burning winds, but seemed a little more open and therefore less likely to be affected by collapsing masonry. He stumbled over to it and found five other comrades who had escaped the Vitzthum High School.

The half-dozen survivors formed into a circle, standing with hands on each other's shoulders. They had wet cloths over their faces, which helped them to breathe, but their clothes were bone-dry and liable to be set alight by the host of sparks flying around the area. It was the task of each man to beat out any small fires that started on the clothing of the man in front of him. They did this for six hours, until the storm began to die down, and with it the fires. So they were among the few who survived the immolation of the shelter at the Vitzthum High School Reserve Military Hospital.

AT THE HAUPTBAHNHOF, the second wave caused terrible damage. The entire area was subjected to a rain of bombs—the Bismarckstrasse, the Wiener Platz, leading into the first buildings of the elegant shops and cafés of the Prager Strasse—and the station itself. Its high glass roof was shattered. Trains still waiting at the platforms were blasted and burned. In the network of cellars underneath, where the vast majority of travelers had taken refuge when the sirens had first sounded, the scenes were appalling.

It was not, for the most part, the fire or the blasts that killed them. At least not directly. It was the air. Or the lack of it. With no filtered air supply and few emergency exits—in any case mostly blocked by human bodies and belongings—as the firestorm sucked the oxygen out of the air in the vastly overcrowded underground complex, many hundreds died of simple asphyxiation. A few managed to find their way out of the labyrinth and stagger back out to the surface, but not many.

One survivor, a woman refugee from Silesia, passed through some hours later, helped out by an army officer "through a long passage." She added: "We passed through the basement. There must have been several thousand people there, all lying very still." Lack of oxygen, smoke poisoning, and carbon monoxide poisoning had once again done their work, as so often in Dresden that night.

Outside the station, the situation was unpredictable. Gertraud Freundel's family lived in the Reichstrasse, a wide street that ran south from the station toward the Technical University and the Münchner Platz law court complex, and from there into the higher ground of the southern suburbs.

The family's apartment was in an old building that abutted the Bismarckplatz, just in front of the Hauptbahnhof. This large square was planted with trees, giving the whole area an airy, spacious feel. For this reason, after they had survived both raids in the sturdy cellar, many in the shelter argued there would be no problems with oxygen shortage—so it would be safer to stay there until they could be sure the danger had passed. Similar decisions were being made all over Dresden in the small hours of February 14. For some, tragically, it had already been made by the bombs—buildings had collapsed, burying alive those sheltering beneath. But for the survivors, with the disappearance of the bombers came a testing time of courage and imagination. Gertraud's father showed both those qualities that night. He said firmly to his daughter that they must go up and out. Even as they left the cellar, others lingered uncertainly in the entrance.

Pets were not permitted in shelters. To Gertraud's surprise and delight, the family's pet dachshund, Jockely, had survived both raids shut inside their apartment. He had somehow escaped from the building—probably through a blown-out window or door—and enthusiastically greeted Gertraud and her father as they emerged. They dipped their hats and scarves in one of the buckets of water provided at the basement doorway and splashed some over their coats for good measure. Gertraud wet her handkerchief and wrapped it around the little dog's nose. She scooped up the animal and cradled him in her arms as they prepared to flee.

Before being called up for the Volkssturm, Gertraud's father ran a state-licensed lottery ticket shop near Postplatz—during his absence, Gertraud and her mother had taken over the business. After the first raid, they had tried to make their way through the underpass to the other side of the Hauptbahnhof, from where they could walk to the shop and check it for damage. Ominously, they had immediately come across burned corpses, blazing buildings, and just a little farther on a police patrol, which turned them back.

So now, three hours and a second devastating hail of bombs later, they knew that the only viable escape route was one that led out of the city. Their agreed destination was the semirural southern suburb of Mockritz. At Mockritz there would be higher ground, fresh air, and, with luck, no bombs. It was about a mile and a half distant. But first they had to get across the Reichstrasse.

This was where they encountered the horror of the firestorm. Here it ruled, as it ruled all the wide streets radiating from the Altstadt, the fire's core. It seemed to claw jealously at those who tried to escape the conflagration:

> Outside the firestorm howled. It was blowing furiously, and the draft was pulling us into the city . . . Father held me tight by the arm and I held the dog with the other. We had to cross this infernal tempest that was raging down Reichstrasse with storm and fire. I was terrified and held back, but father pulled me and implored, shouting through the wind's roar: "We must get through this!"

Bent forward to avoid the suck of the storm, they made their way southward, past a group of half-intact buildings, which provided some shelter. Progress was slow but sure. Then, just as they fancied they were putting distance between themselves and the firestorm, came a shock.

> Turning into Sedanstrasse, we saw people coming back down towards us, from the direction we planned to follow, shouting: "You

can't go on up the hill!" The street was blocked by fallen trees, which were on fire. But Father firmly stuck to his intention and we scrambled over the burning tree trunks. The air was full of fire. Great chunks of burning wood sailed through the air, and sparks and flame rained from all directions. It was lucky we had soaked our clothes.

We passed my old primary school. It was full ablaze. We had to keep to the middle of the street, to avoid the searing flames from buildings on either side. All the time I clung to father and held onto my dog. We safely reached Reichenbachstrasse, which had houses on only one side, and made our way up the hill. We finally settled on a slope by Zellscher Weg, a wide arterial road with hardly any buildings. The air was filled with ash, but it was not at all cold . . .

They never went on to Mockritz. They had found what they needed: an open place where there was less danger from collapsing buildings and greedily spreading fires. The air quality was just adequate for survival. There, with a large group of other exhausted and terrified survivors, they spent the rest of the night. As the cloth over her mouth dried out, Gertraud began to find breathing more difficult. A lady matter-of-factly offered to have her little boy urinate on the cloth to restore its moistness. Gertraud summoned the dignity to deliver a polite refusal.

At first light Gertraud, her father, and the little dachshund made for the southeastern district of Strehlen, where they had relatives. The family had long ago agreed to meet there if they got separated in an air raid. It was a plan they had never imagined would be necessary.

Many of the people who took shelter in the center of the city burned to death. Those who somehow made it out alive will never forget the smell of burning flesh rising from the basements and cellars of the old town. Nevertheless, the overwhelming recurring factor in the stories of such survivors (and being survivors, they are naturally a self-selecting group) is the fierce, concentrated quality of the fight for survival; a fight first against suffocation, and then against the tornadoes of burning sparks and debris that threatened to blind and disfigure them as they struggled to find their way out of the maze of the Altstadt— usually, by some almost animal instinct, heading for the river.

Escape from the central districts of Dresden was a Darwinian business. Success usually fell to the young, and the mentally and physically strong. Berthold Meyer, a twenty-one-year-old engineering student, was determined to reach the Elbe after narrowly escaping the basement of a burning house in the Blochmannstrasse, east of the city center:

> Only someone who has been in such a sea of flame can judge what it means to breathe such an oxygen-deficient atmosphere . . . while battling against terribly hot, constantly changing currents of fire and air. My lungs were heaving. My knees began to turn weak. It was horrifying. Some individuals, especially the older people, started to hang back. They would sit down apathetically on the street, or on piles of rubble, and just perish from asphyxiation. . . .

Anita Kurz and her parents survived the second wave of the raid, though there were no happy endings in view for this close little family.

> After the second raid was over, and things had quietened down somewhat, my parents and other people from the building tried to get out of the basement. My father took me by the arm, and we got through the door, and then once we were out of the basement we had to go for a stretch onto the street. And on this street, at that moment, there was a firestorm, impossible to get through. My father dragged me and my mother back . . . it was like a boiling cauldron. So we went back into the basement. And in the big blocks of buildings the basements all joined together. They had been walled off, but the air raid regulations meant that they were opened up. It might have been possible to get through into the neighboring basements and so escape, but people had tried this and said it was impossible. This was bad. Because there was a shortage of oxygen.
>
> But you know, people were exhausted, physically and mentally. It was hard for them to gather their thoughts. Anyway, I had my bathrobe. I had put water on it, made it really wet. My parents did the same. Everyone knew they had to get out, but how? I'm not sure if there were people in front of me, but . . . anyway, my parents took me into this coal cellar, and I fell onto a pile of coal. I'm not sure how. Luckily I had buried my face in my bathrobe.

The provision of these "breakthroughs" (*Durchbrüche*) was one of the peculiarities of Dresden's response to the bombing danger. It was a function of its densely populated center, the lack of large public shelters with proper air-filtration systems, sealed doors, and firewalls. The idea of knocking crawl-through holes in the walls of all the cellars in a given street—and sometimes a whole district—was that if you had a problem getting out of your particular building's basement—if rubble had blocked the exits, or the upstairs was on fire—you could find your way out of your unventilated basement and through to some other building from which escape might be possible.

Like almost everything else in Gauleiter Mutschmann's capital (except his own private bunker), this system had been designed to deal with a "normal" air raid, in which there would be damage to individual buildings and some fires, but nothing apocalyptic except for those unlucky people who suffered direct hits. The system was cheap, easy to arrange, and seemed logical. In the northern part of the Altstadt, there had been some extra tunneling work. This provided an extensive supplementary network, which moreover emerged via purpose-built concrete exit shafts close to the banks of the Elbe at the Neumarkt (by the Frauenkirche), the Postplatz (the Altstadt's main tram terminal), and the Adolf-HitlerPlatz.

Even where fairly sophisticated in construction, the system proved totally unsuitable for dealing with a lengthy, heavy raid or series of raids. In the case of a firestorm it was actually counterproductive. The policy of creating "breakthroughs" turned each arrangement of connected tunnels into a potential death trap—a trap with a multitude of chambers.

Instead of being limited to sucking the oxygen out of one basement at a time, the firestorm could suck it out of the entire network. And smoke and flame could likewise spread through the underground labyrinth with rapid abandon.

Margret Freyer, then a young woman in her twenties, was one of the few who did escape from the Altstadt after the second raid. She found herself that night with a female friend, Cenci, in the local shelter on the residential Struvestrasse, extending east from the elegant shops of the Prager Strasse:

Three women went up the stairs in front of us, only to come rushing down again, wringing their hands. "We can't get out of here! Everything outside is burning!" they cried. Cenci and I went up to make sure. It was true.

Then we tried the "break-through," which had been installed in each cellar, so people could exit from one basement to the other. But here we met only thick smoke, which made it impossible to breathe.

So we went upstairs. The back door . . . was completely on fire. It would have been madness to touch it. And at the front entrance, flames a metre and a half high came licking at short intervals into the hall.

In spite of this, it was clear that we could not stay in the building unless we wanted to suffocate. So we went downstairs again and picked up our suitcases. I put two handfuls of handkerchiefs into a water tub and stuffed them soaking wet into my coat pocket. They probably saved my life later on.

But as we went up the stairs out of the cellar, Cenci's husband came up and said: "Cenci, please stay here, you must help my sister. She's ill."

I made a last attempt to convince everyone in the cellar to leave, because they would suffocate if they did not; but they didn't want to. Most died down there, but three women were found outside the door, among them Cenci.

The center of Dresden in the small hours of Wednesday, February 14, was a tragic arena of human suffering. In that terrible confusion, unspeakable consequences could spring from a moment of carelessness or miscalculation; not to mention a judgment arrived at amid the burning fog of smoke and sparks and hot air that stripped and burst the lungs. Destiny could be determined by such an apparently trivial matter as choice of clothing.

Margret Freyer, for instance, ascribed her survival once she had left the doomed cellar on the Struvestrasse—the streets were already like ovens—to the fact that she had chosen to wear knee-boots when she went out that winter night to visit her friend. In the heat, the tar on the streets melted. Others who tried to flee through this viscous quagmire rapidly lost their slip-on shoes, even their lace-ups, which stuck in the tar. Their bare feet were quickly burned so badly that they could no longer move. They died. Margret Freyer's snug boots stayed stubbornly on, and she lived.

One man who lost everything and everyone he cared about that night wrote a letter of condolence a few months later to a woman whose parents' deaths he had witnessed. They had all lived in the same block on the Marienstrasse, on the edge of the Altstadt. Hans Schröter's letter conveys the roles of chance and courage in survival with a kind of still, desperate vividness:

> We had got through both raids and thought that we would now survive. This unfortunately was not to be the case. The door of the basement of No. 38 was buried under rubble, so the only option was the emergency exit to No. 40 and 42. When we got through to No. 40, flames were already pouring down the steps, so that to save our lives we had to act with the utmost haste. Everyone behaved very calmly. The electric light had gone, we had electric torches and oil lamps with us. To push through the exit required enormous courage, and many could not summon it—including, perhaps, your beloved parents. They may have thought, we'll be all right in the basement, but they had not reckoned with oxygen shortage. As I emerged, I saw my wife and son standing by the security post on the *parterre* of No. 42. They looked so helpless, but since I had an elderly aunt from Liegnitz staying and I wanted to get her out, I said to my wife, I'll be back in two minutes. When we got back after this time, however, my loved ones had disappeared. I checked every shelter and basement on the street. Nowhere were they to be seen, everything wreathed in flame, no entry possible. Unable to find my family, I summoned my last instincts for survival, got as far as the Bismarck Memorial. I stood for an hour by the little building there, until its roof also began to burn. I walked thirty meters further along the Ringstrasse and stayed there until it got light.

As the new day dawned over Dresden, Schröter started to pick his way back through the shattered streets to what had been his home.

> The sight that greeted my eyes was appalling . . . Everywhere charred corpses. I quickly headed home, hoping to find my loved ones alive, but unfortunately this was not so. They lay on the street in front of No. 38, as peacefully as if they were asleep. What I went through at that point you can easily imagine. Now I had to find out if my parents-in-law or other friends could be rescued from our basement alive. For this I summoned two men from the Wehrmacht . . . As we opened up the emergency exit from No. 38, the heat that came out was so intense that we could not go down there. So we

had to remove the boot-scraper from the entrance of No. 40 so that we could get into the bathroom, and through there into the basement of 40 and 42. The basement of No. 42 was full of bodies. I counted about fifty. Eulitz was among them. I could not see your parents, as everyone was piled on top of each other . . .

IN THE WINTER OF 1944 prisoners of war had been drafted in to sink concrete-lined reservoirs into open spaces in Dresden. The object was to ensure that fire-fighting teams had reliable and plentiful sources of water other than the mains and the river. The largest of these tanks was dug on the wide expanses of the Altmarkt, where the victorious troops had paraded in those far-off days of 1940. Two others, on the Sidonienstrasse, near the Hauptbahnhof, and the Seidlitzer Platz, just north of the Grosser Garten, were more modest, but nevertheless held large quantities of potentially life-saving water. There were smaller upright tanks in other areas of the city. Like the "breakthroughs," in the near-apocalyptic conditions of February 13–14, their dangers turned out to exceed their usefulness.

At the height of the firestorm, Dresdeners fleeing the blazing Altstadt made for the massive water reservoir in the Altmarkt. This container-cum-lake, 130 feet long, 65 feet across, and some 10 feet deep, was surrounded by a concrete safety wall that rose two feet or so above ground level. It dominated the old city square in the last months of the war, the subject of constant comment and discussion. Anyone *in extremis* around the center of Dresden would have known it was there.

That night the water reservoir had been of little use to the fire-fighting teams. By the time the second raid was over, most routes to it were blocked to vehicles by collapsed buildings, or made impassable by deep bomb craters. In any case, most of the surrounding Baroque apartment houses and public buildings were already beyond saving. The vast tank now had only one possible use, and that was the protection its depths provided from water's traditional enemy, fire.

Hundreds of desperate human beings, some already on fire, found their way through the burning streets to the Altmarkt. They plunged gratefully into the apparent safety of the cool, plentiful water. As the night wore on, however, the searing air from the surrounding conflagrations and the accumulated effect of all the burning human beings who had crowded into the reservoir began to have an effect. The heat within became intolerable, the air unbreathable. In the tank, hosts of survivors, many injured, many poor or nonswimmers, tried to clamber out again, only to find out that the Altmarkt emergency water reservoir tank had not been built as a swimming pool. There were no bars or handles, no ladders. On the contrary, the sides of the reservoir were smooth cement, on which it proved almost impossible to obtain a purchase.

The weak scrabbled hopelessly until they drowned. In many cases they seized hold of the strong swimmers and dragged them down to the bottom with them. A very few of the strongest swimmers and nimblest climbers managed to get back out. The great reservoir in the Altmarkt was both a terrible place of struggle that night and, with bitter irony amid a city on fire, a watery graveyard for hundreds of unlucky souls who lost that struggle.

The next day, when rescue gangs cleared their way through to the square, half the huge quantity of water had evaporated. All the people left in the great reservoir were dead. A macabre ring of charred corpses surrounded the walls of the reservoir; these were the bodies of those who had not quite made it to the water before they burned to death, or were overcome by fumes. In the Seidlitzer Platz tank, which was about fifty feet square, would-be survivors had crowded into the water up to the rim—it was shallow enough to stand upright—until it could take no more. The next day, they were still there, most still packed next to one another in an orderly fashion. All dead of asphyxiation.

In smaller tanks, the water became so hot that the people who had taken refuge in them were literally cooked. Only in one case did an officer observe two men clambering out of a water container, having sat out the first wave of the attack successfully. This was, admittedly, quite early in the night, before the air became truly toxic.

Margret Freyer passed some tanks the next day, as she searched for her fiancé in the smoking ruins of the city:

> From some of the debris poked arms, heads, legs, and shattered skulls. The static water-tanks were filled up to the top with dead human beings, with large pieces of masonry lying on top of that again. Most people looked as if they had been inflated, with large yellow and brown stains on their bodies. People whose clothes were still glowing . . .

Most firefighters caught in the city that night died. An exception was the commander of the Third Company of the Ninth Unit of the Dresden "Fire Police," drafted into the city center from Neustadt. He survived to write a report for his superiors. There were heroes that night, irrespective of politics or war.

The commander and his team had taken refuge from the second British attack in an official shelter near the New Town Hall, which was subjected to a direct hit by a high-explosive bomb. The commander's report laconically expresses his agony of frustration that he could do so little:

I immediately made my way out of the cellar with my men through the emergency exit on Kreuzstrasse 4. Since a large-scale conflagration had meanwhile developed, it was impossible to get through with our vehicles. Hundreds of people were on the Town Hall Square, with further streams of people coming from all directions and moving across the Georgplatz, heading for the Elbe. To save lives, I tried to create a way through with our hoses, using supplies of water still available from the tank there. My intention was to open an escape route through the Ringstrasse in the direction of the Elbe. My Fire Truck 8 had been destroyed by enemy aircraft fire. My fire truck 15 had been knocked out by a bomb. The pump store had been hit. The hose truck could not be used, which meant that I had to abandon the intention of creating an escape route. I was forced to restrict my plans to saving my own men. While, the firestorm in the Town Hall Square had reached such an intensity that it was hurling people to the ground and annihilating them. I gave my men a final order. They were to follow me, and rush through the wall of fire to the Elbe a hundred meters away. I went ahead and cleared the way. I brought all my squad to safety, except two men. The two firemen Rietsche and Kaufmann do not seem to have followed me. I myself and six of our men received second degree burns, but survived . . . I personally reported the withdrawal of my squad to Hauptmann Thieme in the Kaiser Wilhelmplatz at 4:15 A.M. on 14 February.

SOME DRESDENERS who escaped from the inner city survived, like Hans Schröter, by finding islands of space—air pockets. For most, though, unless they managed to get out of the city altogether, there were only two sure refuges: the green spaces of the Grosser Garten and, north of this, the terraces and water meadows beside the Elbe.

Often it was only a few hundred yards, but with the firestorm raging, narrow streets blocked by collapsed buildings, and many familiar landmarks obliterated, the journey was often epically, sometimes fatally, tortuous and slow. Berthold Meyer had begun by escaping from a burning building in the Blochmannstrasse, around a third of a mile from the Elbe as the crow flies.

Clutching a briefcase packed with his most important possessions—he had already been bombed out of his home in Bremen earlier in the war—and with his free hand holding a plank in front of him to protect himself from burning debris, the twenty-one-year-old had not gotten far when he lost his way and was tempted to give up, to sink into deoxygenated oblivion.

> I feel my strength being sapped—my lungs are tensed to bursting. Breathing in short, panting gasps I stumble and weave through the sea of flame . . . the hot wind presses so hard against the plank I am holding out in front of me that it feels like I'm pushing a ton's weight. If something doesn't happen soon, I'm finished! Is this the end? I straighten up, summon up my last reserves, and the will to live somehow gives me a giant's strength. There! What should loom up but a steel-walled *pissoir* [lavatory] on the corner of the street. A few people are already crammed in there and taken cover—truly salvation at a time of great need! It offers a moment to grab a breather, and actually the heat in there is not so bad. Above all, there is some protection from the firestorm and the tornado of sparks.

It was a miraculous place of safety, but the only way to be sure of survival was, as Meyer instinctively knew, to keep moving, heading toward the water:

> Even here it is too dangerous to stay for long, for the fire is continuing to spread. Meanwhile I have made inquiries about the way to the Elbe. There are still about 150 meters to go. I am at the Güntzplatz, I hear. I give up my place and go outside, heading in the direction of the river, still with my briefcase and my plank. I lurch down the Sachsenallee and I start to feel—O greatest blessing of my life!—the

first fresh drafts of air coming from the Elbe. I greedily fill my heav-
ing lungs—what a gift!—and I see that I am about ten meters from
the Albertbrücke. I must, I think, get down to the river to wash my
eyes, for I can scarcely see. Beside the bridge I follow the stream of
humanity down to the Elbe, past an unexploded bomb that people
warn me about. Then I reach the lifesaving strand of the river. It is
so wonderful to breathe this glorious, fresh air. I wade ankle-deep
into the water and, taking my dirty handkerchief, which I had used
in the shelter to protect my face and eyes, I wash my eyes as best I
can. They are swollen and totally covered in filth. Then I lower
myself onto a rock. There I sit for about an hour, with the burning
city of Dresden in front of me.

Farther to the east, at the Johannstadt hospital, young Günther
Kannegiesser and his charges had arrived just as the second raid began.
Unable to find the hospital's shelter, he had left the women and children
he had led here in a corner of the entrance hall, hoping they would be
protected by the heavy staircase, and set off to look for a place of safety.
There was none. Günther finally just lay down in the corridor near the
porter's lodge, on a landing between the ground and first floors.

> Because the explosions went on so long and seemed to get louder
> all the time, I thought that Russian troops had already arrived, and
> these were tanks. They had already advanced deep into Silesia.
> Finally things got quieter and I heard a child crying. Two of the
> women and their children were still standing in the main entrance
> hall. Luckily they had not been injured . . .

Their survival was all the more miraculous, considering that parts
of the hospital had been completely devastated. Especially the
Women's Clinic. According to a later statement by Professor Fischer, a
senior gynecologist, an air mine and two high explosives had hit "B"
wing of the building. Two delivery rooms, an operating theater, the
nursing mothers' department, as well as various wards and steriliza-
tion facilities had been destroyed during the raid itself. As staff tried to
move patients from the endangered "B" wing to "A" wing, there were
fires and explosions everywhere. "A" wing now caught fire. The

patients had to be moved yet again. They were led into the street, where transport was available for the most needy.

Some two hundred nursing mothers, patients, and others still lay beneath the ruins of "B" wing. As staff and fire-fighting teams fought to contain the blaze and reach those underground, a boiler exploded. Within minutes the rescue teams stood knee-deep in water. This was not, they could be thankful, scalding hot, but its dispersal was disastrous in another way. It left no water for the firefighters' hoses. A few women were found beneath shallow rubble and saved, but there seemed no way through to the cellar. The would-be rescuers could only listen helplessly to the victims' faint, desperate cries for aid. Then the walls of the remaining structure began to tremble. The place had to be evacuated, leaving what was left of the building to disintegrate and the trapped women to be buried alive.

Günther Kannegiesser knew nothing of the horrendous scenes in other parts of the hospital. The plucky fourteen-year-old's job was to save himself and his charges, and that was what he did:

> We had to press on toward the Elbe! In 1945, at the place where the Fürstenstrasse meets the water meadows by the river, they used to store planks. These had been taken out of the upper floors of buildings to make it easier to combat incendiary bombs. They were nailed together, forming alleyways of planks, ready to be reinstalled after the war. We crawled in among these planks to get a little rest.
>
> A way toward the Elbe, a wooden building was on fire. It was probably the headquarters of the rifle club that met in the Vogelwiese. One of the women kept crying. I promised I would ask my mother if she could come and live with us. Then I fell asleep.

Nora Lang, her little brother, and the determined young woman with the baby had ducked into a cellar on the Dürerstrasse when the bombs began to fall on Dresden once more. Like Günther Kannegiesser, they had been trying to reach the Fürstenstrasse, and via that the Elbe, but the storm proved too much for them. The cellar was dark, with only the occasional flickering of flames visible through the tiny street-level windows, but she could see that it was full of the vulnerable and the old.

That was much more shocking. During the first attack, I still had my parents with me, but now suddenly I was alone with my little brother. And the misery in the basement, I can still feel it. They were mostly elderly people, fragile, sick people and women with children. There was no beginning and no end to the bombing, it seemed, just these endless explosions everywhere.

The noise level grew so loud and so constant that for her, as a child, it felt like the Bible descriptions of the end of the world. At some point toward the end of their ordeal those sheltering in the cellar realized with a sickening certainty that the building on top of them had collapsed. There would be no exit through the door they had used to find refuge from the naked peril of the street.

When the sound of bombs and aircraft finally died away, they decided they had no choice but to use the "breakthrough," though it was small and narrow. For the older people and the sick, this escape route was not an option.

We slipped through from one cellar into the next. First I had to hand my bags through, then my little brother. But even in the next building there was no way out either. But somehow we succeeded in getting out through some basement or other. How, I couldn't tell you now. Actually I only made it because the young woman with the baby in her arms kept urging me on.

Back out on the Dürerstrasse, they once more did battle with the firestorm, which seemed to have grown even further in intensity. Nora's eyes were now so affected by the rank smoke that she could scarcely see. A few yards short of the Fürstenstrasse, they had to take refuge in the doorway of a shop. This provided a few moments' respite from the maelstrom, but the shop turned out to be a pharmacy, and fellow survivors hurrying past warned them about drums of inflammable chemicals stored there, which could blow up at any moment. Screwing up their courage, they stumbled on: the woman and her baby, thirteen-year-old Nora and her little brother, with their few possessions.

They finally staggered into the Fürstenstrasse. It was a much wider street. Somewhere within the suffocating swirl of smoke and burning debris there was that life-saving current of fresh air, blowing up from the river. They could breathe almost normally. At last.

A complete matter of luck . . . that we didn't all die was pure luck. Some houses were still standing. And there was a truck trailer there. We crept under it and just lay down. We were so exhausted. And . . . it was so cruel . . . there was this man there who had gone mad. He just stood there and bawled into the night, over and over again: *Auto! Auto!* [Car! Car!] My brother was five years old at the time. He can't remember much of what happened that night, except for that man's voice, and the *Auto!* He can never forget it. When we had recovered a bit we went on further to the Johannstadt hospital. They had also been hit there, but one building was more or less intact, and there was a big basement. We went in. There were a lot of people inside.

And then, the next morning, we continued our journey . . . we went down to the Elbe, and there were so many people on the road that we lost the young woman and her baby. Some people were heading upriver, some down. There was such a turmoil and we went down onto the water meadows by the Elbe because there was such a crush. But I still couldn't see well enough to spot the dead bodies there, and the body parts, and so my brother told me, be careful here, here you have to make a detour, and so on . . .

Unlike Nora Lang's family, Christoph Adam, his parents, and his small brother had managed to stay together. They too fought their way eastward, away from the center of the city, all the time resisting the burning suck of the firestorm. Between the first and second attacks they made their way into an area that was as yet undamaged and where at last they began to feel safe. As for so many others, that very feeling was treacherous. They were caught in the open by the British bombers.

We experienced the second attack on the Wallotplatz—not far from the Grosser Garten—where at that time nothing had happened. Suddenly the sky lit up—bright as day, in fact. All the houses were closed up, there was nowhere we could take cover.

There was nowhere else to go, so the family just stood there in the square, fully exposed to the bombers, and took their chance. Terrifying as the experience was, the only near disaster occurred when Christoph suddenly felt "things get hot." An incendiary had landed right by him, and he was actually on fire. Luckily the quilt he wore wrapped around him was still wet. They managed to smother the bomb in time.

Having survived in the open, on the edge of the area of greatest destruction, Christoph Adam and his family at least faced no struggle to escape from a collapsed building or a burning basement. Still in shock, they stumbled the short distance to the Grosser Garten. The eastern section of the great royal park was somewhat less badly ravaged and also less crowded than its western and central areas, though he recalls that "every five meters or so, something was on fire." There Christoph fell asleep, among the scattered groups of equally ash-encrusted, blister-covered refugees from the city. He was still wrapped in the quilt that had saved him from serious burns or worse.

Sometime toward morning it began to rain.

HENNY WOLF and her parents were still in the area of the Hauptbahnhof when they heard a distant alarm siren and sought out the nearest shelter. Someone had a battery-operated radio, so they were able to hear announcements of the incoming attack.

There they sat out the bombs. The shelter, as usual a converted basement, was—also as usual—filled with women and children, plus an illegal German shepherd dog, which howled throughout the entire raid. Herr Wolf, as the solitary fit adult male, took charge of keeping everyone calm. He also quietly noted the location of the emergency exit, which had been fortified with sandbags against bomb blasts. He aimed to get himself and his family out as quickly as possible once the raid was over, before anyone had a chance to ask them questions.

When they hurried from the shelter, Henny said to her parents that they had to seek out fresh air. They must follow other survivors down to the river. This meant going north through the Altstadt.

> However, soon we realized that progress was unthinkable. As we passed a narrow street behind the Altmarkt—the Webergasse, I think—we found ourselves pulled into it by a fierce undertow of fire. Suddenly my mother looked like she was flying away. Father said, "We have to get out of here!" We made it to through to the

Zeughausstrasse. The Jewish community house was in flames. We had been due to report there two days later for "transportation." What an irony that we were now standing in front of the blazing building with that deportation order in our knapsack!

They looked for Henny's friend Werner Lang, another *Mischling* who had worked at Zeiss-Ikon and at Bauer, but without success. Nearby they saw a wall covered in contact messages, left by those seeking loved ones, but the Wolfs dared not leave any such traces of themselves, for fear of the Gestapo. Discussing their next move, Herr Wolf considered crossing over to Neustadt to see if his property in the Alaunstrasse was still intact, but soon abandoned the idea. Finally they followed thousands of other survivors down to the Elbe water meadows to wait for daylight.

Victor Klemperer and his wife, Eva, had been at the Zeughausstrasse earlier that night. They had lived for some years in one of the "Jew houses" there. They survived the first raid relatively easily. Some bombs had fallen nearby, but they had suffered no direct hit. A couple of incendiaries were found but successfully extinguished. All the same, most of the windows had been blown in, so the house was a mass of broken glass. Sometime after midnight, they took to their beds—"Just sleep, we're alive!"—Eva Klemperer first having to clear stray splinters from her mattress.

Victor dropped off for a while, despite the "terrible strong wind" that had started to blow. Then at around one, Eva said suddenly, "Air raid warning!" Her husband had heard nothing. She had somehow picked up the sound of sirens from the other side of the river, or perhaps—as her husband thought—there were hand-operated sirens being carried around the streets. A fellow resident of the Jew house knocked on their door to confirm the warning. The Klemperers in turn woke the woman in the next room. Then they hurried downstairs.

The street was bright as day and almost empty, fires were burning, and the storm was blowing as before. As usual there was a steel-helmeted sentry in front of the wall between the two Zeughausstrasse houses (the wall of the former synagogue with the barracks behind it). In passing I asked him whether there was a warning.—"Yes."—Eva was two steps ahead of me.

We came to the entrance hall of no. 3. At that moment a big explosion nearby. I knelt, pressing myself up against the wall, close to the courtyard door. When I looked up, Eva had disappeared. I thought she was in our cellar. It was quiet, I ran across the yard to our Jews' cellar. The door was wide open. A group of people cowered whimpering to the right of the door. Big explosions . . . the window in the wall opposite burst open . . . Then an explosion at the window close to me. Something hard and glowing hot struck the right side of my face. I put my hand up, it was covered in blood. I felt for my eye, it was still there. A group of Russians—where had they come from?—pushed out of the door. I jumped over to them. I had the rucksack on my back, the grey bag with our manuscripts and Eva's jewellery in my hand, my old hat had fallen off. I stumbled and fell. A Russian lifted me up. To the side there was a vaulting, God knows of what already half-destroyed cellar. We crowded in. It was hot. The Russians ran on in some other direction, I with them. Now we stood in an open passageway, heads down, crowded together. In front of me lay a large unrecognisable open space, in the middle of it an enormous crater. Bangs, as light as day, explosions. I had no thoughts, I was not even afraid, I was simply tremendously exhausted, I think I was expecting the end.

Astonishingly, the professor, with his heart trouble—and his wife, Eva, now missing—showed great determination and presence of mind. Exhausted he may have been, and "expecting the end," but after a moment's rest he scrambled over a parapet into the open air, threw himself into a bomb crater, and lay there for some time.

Eventually clambering out of the hole, Klemperer sheltered in a telephone kiosk. Then he saw another man, Herr Eisenmann, from the Jew house. The man was carrying his small daughter in his arms. Together they took refuge in what was left of the Reich bank's regional office building. It was largely shrouded in flames, but solid, with thick walls. No bombs fell on them. Later Eisenmann suggested they take a risk and run the relatively short distance down to the river. The younger man quickly left Klemperer behind—running was too much for him, especially in the oxygen-poor air.

Instead the professor joined a group clambering up the slopes of the ornamental gardens that led to the Brühl Terrace. With difficulty, he reached the high terrace. There he felt a cooler breeze. Still the stinging sirocco of sparks and hot air, but an atmosphere that was breathable.

High atop the riverside terrace, he had become one of the select band who had escaped the blind, infernal maze of the Altstadt. Klemperer's vantage point afforded him a matchless view of a terrible panorama of destruction.

> Within a wider radius, nothing but fires. Standing out like a torch on this side of the Elbe, the tall building at Pirnaischer Platz, glowing white; as bright as day on the other side, the roof of the Finance Ministry. Slowly thoughts came to me. Was Eva lost, had she been able to save herself, had I thought too little about her? I had wrapped the woollen blanket—I had probably lost the other with my hat—around my head and shoulders. It also covered the star. In my hands I held the precious bag and—yes, also the small leather case with Eva's woollen things . . . the storm again and again tore at my blanket, hurt my head. It had begun to rain, the ground was soft and wet, I did not want to put anything down, so there was serious physical strain, and that probably stupefied and distracted me. But in between there was constantly present, as dull pressure and pang of conscience, what had happened to Eva, why had I not thought enough about her? Sometimes I thought: She is more capable and courageous than I am, she will have got to safety; sometimes: If at least she didn't suffer! Then again simply: If only the night were over! Once I asked some people if I could put my things on their box for a moment, so as to be able to adjust my blanket. Once a man addressed me: "You're also a Jew, aren't you? I've been living in your house since yesterday." Löwenstamm. His wife handed me a napkin with which I was supposed to bandage my face. The bandage didn't hold, I then used the serviette as a handkerchief. Another time a young man, who was holding up his trousers with his hand, came up to me. In broken German: Dutch, imprisoned (hence without braces) at police headquarters. "Ran for it—the others are burning in the prison."

Another whose fate had been transformed by the intervention of the RAF.

The Dutchman's fellow prisoners had perished, and so had some Jews too. Friends of Henny Wolf lived in the Jew house in the

Sporergasse, from which many had been removed to the Hellerberg camp. They always used to say, if there was an air raid the Wolfs should take refuge with them. It was an old and shabby house, but with sturdy walls and foundations, and on one side the massive remains of Dresden's ancient fortifications. Those walls would hold! And so they did. All too well. In the night of the firestorm, a bomb hit the house. It collapsed over the cellar where the Jews had taken shelter.

As Henny Wolf sadly recounted:

> No one could get them out, although for hours knocking signals could be heard from the interior of the house. In this inferno, there were no rescue workers, let alone an excavator or something of the sort, which could have broken through the old walls. A doctor, Werner Lang's brother, was among those buried under rubble. We hoped that he had enough cyanide to enable them to be spared the horrors of death by suffocation. Around 40 of the 170 or so Jews still living in Dresden died there, at the hands of their liberators, and so close to the end. For us, however, macabre as it may sound, the air raid was our salvation, and that was exactly how we understood it.

It was different for the concentration camp Jews. Like the prisoners who had come to Dresden from the Lodz ghetto in October 1944, the Jews had been moved steadily west, to keep them from being liberated by the Russian advance. That night when the Klemperers and the Wolfs found their salvation, the five hundred Jewish workers at the Bernsdorf & Co. factory in Striesen, northeast of the Grosser Garten, were locked into what passed for a shelter.

Ilana Turner had just turned seventeen when Dresden was bombed. She had been working twelve-hour shifts making things for the German war effort since the age of thirteen. This had saved her, so far, from joining the 1.5 million children who died in the Nazi regime's concentration camps. She was with the other slave-laborers in the Bernsdorf factory when the sirens sounded that evening.

> It was after nine o'clock. We had alarms all the time, so we didn't pay it much attention, but when the sirens sounded and bombs began falling, we had to take refuge. It was not really a shelter but a

kind of half-shelter. Half of the window was on the street, so it was a
semibasement. There we spent the whole night and the bombs were
falling all around us . . . And the funny thing was, the Germans—the
SS and all the others—came to us about twelve o'clock, and they
said, we came to stay with you because we have heard that the Jews
are lucky . . .

That night, at least, they were. They suffered no direct hits. After
the second raid, however, it became clear that incendiary bombs had
penetrated the factory roof. The building was on fire. They were evacuated and marched down to the river Elbe about a kilometer away. It
was a strange, almost surreal experience as they followed the route
taken by many German survivors. The shaven-headed Jewish prisoners and their superstitious SS minders, shuffling through the burning
streets. Ilana recalled, "It was a terrible sight . . . there were electric
tramways that had been set ablaze. There were people sitting inside,
on fire."

They reached the river. Ilana is not sure where—the nearest place
would be the waterside suburb of Tollkewitz. Unlike most of the
shore, Tollkewitz is served by two tramlines, which would account for
those horrifying, spectral figures inside the streetcars.

It was quite isolated, but full of Germans . . . the whole population
went to the river because probably they thought they could duck
into it if something happened. It was still dark, I think four, five
o'clock in the morning. The bombing had stopped some time ago.
We were there for three or four hours.

"There was one German soldier who went crazy," Ilana added.
"But otherwise everyone else survived." The factory was quite badly
damaged by the incendiary bombs. The Jews could not return. At first
light they were marched off along the river to Pirna, a picturesque
small town twelve miles southeast of Dresden. Early in the war, Pirna
had witnessed the experimental mass gassings of the mentally ill (followed later by Jews), carried out under the notorious "T4" program at
the asylum in the suburb of Sonnenstein. It was currently the site of a
small concentration camp.

We stayed there for six days to one week . . . they had Russian prisoners of war there. They were in terrible condition. We had nothing, but what little we had we shared with the Russians. These
young men looked like skeletons.

Then the Jews were returned to Bernsdorf & Co. Sufficient
repairs had been accomplished for them to start work once more,
making armaments in what was left of Dresden for what was left of the
Greater German Reich. The difference was, they now had to sleep on
the floor of the machine shop.

THE CIRCUS SARRASANI, in its two-thousand-seater permanent "big top"
at the Carolaplatz, on the Neustadt side of the Elbe, was one of the few
entertainments still permitted in Dresden in February 1945. There
were cinemas, the occasional church or military concert, but the traditional status of the city as a center for the performing arts had been
suspended for the duration of the war.

Since September 1944, as part of the enhanced "total war" program
following the attempt on Hitler's life, all other places of amusement or
artistic activity had been closed by the regime, including theaters, opera,
dance, and art schools. The Dresden Philharmonia orchestra and the
choir of the opera house were disbanded. Their members, along with the
actors from the city's famous theaters, had been drafted into the
Wehrmacht and to the armaments workshops. At the Universelle factory,
which had once produced typewriters and now made parts for the military, there was an entire section made up of thespians.

Perhaps because it was peculiarly suitable for entertaining troops
and armaments workers, Dresden's famous circus had been selected
as "vital to the war effort." The management, under owner's wife/per-
former Trude Sarrasani and her Hungarian artistic director, Gabor
Nemedi, continued to mount a lavish spectacle.

On the evening of February 13, 1945, the Circus Sarrasani was
filled to overflowing. This being Fasching Tuesday, the Cavallini
clown troupe was bringing gales of laughter with its special carnival
routine. Many of the seats were filled with refugees, seizing a rare
opportunity for an interlude of glamour and light relief during their

wretched trek westward. In their programs, as they settled into their seats, they would have read a carefully worded warning:

> The Sarrasani management announces: In case of air raid warning, we ask our guests first of all to remain seated and to follow the instructions of the circus staff. The cloakroom cannot be opened. Do not walk quickly, or crowd together—everything has been arranged. Do not smoke. Do not leave the building. There is no need for concern! Remain calm!

The program also assured the audience that the circus, as well as being thoroughly blacked-out, was "camouflaged from the air." It also possessed "exemplary air raid shelters"—including the cellar bar. On the evening of February 13, the precautions were put to the test. Sarrasani's own air raid warden, a certain Herr Curt Sonntag, reported later on the procedure:

> At 21:20 the circus received a preliminary warning . . . a little later the sirens sounded for the full-scale warning. Our show continued according to schedule until 22:00 hours. I ordered it to be curtailed immediately. The last act that had just finished their performance were the Lindströms. They had been followed by Preto's Unrideable Donkey. This performance was abandoned. The Cavallinis and the Hungarian post-horses had already stopped. I had the seats evacuated according to the pre-established sequence, beginning with the 2nd tier, the standing room and the balcony, then the 1st tier, the middle circle, and so on. The circus possessed three air raid shelters: the tunnel bar, the hunting animal cellar, and the scenery stores. The evacuation proceeded according to plan and without panic. Only a few people, contrary to instructions, left the circus building. In the corridor leading to the basement, a woman became hysterical, and a naval officer drew his revolver. I stepped between them. We were trained for such eventualities. The last guests had not yet been brought to safety when the first bombs began to fall. Some anxiety was understandable but was kept within bounds . . .

When the all-clear was sounded at about 11 P.M.—there was still a power supply in the neighborhood—the audience was allowed to go home. It was, all in all, an exemplary performance by both the Sarrasani's staff and those in their care. They had suffered one near-direct hit, which had damaged rooms over the main entrance, and some incendiaries were found burning in the hay and straw stores. Staff got to work extinguishing these, and succeeded. All the same, the built-up area around them was on fire. There were suggestions of a rising firestorm even on the Neustadt side. Locals who had been driven from the homes came with their belongings to the Sarrasani, which seemed almost entirely undamaged.

> People were running crazily hither and thither, between them you could hear the fearful screaming of the beasts that had been tied to fences and trees after being freed from their stables. Nevertheless, we thought the worst was over. Against my instructions, Trude Sarrasani had her valuable Lipizzaner horses and other animals taken down to the banks of the Elbe. She herself rushed off to her apartment, to salvage her personal effects. Then the second attack came: more unexpected, more intense and more terrible than the first. In the short time available, it was not possible to put all the animals back in their stalls, and we could give no thought to the ones down by the river. Everyone rushed headlong into the air raid shelter.

The raid passed in din and impact tremors, dust and smoke. Then it was over. But soon the flames from the inevitable incendiary fires found their way down into the underground rooms used as shelters. The paneling in the cellar bar caught alight. Fortunately the cellar had alternative exits, and circus staff and sheltering neighbors were able to leave. All who took to the circus's air raid shelter, in either the first or second raid, survived. Once outside, however, they found the building on fire. An entire canister of incendiaries had crashed through the roof of the dome that surmounted the big top. By the time the staff and their helpers emerged from the shelter, the four-pound fire raisers had distributed themselves and caused many small blazes. The building was full of curtains and wood paneling, and the circus ring itself was carpeted with dry coconut matting; there were storerooms for saddles, harnesses, and costumes—all eminently combustible.

Curt Sonntag, the circus's air raid warden, recounts the grim tale of what came next:

Attempts to extinguish the blaze were hopeless. We could only save what we could. The beasts that we had been able to return to their stalls before the second raid had remained uninjured, but now they had to be rescued from the rapidly spreading fire. This was only partly possible. The tigers, in their traveling cages, died pitifully in the flames; we had to abandon our attempts to save them because of the strength of the raging conflagration. Horses and other animals that had survived the air raid in the open were bleeding from many wounds.

The cupola of the big top finally and noisily collapsed at around 4 A.M. Sonntag could nevertheless proudly report that no one who stayed in the circus building had died as a result of enemy action. Of the staff who perished, all were either with the doomed animals on the banks of the Elbe during the second attack—these included the star rider, Regina Beer, her body pierced by seventeen bomb splinters—or had set off for home after the first raid. They died at home or were killed in the streets, like so many other Dresdeners. One of the latter was the Chinese acrobat who, to the disgust of the racist Nazi authorities, had married a Dresden woman. He kissed his wife good-bye at a tram stop that afternoon before going off to work at the Sarrasani and she never saw him again.

On the morning of 14 February, we collected the five dead from the Elbe meadows and laid them out on the elephant podium. Since at this point the entire circus building was open to all, strangers—people from the neighborhood—brought their own dead and likewise laid them out on the podium. Thus the rumor arose that during the night's bombing numerous people had died at the Sarrasani.

Figures of at least a hundred dead at the circus continue to circulate. In fact, the main casualties seem to have been the helpless animals whose safety could not be secured before the fire took hold. A Reich labor service officer witnessed the scene:

In the middle of the square was the round circus building; I believe there had been a special Carnival night performance. The building was burning fiercely, and was collapsing even as we watched. In a nearby street I saw a terrified group of dappled circus horses with brightly coloured trappings standing in a circle close to each other.

HIGH ART AND LOW, Dresden lost everything during those desperate hours.

By the time the Circus Sarrasani's cupola collapsed during the small hours of the morning of February 14, 1945, all Dresden's other famous performance spaces were also either destroyed or burning inexorably to destruction: the opera house, the state theater, the Zwinger. As far as the galleries and the collections of artistic treasures were concerned, the really valuable or important works and artifacts had mostly been spirited away to places of safety.

On the night of the British raid, forty-two large paintings (too large to be moved) still hung on the walls of the royal *Schloss* in the center of Dresden. The building was still full of antique furniture. By chance, a truck full of more than a hundred paintings and other precious items was also parked in one of the inner yards, resting over night in transit from its former place of safety, now endangered by the Russian advance. It had been due to proceed next morning to Meissen. Everything burned.

In the royal Catholic Court Church (Hofkirche) stood beautiful wooden pews and carved panels, and, beneath the church, the crypt containing the coffins of members of the Saxon royal family going back hundreds of years. The last of the Wettin dynasty to be laid to rest there was Georg, the eldest brother of Prince Ernst Heinrich. Georg would have inherited the throne had the monarchy not been overthrown in 1918. Relieved of that responsibility, he became Father Georg, S.J., a Jesuit priest. His funeral in 1943 had brought the family together for the last time.

As for Prince Ernst Heinrich, the youngest son of the last king, in 1945 he found himself on February 13 back in the city where he was born. He had somehow survived the past eleven years—despite his democratic sympathies, which had almost seen him "liquidated" in the Night of the Long Knives in June 1934. His most recent brush with death at the hands of the regime had come on the turn of 1943 when the prince, serving as an officer with the German Intelligence Service, the Abwehr, had suddenly been ordered into the doomed German cauldron of Stalingrad.

The Sixth Army had already been surrounded there for weeks—it amounted to a deliberate death sentence. The order must have come from right at the top of the Nazi hierarchy, because a special law generally barred members of German royal houses from active service. During the last few hours before he was due to fly into the beleaguered city on the Volga, Ernst Heinrich had seen himself as a "dead man walking." He said his farewells to his children (his first wife had died young of a blood disease two years before). Then came an equally mysterious phone call from the Army High Command, canceling his posting to Stalingrad and releasing him from obligations to serve in the Wehrmacht. Someone equally highly placed must have stepped in.

The widowed prince had been living for years along with his children at the old royal hunting lodge at Moritzburg (where he had been arrested in 1934). He was now engaged to be married once more, to an actress—daughter of a distinguished cavalry officer—who lived in Dresden. This, and routine commitments to family business in the old capital, brought him into the city on February 13, 1945. Ernst Heinrich spent the morning at one of the royal palaces still controlled by the family, completing some administrative tasks, before a dentist's appointment. Only then was he free to visit his fiancée, Virginia (Gina) Dulon, at her apartment in the well-to-do "Swiss Quarter" of Dresden, south of the Hauptbahnhof.

After dinner Ernst Heinrich was considering setting off back to Moritzburg (about twelve miles north of the city) when the air raid warning sounded. He, his fiancée, and her sister took to the cellar of the villa, and emerged to find the house intact except for a few broken windows. The area had escaped lightly from 5 Group's attack. A glance in the direction of the city center almost two miles away revealed fires, but at this distance the effects of the raid appeared limited.

"This was," as the prince remarked in typically understated fashion, "a great delusion."

They decided to drive the short distance over to the Reichsstrasse—where Gertraud Freundel and her father had first encountered the firestorm—to see Gina's parents, who lived in a boardinghouse there. The elderly couple had survived the raid. They took them back to the apartment and began tidying up. It was then they heard distant sirens from the edge of town, heralding the approach of the second wave of British bombers. After the first raid, as the fires seemed to increase in the

Altstadt, they had decided against trying to drive back to Moritzburg. Now Ernst Heinrich took charge. He told everyone—Gina, her parents, her sister, and a girlfriend—to get into his car. They must be out of the city before the new attack began.

The first bombs had started to fall as the group, crammed into the prince's little DKW sedan, reached the southern limit of the city, and took the main road toward the town of Dippoldiswalde. Ernst Heinrich was concerned about getting caught on the road, but could see no alternative except to proceed. Then Gina's girlfriend told him to stop. She knew there was tunneling going on beneath the road here; on a country stroll a short while before she had seen a huge pipe being laid, covered with twenty feet or so of earth, and had thought at the time that it would make a good shelter if there were bombing or fighting in the area. The prince stopped his car, and saw that the young woman was right.

> We had scarcely got into the hiding place when the witches' Sabbath began. There was the scream and the whistle of bombs falling. It seemed to be raining, with a strong wind. I was standing at the entrance to the pipe. Gina stood behind me. With us were three French prisoners of war. We could see a fiery phenomenon moving toward us along the upper edge of the valley. This was revealed as a wall of phosphor, eighty meters wide and three meters high, being pushed by the wind in our direction. When it got into the calmer valley, however, it lost speed, came to a standstill, and then collapsed, leaving burning brands littering the ground for some time afterward. Soon there was a huge explosion, and a pillar of flame shot into the air. About three hundred meters from us, a bomber and its load had crashed. Soon after I saw a man floating down to the ground on a parachute; he was part of the crew of the bomber and evidently the only one who had survived.

Afterward they returned to the car. All its windows were shattered, but it started smoothly. They drove to the town of Dippoldiswalde and found temporary shelter for their passengers in the nearby village of Bannewitz.

From Bannewitz, standing on a hill in the small hours, looking through the darkness toward Dresden, Prince Ernst Heinrich of

Saxony watched as the capital his ancestors had built burned to the ground.

> The entire city was a sea of flame. This was the end! Glorious Dresden was burning, our Florence on the Elbe, in which my family had resided for almost four hundred years. The art and tradition and beauty of centuries had been destroyed in a single night! I stood as if turned to stone.

The prince spent the remaining part of the night in a bed made up on the billiard table of the crowded local inn. He hardly slept. All through the night, streams of distraught refugees arrived, some almost zombielike in their passivity, some powered by a fragile hysterical energy. Everyone reported the terrible scenes, the appalling loss of life, the complete annihilation of the city's historic center.

The prince was up before dawn the next day. Once more he surveyed distant Dresden. If anything, the fires seemed worse, the huge shroud of smoke more vast and billowing. He got into his car and headed back to Moritzburg, making a long detour to the east to cross the Elbe at Pirna before looping back downriver.

When Prince Ernst Heinrich arrived back at the royal hunting lodge, with its ornamental lakes and parklands and woods, every stone and tile, every sculpture and painting, stood intact. It was once again as if nothing had happened, nothing had changed, and nothing ever would.

23
Ash Wednesday

JUST ABOUT THE TIME that Prince Ernst Heinrich and his companions reached the primitive safety of the inn at Bannewitz, in England more aircrew were being wakened from their chilly beds. This time not British and Commonwealth, but American.

The Eighth U.S. Army Air Force had originally been scheduled to go in ahead of the British against Dresden on February 13. The whims of the weather had changed that. Now they were going in after the British. This was not unusual. What was unusual was how much of the target city had already been wrecked—in the course of a single night—before the Americans even took to the air.

Not that their people could be expected to know that. Around 4 A.M., when the wake-up calls started to go out to the big daylight bomber bases in East Anglia, the British aircraft of the second wave were still on their way back. No report or narrative had yet been drafted; the only thing that might have been known—through telephone calls or personal conversations—was that the RAF had hit Dresden hard. Harder than Hell's Angels or the other outfits that had attacked the place as a secondary in October and January had ever managed or been intended to do.

Sergeant William Stewart, a ball turret gunner of 325th Squadron, 92nd Group of the U.S. First Air Division, was awakened before dawn on February 14, 1945:

> I rose, dressed and made my way to the mess hall and ate a breakfast of bacon and eggs. From there, I took a ride in a truck to the building on the flight line where the briefing was to be held. As I walked

in, I could see the map of Europe and the British Isles covering the
front wall. On the map was a red ribbon, showing the route to be
flown in and out of Germany, held by pins. The target on the map
was Dresden. To me, Dresden was just another city in Germany and
I didn't recall having heard of it before that time. The briefing officer went over the schedule for the mission and the possible conditions of weather, flak and opposition we might encounter. Take-off
was to be at 0700 hours that morning.

A total of 431 B-17s were under orders to bomb Dresden around
noon that day. The force consisted of the entire First Bombardment
Division, composed of twelve groups, which were divided into four
combat wings. The Third Division was to follow the First. Its force
(also twelve groups) was to attack marshaling yards in nearby
Chemnitz. Second Division would bomb the hydrogenation works in
Magdeburg, making a total of just over thirteen hundred American
bombers operating against major German targets in the daylight hours
of February 14.

There would be an awful lot of American aircraft in Saxon airspace this Ash Wednesday. Even more than the bomber figures
imply—for the three divisions were to be escorted all the way by 784
P-51 Mustangs of Eighth Fighter Command. These were the long-
distance fighters, with their drop tanks and high performance, which
had transformed Allied fortunes in the spring of 1944 by providing,
for the first time, armed escort to the vulnerable bombers all the way to
their targets.

Protection for the First Division came from Twentieth Fighter
Group, 352nd, 356th, 359th, and 364th. The Twentieth FG was to
escort the spearhead units of the bomber force, the First Bombardment
Wing, consisting of the 398th Bomber Group, followed by the 91st and
381st.

So, including the fighters, almost twenty-one hundred American
aircraft were to be found over central and eastern Germany around the
middle of that day. For the German population on the ground, it must
have seemed that the sky was black with machines that meant them
harm.

What kind of harm is a matter of discussion, but at 9:30 P.M. the
previous evening teleprinters rattled out the orders for the raid to the

American bomber bases (just as the RAF's 5 Group was making its turn down toward Dresden), defining the composition of the combat groups into which the force would be divided and stipulating the bomb loads to be carried. These were unusual—for American units. The entire First Division would deliver 678.3 tons of HE ("general purpose") bombs and 400 tons of incendiaries. As things turned out, not all these would be dropped on Dresden, but the correlation remained the same. This was more like the proportion employed on British "city-busting" raids than he usual "precision" missions fastidiously undertaken by the USAAF. It was certainly more appropriate for attacking population and industrial centers rather than railways or communications.

There is some confusion as to the exact nature of the orders given to the First Division of the Eighth Air Force against Dresden.

The initial report, dated February 14, 1945, and distributed by teleprinter to senior officers, states the target as "Dresden Marshaling Yard." The individual bomber groups' reports state it as mostly simply "Dresden," or in one case "military objectives in Dresden" (303rd Bomber Group—which at the same time specifically refers to "Marshaling yard in Chemnitz"). On the other hand, the definitive report from the commander of the First Division, Brigadier General Turner, to Headquarters Eighth Air Force on the Dresden raid, dated February 25, states quite baldly:

> Primary Target—visual—Centre of built up area Dresden.
>
> Secondary Target—Visual—M/Y Chemnitz
>
> H2X—Centre of Dresden
>
> Last resort—Any military objective positively identified as being in Germany and east of the current bomb line.

In other words, if visual bombing is possible, the bombers are to attack the center of Dresden. If Dresden is swathed in cloud, but Chemnitz—a few minutes' flying time distant—is clear, then the smaller city's marshaling yards are to be attacked. If all are under cloud, then H2X (the equivalent of the RAF's H2S, or "radar bombing") is to be carried out, once again on the center of Dresden. The "current bomb line" refers not to the line agreed at Yalta (operative east of Dresden to avoid the Russians), but the security line protecting the Anglo-American forces on the western front from accidental bombardment. This begins to look like the other allegedly Thunderclap-style attack on the center of Berlin eleven days earlier, on February 3— except that the ratio of high-explosive to incendiary bombs in the Berlin raid was 90 percent to 10 percent, not 60 percent to 40 percent as it will be against Dresden.

Top Sergeant Harold W. Hall, a radio operator with a B-17 Flying Fortress of 527th Squadron, 379th Bomber Group, recalled the briefing for the mission:

> The reason I remember that mission is that during our briefing the officer pointed to a small building located on the map as in the centre of Dresden . . . I felt it was indiscriminate bombing of all the refugees fleeing the Russians. I have to say that I felt ashamed we had levelled ourselves to the Krauts. During the briefing, no mention of refugees in the city was made, but it wasn't needed, and the implication of "foul play" was strong. Incidentally, that was the only time I ever felt (and others) that the mission was unusual.

William Stewart, ball turret gunner, does not record feeling any such compunction as he ate his breakfast and took a truck out to where his aircraft, a Flying Fortress, was waiting. It was dark and cold, a real East Anglian February morning. His squadron, based at Podington in the flatlands of the Northamptonshire/Bedfordshire border, had been due to take off at 7 A.M., but as he perched in his bubble, testing his twin .50 caliber guns and doing his assigned checks, a voice over the intercom said they would be leaving an hour later. He was flying with a crew he did not know, which added to his apprehension. When they finally taxied out with the other Fortresses, it was getting light. A roar of the aircraft's four Wright Cyclone engines, a series of bounces, and they were airborne. Stewart wrote that he headed for Dresden that day huddled in his bubble "like an embryo in an egg."

The swarm of fortresses—all three divisions together at this point—headed as usual over the coastal town of Felixstowe, then to the Zuider Zee in Holland. They proceeded eastward to 8 degrees of longitude somewhere northeast of Münster. There the Second Division left the stream and continued straight toward Magdeburg. The First and Third Divisions turned southeast, a course they were due to maintain for about two hundred miles before orienting themselves toward their final destinations.

Some of the force of escort fighters had been with the bombers since they left the coast of England behind them. Some joined them only once they were across the North Sea. A number of these units had been stationed, since the end of 1944, on the continent. These included the 352nd Fighter Group (the "Blue-Nosed Bastards of Bodney"), who at the time of the Battle of the Bulge were transferred from Cambridgeshire to Asche in Belgium, and then a few weeks later to Chievres, west of Charleroi. It was from Chievres that, on February 14, 1945, they took to the air to act as escorts to the bomber stream heading for Dresden. Lieutenant Alden Rigby, from Utah, a faithful member of his church and, though only twenty-two, married with a baby daughter, was a Mustang pilot with 487th Squadron, 352nd Fighter Group.

Al Rigby had distinguished himself on January 1, when the Luftwaffe, in its final attempt at self-assertion, had launched Operation Bodenplatte ("floor slab"), involving surprise attacks by twelve hundred aircraft against Allied airfields in liberated France and Belgium. The German raid was quite successful. Scores of Allied aircraft were destroyed on the ground. A dozen Mustangs of Rigby's unit had been among the few that managed to fight back, taking off as the Germans strafed the airfield and downing more than a score of enemy FW 190s in the dogfights that ensued. Rigby shot down four Germans, which along with another "kill" back in November made him an "ace." His P51-D Mustang was emblazoned with the names of his wife and baby daughter—"Eleen and Jerry." Now, six weeks later, he and his squadron took to the wintry skies to protect the big bomber stream heading into Germany.

Al Rigby recalls it as being late morning when they began their duty that day. The trip was routine, and as far as he is concerned, it stayed that way. It is in his logbook as simply "Escort—Eastern Germany—Dresden—5 hrs 10 mins."

We were not briefed on what kind of mission it was, other than escort. I didn't really know anything about the bombing—the fire-bombing—until sometime later. I don't recall it was any big thing at all . . . my main emphasis was that it was a long haul . . . five hours was quite a long deal.

The 67 P-51s of the 352nd Group slotted themselves at the back of the stream. They would cover the hindmost bomber groups all the way to Dresden and back. It was the "A" and "B" flights of the Twentieth Fighter Group, based in King's Cliffe, Northamptonshire, that flew in the vanguard all the way from England, accompanying the lead bombers to Dresden. Or that was the plan. Not for the first time when it came to bombing Dresden, the weather took a hand.

Over Holland they had run into a big weather front, bringing thick cloud. The lead 398th Bombardment Group had already gone somewhat off course around then. The division leader and his deputy found their Gee equipment faulty, and suffered from persistent German jamming activity. The division leader—call sign "Swordfish Able"—informed the Ninety-first Bombardment Group, just behind him, that he had decided to try to avoid the cloud by taking a more southerly course (a thirty-mile detour). The navigator of the Ninety-first's lead aircraft objected that this would mean overflying the flak hot spot around the city of Münster, but his superior decided to follow the 398th, as did the 381st in third position. So the entire First Bombardment Wing—three groups amounting to 137 aircraft, almost a third of the First Division—went on a supposedly temporary detour to avoid the cloud.

The rest of the bomber stream decided to fly over the cloud rather than around it, even though this meant increasing their altitude and thereby their fuel consumption. The 379th Bombardment Group was now leader of the nine remaining groups, at least until First Wing and its escorts, the P-51s of Twentieth Fighter Group, rejoined the stream. Flying at around thirty thousand feet (five thousand feet above their assigned altitude), the rest of the bombers proceeded to Dresden.

Meanwhile, the division leader and his three errant groups were getting into ever worse trouble. Over Münster the Ninety-first was shot up by flak. There were wounded, damage to aircraft, plus a further deviation from the assigned route. They were by now almost fifty miles off course. This found them liable to pay an undesirable visit to the heavily defended city of Schweinfurt, so they looped farther south to avoid this. They—and their fighter escort from the Twentieth

Fighter Group—were now sixty-five miles or so off course, and the bombers were having problems orienting themselves by radar. The bandwidth was not what it should be. Soon the lead aircraft's radar gave out altogether, and command had to be transferred to the deputy leader, whose radar seemed to be working. There were discussions with the weather pathfinders, who said it should be possible to find the main target. There was confusion between the secondary and the primary targets, and there were misunderstandings between the acting commander and his radar observer.

Suddenly, miraculously, they were presented with a break in the cloud. An urban area became visible, apparently the main target (Dresden). The deputy leader ordered an attack. With three minutes to go before the first bombs were due to be dropped, the deputy leader's radar also went completely out of commission. However, a short while later, to everyone's relief, the bombardier in the lead aircraft got a visual fix on the urban area, which lay to his right. A good-looking city with a river running through it. The aircraft banked around and started its run on the target.

The drop, forty seconds later, was successful. There was "meager to moderate flak" and no sign of enemy aircraft to trouble the fighter escort. The other Flying Fortresses of 398th Group followed in tidy sequence over the following few minutes.

Unfortunately, the city they had just bombed was not Dresden but Prague.

MEANWHILE, eighty miles to the north, morning had finally come to Dresden.

Dirty gray smoke drifted over what remained of the city area, blown by the same northwesterly that had carried 5 Group's first wave of bombers over its target with such fatal ease the previous night. In many parts of the city, survivors had not realized that it was light, so dense was the haze blocking out the winter-thin sun as it rose from the east. The firestorm had subsided from its height, but there were still a thousand individual fires burning in the ruins. A few survivors and rescue teams were starting to pick their way across the rubble, though many places were still too hot to enter. It was an unbearably bleak, inhuman prospect. Ash Wednesday.

Fourteen-year-old Günther Kannegiesser, in the refuge among the planks by the Elbe that he shared with the young women and their babies, regained consciousness with a start. It was still early, he recalls.

> I was awoken by the sound of an aircraft. A Fieseler Stork [a German light aircraft, often used as a spotter plane] was flying low over the meadows by the Elbe. It was already somewhat light. So I said my farewells and promised to return soon. I headed as far as the Blasewitzer Strasse, via the Fürstenstrasse. I saw my first dead person lying there in the middle of the street. There was no prospect of getting through that way.

Günther's thoughts were for his mother, sister, and little brother, whom he had last seen between the two night raids. He had to find his way to the Zöllner Strasse, to their flat. The streets were blocked with rubble, thick with smoke. The fourteen-year-old realized he was entering a city of the dead.

> I got through to the 20th Grade School on the Zöllnerplatz. There I saw several people sitting on benches. As I approached, intending to ask them for help, I saw that they were all dead. I now went through the Stefanienstrasse to the Striesener Strasse. It was possible to make progress on the right-hand side of the street. The houses were somewhat set back. On the corner of Stefanienstrasse there was a pharmacy. This was burning especially fiercely. Between Stefanienstrasse and Zöllnerstrasse there lay many completely burned bodies. All you could distinguish was, say, a shoe and a stocking.

Finally, Günther found the building where his family lived. Or had once lived.

> The section on the Zöllnerstrasse had been razed to the ground except for a few remnants of wall. On the Striesener Strasse side, the walls and chimneys were still standing as far as the fourth story. I clambered over the rubble and tried to climb in through the emergency exit of the air raid shelter. It was on fire in one corner. So I had to go back onto the street.

The fire "in one corner" was where the stairs up from cellar had been. As young Günther stood atop the rubble, peering down into the gloom, he called out to his mother and the younger children. He got no reply. So he retreated to the street. He would never see his mother or his sister or little brother again. And he would never quite forgive himself for not having pressed on, despite the danger, into the burning air raid shelter.

Günther had agreed with his mother that if the family ever became separated, they would meet at her sister's house, in a village near Meissen. For now, though, he decided he would go to the agreed meeting point for local residents that the party's local group had stipulated. No sign of his family. He recalled his promise to the women in the plank city down by the river. When he returned they were still there. He escorted them eastward along the Elbe toward the waterside suburb of Laubegast, where one of them had relatives. It was still only midmorning.

At about this time, south of the Hauptbahnhof, Hannelore Kuhn and her family were still clearing up the mess. The villa had suffered broken windows and incendiary damage, but they had put all the fires out. They were in much better shape than most families in that part of Dresden. During the night, friends whose homes had been destroyed began to appear, seeking shelter. The family took them in. Even the parents of Hannelore's husband-to-be Fritz (at that time a prisoner of war in Russia), who were family friends, arrived. Their house in the Nürnberger Strasse, two blocks closer in toward the center, had been destroyed. No one slept much. They worked through the night, clearing up and repairing. "We put all this rubbish right out into the street. And I saw this bucket of rubbish out there and right in the middle, a bust of Hitler! Someone had chucked it out from sheer anger!"

In the morning they realized there was no running water. They needed to clean, to wash, and to cook for themselves and all their unexpected guests. "We heard that a few hundred meters away on the street there is a natural spring . . . So my mother got a little cart and put two buckets on it, and set off up there to get some water."

WHILE "SWORDFISH ABLE" and the three groups that had followed him were on their mistaken way to the Czech capital, the division leader had stayed in radio contact with the nine groups that had kept course for Dresden. Still believing he was headed for the target himself, if somewhat tardily and indirectly, he nevertheless gave 379th permission to start the first bombing run on Dresden.

The Flying Fortresses began bombing sometime after noon.

At 12:17 P.M., the thirty-seven aircraft of the 379th approached on a course east-northeast, exactly as ordered, at between twenty-seven thousand and twenty-eight thousand feet, and stayed at that height to bomb. Conditions were 7/10 cloud, with fires from the night raids continuing to burn. Smoke drifting southeast across the target (encouraged by the continuing northwesterly wind) was causing problems closer to the ground. In the end, the 379th bombed mostly where there was less smoke—the Friedrichstadt marshaling yards and industrial and residential areas to the west of the city center, which had been much less affected by the British attack than the Altstadt and the eastern suburbs.

The situation for the 303rd Group ("Hell's Angels") was quite different. Though its aircraft began their runs only two minutes later, strong winds at high altitude had suddenly blown near 10/10 cloud over the city. This is reminiscent of the big February 3 raid on Berlin, where an equally sudden and dramatic obscuring of the target area between two waves of the attack may well have saved the heart of the German capital from a firestorm comparable to Dresden's. So, because of the sudden cloud, the 303rd and subsequent groups went in on H2X—radar bombing. The group's commander wrote in his report:

> Thirty-six A/C dropped a total of 210 x 500 G.P. bombs, 140 x 500 M17 incendiary bombs and ten units of T-298 leaflets on Dresden. Bombing was PFF through practically solid undercast with results generally unobserved, although there are a few reports of bombs hitting in the city.

The general area in which these bombs fell was thought to have been southeast of the aiming point, which could have meant the Südvorstadt.

Some confusion ensued that would feature prominently in the reports submitted after the raid. The 384th, 92nd, 306th, and 401st Bombardment Groups also bombed by radar, though not always in correct squadron order. The 457th overshot the target area. Some bombs fell in Neustadt up to one and a half miles from the aiming point. Some visual confirmation was claimed for the report that bombs damaged "an industrial area in Dresden-Neustadt and parts of a marshaling yard."

The 305th Bombardment Group had the worst time of all. Shortly after crossing the coast onto the continent, the low squadron was separated from the other two (lead and high squadrons) "due to dense contrails." Now, as they began their bombing runs, the two remaining squadrons found themselves flying a collision course with aircraft of the 379th. They separated rather hastily. Finally the high squadron dropped its loads on the "fresh smoke bombs of the 379th . . . photos show only clouds, but crews report that bombs hit in the built-up area of Dresden." Meanwhile the scattered lead squadron reassembled in part, and on its commander's initiative, five aircraft flew south and attacked the Brüx synthetic oil plant, just across the border in the protectorate of Bohemia and Moravia. The rest of the lead squadron bombed Dresden on "unidentified smoke bombs."

Less than ten minutes after the first bombs had fallen, the reduced daylight raid was almost at an end. It was extended only because the low squadron of the 306th, on its first approach, missed the aiming point. The Fortresses continued east for about thirty miles, turned near Bischofswerda, selected a new bombing route, and finally bombed Dresden at 12:30 P.M., more than ten minutes after beginning their initial run, and several minutes after the rest of the force had left the area.

The American daylight-bombing raid on Dresden was over. The subsequent intelligence report spoke of "unobserved to fair results."

"UNOBSERVED" and only "fair" as their effects might have been, beneath these bombs was the young Götz Bergander, along with his family and other residents of the "industrial area" by the Friedrichstadt marshaling yards.

Given its position in the extreme west of 5 Group's bombing sector the previous night, the Bramsch & Co. distillery had experienced serious bombing only during the first part of the first wave. During the second British wave, at 1:30 A.M., as the rest of Dresden burned, everyone had once more taken to the distillery's state-of-the-art shelter—now crowded with bombed-out refugees from the city center. The shelter held.

After the second British raid, the Berganders emerged and went to the family apartment for some sleep. The next morning they were up inspecting the property:

> I was standing in the factory yard the next day around noon, and there were all those people there who didn't know what to do. They were all thinking, we have to get out of here, out of the city . . . and I heard from far away a siren . . . maybe from Freital or somewhere . . . We all said, "Another warning, how can that be?" but we went down into the cellar. Then came the daylight raid. This affected us much more. I mean, our neighborhood was close to the marshaling yards. So we ended up with half a dozen high-explosive bomb craters in the yard and the building and the cemetery next door. And our house . . . there was such a crashing and banging and rattling, I thought it must be gone. Anyway, there was the din of the bombs dropping and then finally it was over. The air raid warden said, we must go up and check for incendiary bombs. I went with him and . . . to my amazement . . . the house was still standing! It was built at the beginning of the century, when they still constructed walls a meter thick . . .

No one in the shelter had been injured. But as far as Friedrichstadt was concerned, this counted as a more serious raid than the cataclysm that had struck central Dresden the previous night. The house and factory had now lost all their windows. The Ostragut area, not far to the west, was fully ablaze, and sparks and burning debris had begun to float toward the distillery and the Berganders' home. The danger was that these sparks would be blown into the buildings and start fires there—many homes, initially undamaged, had been lost that way in the hours following the air raids. The next few hours were spent ensuring that the buildings survived. The family's apartment was now, however, uninhabitable. They, and other staff and neighbors, had to move down into cramped bunk beds in the shelter, where they would spend their nights until the end of the war.

In the small hours of February 13–14, Günter Jäckel had lain in his meadow in his pajamas and slippers and watched the second wave of the British attack on Dresden. When it was over, he walked back up to the street and saw his native city burning in its aftermath:

> This biblical pillar of fire. And there was this woman, and she had nothing better to say than, "Look, look! A parachute!" This parachute they used in order to drop the marker flares was still hanging there where it had caught in the branches of a chestnut tree. Another woman was shrieking like a lunatic in the meadow, she had lost her child. And this hissing and roaring and she was calling out some name or other . . . yes, this strange hissing and roaring and then began an explosion. A delayed-action bomb.

And then, astonishingly, they all went back to bed. The fever he had suffered from in the eight weeks since his wound became infected had disappeared. Jäckel himself is sure this was a psychosomatic reaction. In the morning he got dressed. The wounded from the schoolhouse were to be evacuated southward, toward the high ground outside the city. This process was under way when the 379th led in the noon attack.

> From up at Dölzschen there were sirens sounding. And then came the planes . . . yes, the Americans . . . Gray and dark. And it was gloomy with smoke, we saw flashes from the aircraft. And the smoke markers . . . we had the impression of enormous precision. Pathfinders. Then the smoke indicators. Finally the bomber formations coming in . . . dropping these heavy crates all over the place. Just down below us, where the courthouse used to be, Münchner Platz. A whole carpet of bombs was laid between us and Münchner Platz. It was bad. This was maybe fifteen to twenty aircraft . . . the first wave was delayed-reaction bombs, with timed fuses that went off about a quarter of an hour later.

The squadron-strength attack that Günter Jäckel describes, between his location close to Plauen and the Münchner Platz about a mile distant, was almost certainly the one that ruined the plans of Hannelore Kuhn's family.

The family had survived the horrors of the night raid, had begun to pick up the pieces, and had made room for those less fortunate. Now, unbelievably, there came more bombs by daylight. Hannelore's mother had gone off to find water so that she could cook for everyone. "But she wasn't properly on her way when the midday attack came," remembers her daughter. She now doesn't recall what nationality the bombers were. Most people did not realize at the time that these were American rather than British aircraft. Conditions were hazy. In any case, aircraft recognition was not a strong suit among Dresden's inhabitants—except the usual select group of militaria-obsessed teenage boys and young returned soldiers such as Günter Jäckel, who the previous autumn had spent a great deal of time in various foxholes in eastern France, carefully watching Allied bombers going overhead and waiting for the next low-flying P-47s or P-51s to make yet another attempt on his eighteen-year-old life.

What the British had not managed to do to Hannelore Kuhn's family home, the American bombers achieved within a matter of minutes. "We had this phosphor coming down the stairs toward me. We had no water or sand prepared anymore. I mean we had prepared buckets of sand everywhere for the previous night, but now we had nothing."

The house had, in fact, probably been hit by an M-17 gasoline bomb canister. Having used up all the fire-fighting materials they had hitherto kept ready, and which had helped them save the house the previous night, the family were powerless. The fires quickly raged out of control and forced them and their guests to leave their home. Hannelore adds stoically, "So I got my bicycle, and we piled our air raid luggage on it, and we set off to a place where my parents had acquaintances who owned a little house. It was still standing. There we were able to stay."

IN THE COURSE of thirteen minutes, the 311 Flying Fortresses that had actually bombed Dresden (excluding the five that decided to divert to Brüx) had dropped almost 1,900 five-hundred-pound high-explosive bombs and 136,800 stick incendiary bombs on the city. The latter amounted to about two-thirds of the weight of incendiaries that had enabled the 235 Lancasters of 5 Group to start the firestorm in the course of the first British wave.

The report that followed collated the immediate impressions of damage from the aircrew:

> Due to cloud cover of 7/10–10/10 density and fires caused by RAF on night of 13/14 February most of bombs dropped on target could not be plotted. Of the 27 concentrations dropped, 11 were posted; however, portions of these are partially obscured. 3 concentrations could be seen bursting in Dresden, Friedrichstadt M/Y. A few bursts could be pin-pointed in forward siding, main weight of bombs fall across central portion of M/Y and into industrial area immediately adjacent to north edge of yard. 2, possibly 3, concentrations could be seen in the Löbtau M/Y, 1 mile SE of the Friedrichstadt yards in central portion of city, hits could not be pin-pointed, however. 2 explosions seen. High possibility that severe damage will result in built-up industrial area adjacent to eastern and western side of the yards. 3 concentrations of GP fell in Kackmitz district 1–1½ miles SW of centre of city.

Three-quarters of an hour after the Flying Fortresses began their return trip, a Mosquito from the RAF's 542nd Intelligence Squadron appeared over the still-smoking ruins and took 108 photographs that were evaluated at Medmenham on the aircraft's return. Evaluation Report K.3742, submitted on February 15, confirmed that the most serious damage done by the Americans had been in the Friedrichstadt marshaling yards and industrial area, west of the city center. Fires were "burning strongly."

The armaments factories along the Hamburger Strasse (including Seidel and Naumann, which had been hit back in October 1944) were largely destroyed. There were many deaths among the foreign workers housed in the Bremer Strasse, a little farther north toward the Elbe. Apartment blocks and factories in this mixed area burned for hours. The area around the Bramsch Distillery was on the edge of this realm of destruction, as was the Friedrichstadt hospital complex, which suffered damage but was not as badly damaged as the hospital in Johannstadt. The damage also extended westward toward the *Alberthafen* river port.

South of this area were Löbtau, where Prince Ernst Heinrich's fiancée had lived, and the areas somewhat farther east where Günter Jäckel's party of wounded were awaiting evacuation, and the Südvorstadt, where the showers of American bombs proved too much for Hannelore Kuhn's doughty defenders of the family property. The Altstadt marshaling yards, west of the Hauptbahnhof, were also bombed. There was, however, not much left of them after the RAF's visit the previous night.

"North of the Friedrichstadt marshaling yards" was where the American bombs found Nora Lang and her little brother. Having followed the stunned crowds swarming along the Elbe meadows, avoiding the still-burning Altstadt by crossing the river on the damaged but still-usable Albertbrücke Bridge. Nora was still carrying the family suitcase with one hand and holding on to her five-year-old brother with the other. The agreed family rendezvous point, in case they got separated, was their grandfather's little patch of land and weekend cabin at Wilschdorf, several miles northwest of Dresden. So they found their way around in a loop that brought them into the area just north of Friedrichstadt. At first this seemed like a blessing, for here not all houses had been destroyed, and there were some intact welfare facilities:

> Finally we got to a square and we were tired, so we sat down, and a woman came and said, oh children, you must go to the hospital. So there was a hospital, a temporary one in a school, and they told us to eat and drink something, which we did. Something to drink and a bit of chocolate. And it was then that the third raid happened, the one at midday. And actually some incendiary bombs fell in the Neustadt quite close. There were some railway yards. And there was a house in the vicinity which got hit by an explosive bomb. And there were all these amputees and so on, and people in wheelchairs, and no nurses to look after them. They rolled down the stairs into the cellar and . . . it was bad . . .

Fires spread from the nearby house. The children fled the basement shelter, leaving the belongings they had held on to so firmly all through the firestorm night. When the aircraft had gone, some soldiers went back into the shelter, to try to help them find their things, but inside everything was burning, precious suitcase and all. Almost too weary to feel more than the dullest sense of fear or loss, they salvaged what they could and pressed on.

The American midday raid was far less destructive than either of the British ones. Partly this was because it ended up smaller than planned, owing to the absence of the three bomber groups that found their way to Prague; the fact that so much of Dresden's built-up area had already been destroyed; and finally, the unusual preponderance of incendiaries. A four-pound incendiary dropped amid rubble can do little more than strike and flare ineffectually, no matter the quantities.

The people of Dresden were by this point overwhelmed by the unpredictable violence being done to them from the air.

"The people who had been made homeless during the night, and who had fled into the western part of the city," Götz Bergander observed, "really felt as if they were coming in for special persecution in the 14 February noon raid."

Of course they were not—in fact, had the weather not gotten in the way of the original operational plan, the USAAF would have attacked Dresden at noon on February 13, hours *before* the RAF raid. The city would have been still intact, and the course and the effects of the U.S. raid (not to mention the British one) inevitably different. But when an area is subjected to intense attack, terrified individuals understandably take it personally. The sense that the enemy aircraft are "following" you, or have "picked you out" is a strong human instinct. First the British had bombed the Altstadt, then the Grosser Garten where so many escapers from that hell had gathered. Then came the Americans, going for the undamaged areas of the western suburbs. It was as if the enemy had anticipated the Dresdeners' every move, and then killed them like cattle cunningly driven into holding pens.

By noon on February 14, the people of Dresden had suffered three devastating air attacks in just over twelve hours. Each time, the enemy had dealt fresh destruction in that very spot where the survivors had thought themselves safe. Rumors, legends and distortions spread among a terrified populace.*

AS THE SMOKY SKIES faded into dusk at the end of Ash Wednesday, and the last American aircraft returned to its base, Dresden's destruction was becoming known to the outside world. At first it was just another news item. Another routine story breaking on a war-weary world where thousands of human beings still died violently every day.

* See Appendix A.

24
Aftermath

ANITA KURZ was discovered alive late on the afternoon of Ash Wednesday. Sixteen hours before, following the second British raid, she and her parents had tried to escape the city before realizing they were trapped by the firestorm. They had been forced to take shelter once more beneath their apartment house, despite the increasingly bad air in the communal cellar. A little later, the twelve-year-old had huddled down, her face buried in the folds of her water-dampened bathrobe. She felt terribly tired—a result of the carbon monoxide that, though she didn't know it, was starting to fill the room. Her parents were still desperately discussing alternative escape plans. "My father came to me as I lay there and I said, 'Let me lie here'. And he said to Mother, 'If Anita wants to stay here . . .' And Mother nodded and murmured: 'Then let's stay here.'"

For Anita, now a grandmother, this remains her last, indelible memory of her own parents; that short, whispered conversation as she lay exhausted on the cellar floor. Soon after, she must have drifted into unconsciousness.

Anita's savior, a soldier, had come looking for his wife. He knew that she regularly used this same local shelter. The soldier broke in through a small cellar window accessible from the street, allowing healthy air into the room. At some point as he searched the silent cellar, grimly checking the thirteen corpses it contained, he spied a tiny movement from the slight figure in the corner, wrapped in a bathrobe. He realized there was life in the room after all, and within minutes Anita was clasped in his arms, on her way to the nearest medical post. The doctor there said that in the gas-filled cellar, burying her face in the damp robe had been the key to her survival. The water had emitted just enough vital extra oxygen to keep her breathing.

Her parents were not among the dead in the main chamber of the shelter. For a short while, there was hope they might have survived.

> The day after that, a sanitation squad was sent in, and they cleared the shelter. But then they sent another team to check again . . . That was when they found my parents. They had gone off into a little chamber on one side of the main cellar. And the thing is, my mother's body wasn't even stiff. They had to summon a doctor to make sure.

As the soldier's presence in the shelter proved, people were already starting to move around the city, even before the fires had gone out (some would burn for another three days). The soldier had probably come into the city with his unit to help with rescue work, and managed to find his way into his home district.

Others concerned for family members in the city did the same, whether out of courage or of foolhardiness. A young woman from Dresden, who had married and moved out to the nearby town of Hermsdorf, had watched from the roof of her parents-in-law's house as the city burned on the night of the firestorm. Early on the morning of Ash Wednesday, leaving her baby son with her husband's family (he was away in the Wehrmacht), she bicycled into the city, completely against the traffic, and found her way to Johannstadt, where her mother and unmarried sister lived. As she pedaled into the street where their apartment should have been, there were only smoldering ruins. She saw what looked like shop window dummies scattered around the area, then realized that they were dead people. Plunged into despair, she began to wander around, fearing any moment that she would see the mutilated corpse of her mother or her sister. After a while two ghostly figures emerged from the acrid haze, draped in sheets. One of them croaked a greeting. It was her mother.

> They had gone to Kleinschachwitz [an outer suburb] to my grandmother's birthday party! They stayed later than intended, they were all so glad to be together, and by the time they boarded a tram to go back to the city, the bombing had begun. The tram was halted, and

so they could not get back into the city until the raid was over. All their neighbors were dead—and so would they have been if they had not stayed late at grandma's.

Such cases—and such happy endings—were the exception.

Although it would be some time before the authorities could get out a proper newspaper for the Dresden population, by the end of the day Walter Elsner, the *Gau* propaganda leader, had managed to commandeer the services of a printer in the town of Pirna, ten miles upriver from Dresden. Here, late on February 14 —around the time Anita Kurz was carried from the cellar where her parents lay entombed—with ash and singed papers blown from the burning city still descending out of the heavens, Elsner managed to produce a single-sheet "News in Brief for the Population Affected by the Air War" (*Kurznachrichten für die vom Luftkrieg betroffene Bevölkerung*), which was distributed around the accessible parts of the city by party volunteers.

Practical instructions for those bombed out of the city center were laced with off-the-shelf hate propaganda against the British (at that time it was thought that all three raids, including the midday attack, had been carried out by the RAF).

Dresdeners were told to make their way to the edge of the city, where welfare facilities would be provided. As a result, false rumors quickly spread that the entire city was being compulsorily evacuated.

In fact, people scarcely needed to be told to leave the devastated central districts. Instinct, and fear of more attacks, had sent thousands upon thousands of survivors pouring out of Dresden in all directions. The party aid posts on the edge of the city, as promised, offered help, directions, and most importantly the "Provisions Card for Air Damaged Citizens" which was required for obtaining emergency rations and accommodation. The aid offices in just one suburban district registered a thousand homeless survivors a day, every day for two weeks after the raid. The refugees were provided with temporary lodgings in everything from inns to schools and private homes, until they could be shipped on to even more outlying towns and villages. Anywhere, to relieve the pressure on the suddenly overcrowded suburbs.

Günther Kannegiesser went to his aunt's house in Meissen, twelve miles downriver, as arranged with his mother, but although he found a meal and a bed for the night, there was no sign of her or his brother and sister. Typically, in the morning the intrepid fourteen-year-old headed back into the city—and continued to search. He was to wait more than fifty years until he found out what happened to his family.

Thirteen-year-old Nora Lang and her little brother headed north across the river toward their grandfather's small piece of land, where there was also a weekend cabin. This was where they the family had agreed to meet if they ever got separated by air raids or fighting.

Christoph Adam's family, meanwhile, walked south toward Altenburg, a picturesque small town on the Czech border, where relatives had always promised to take them in if there was trouble in Dresden. Within days he would be back at school.

Those for whom the Allied raids on Dresden had been a terrible blessing were also on the move. Henny Wolf and her father and mother had also found their way to the edge of the city. For some days they found refuge with a family friend in Loschwitz before moving into an empty house. Herr Wolf, as the only non-Jew in the household, would from now on be the only member of the family to venture out by day. Victor Klemperer had been reunited with his wife, Eva. He tore off the telltale yellow star on his coat. Together they found their way north to a welfare center near Klotzsche airfield, where they reported themselves as homeless Aryan victims of the bombers who had lost all their papers and belongings. Then, armed with emergency ration cards, they set off on a trek southwest, eager to get away from the place where they were known.

These Dresdeners were part of the exodus that the Allies hoped to catch again on Wednesday, thus ruthlessly sowing further chaos on the supply routes to the eastern front. The target now was Chemnitz, a short journey by road or rail from Dresden. Or such was the Allies' calculation.

DURING THE DAYLIGHT HOURS of February 14, the Eighth Army Air Force had sent its First Division to Dresden for the midday raid, but its Third Division to Chemnitz. The First had been intended to follow up the big British night raid, the Third to prepare the as yet little-damaged smaller city for a British night raid to come. Meanwhile, the British aircrew who made it back to their bases from the Dresden attack—as all but a handful did—had gone straight to bed. They were roused in the afternoon, to be briefed for their next long foray into Saxony.

The planned raid was another "double punch," very similar in conception and almost equivalent in strength to the previous night's devastating blow against Dresden. Chemnitz, their new target, was a city of around four hundred thousand inhabitants (about two-thirds of the population of Dresden), forty miles southwest of the Saxon capital. It was known as the "Manchester of Saxony." Before the Nazis came to power, Chemnitz was a major center for textiles—more than 43 percent of its workforce worked in that industry—and for the manufacture of machinery and vehicles (13.5 percent). By 1944–45 the textile industry had declined somewhat. Chemnitz was now best known, not least to the Allies, for making military vehicles, and especially tanks, which made it a key target.

More than seven hundred British aircraft were earmarked for the night raid planned for February 14–15, 1945, against Chemnitz, made up of roughly two-thirds Lancasters and one-third Halifaxes of 1,3,4, 6 and 8 (Pathfinder) Groups. As for 5 Group, which had set the Dresden firestorm near enough single-handed the previous night, it was sent as a unit to attack the oil refineries at Rositz, south of Leipzig.

It was going to be another crowded night over Saxony.

The configuration was different, and the numbers of attacking aircraft more evenly split between the two attacks, but this was another big raid, planned to produce a similarly destructive outcome to that of the attack on Dresden. If it succeeded, the raid would complete not just the annihilation of two important cities, but it would wipe out the entire industrial, transport and communications system of eastern Saxony just as the Soviets were approaching. The defending Germans would have their backs to a wasteland, and reinforcement would be almost impossible. Or such was the RAF planners' hope.

To most aircrew it was "just another target." At the briefing early that evening Miles Tripp listened, he said, "without any qualms" to the fact that Chemnitz was crammed with refugees, many of who had escaped from Dresden. The payoff, as far as he and other aircrew were concerned, was that if this attack succeeded they were promised no more long-haul visits to the Russian front. Only later did he feel uneasy:

From the abstract application of ethics and morality, as distinct from the practical consideration of helping the Russian advance, the raid on Chemnitz was probably less justifiable, and more inhuman than the Dresden raid.

The raid by the American Third Air Division had not gone well. Only two-thirds of the 441 B-17s that had taken off for Chemnitz by day actually found the target. They bombed by radar through thick cloud and did little damage. The other aircraft missed the city altogether. They bombed targets that became visible through the cloud, including the town of Bamberg in Northern Bavaria and the Eger air base in what is now the Czech Republic.

The RAF's first wave arrived over Chemnitz a few minutes before 9 P.M. that night. Conditions were much poorer than they had been over Dresden the previous night. The pathfinder Lancasters of 8 Group had to confine themselves to "sky marking" with parachute flares, at which the aircraft aimed rather than at markers on the ground. As it turned out, the truly distinctive thing about the Chemnitz raid was that, although carefully planned and accompanied by similar elaborate diversion and deception operations as the attack on Dresden, it was an almost complete failure.

The first wave's efforts were not reckoned a success. By twenty past midnight, when the 350-strong second wave arrived, instead of finding a raging inferno spread out beneath them (as had the second wave at Dresden the previous night), they were faced with conditions of 10/10 cloud. Zero visibility. "Bombing under the circumstance appeared scattered," as the official record put it euphemistically. Despite the poor weather, some crews reported being able to still see the glow of fires burning at Dresden.

Miles Tripp reported the near fiasco:

Over the Continent, layers of cloud stretched from the French coast to East Germany, and over Chemnitz Pathfinder flares disappeared almost as soon as they were dropped. The voice of the Master Bomber, a Canadian, came clearly over the R/T. He kept calling for more flares, but few were forthcoming. He seemed to have little idea where to direct the bomber stream. Eventually he gave up his appeal for flares in disgust. "Oh hell," he said. "I'm going home. See you at breakfast."

Chemnitz, instead of suffering of a crushing hammer blow similar to that inflicted on Dresden, had, by a combination of chance and poor design, escaped almost without a scratch. Imposing as the Chemnitz effort was—717 aircraft involved and more than thirteen hundred tons of bombs dropped—it went almost for nothing.

Much was and is made of the fiendishly efficient planning of the RAF war machine, but that machine could be all too fallible, and the conditions in which it operated all too unpredictable.

THE ONE IMPORTANT LANDMARK of old Dresden that seemed to have withstood both the high-explosive bombing and the firestorm was its most iconic and magnificent: the three-hundred-foot high, domed structure of the eighteenth-century Frauenkirche, the Lutheran cathedral Church of Our Lady.

So safe had the authorities thought the cellars and crypts beneath the Frauenkirche to be that many valuable artifacts and statues from other churches and public buildings in Dresden were stored there during the war years. A great deal of highly inflammable archive film material had also been brought from bomb-endangered Berlin by Göring's Reich Air Ministry and deposited in the crypts beneath the Frauenkirche for safekeeping. Moreover, after the first wave of the British raid on the night of February 13–14, 1945, around three hundred Dresdeners had taken refuge in the adjoining catacombs, attracted by the sturdiness of the building and, perhaps, the hope that God would protect His own.

Then came the second wave, hitting the area of the Neuer Markt, on which the Frauenkirche stood, much harder than the first, and with a preponderance of incendiary bombs. By the end of the attack, most of the buildings in the square surrounding the cathedral were on fire, and the searing, predatory, man-made winds of the firestorm continued to seek out combustible prey among the nearby maze of streets and alleys.

It seems as if no incendiaries or high-explosive bombs succeeded in penetrating the cupola or any weak points in the cathedral's structure during the actual attacks. Just as the Prussian cannonballs had bounced off the copper dome in 1760, so now the British air ordnance was defeated by architecture. All through Ash Wednesday and into the early hours of Thursday, though clearly damaged and with smoke leaking from its dome, the church remained standing

Early on the morning of February 15, Hannelore Kuhn gazed out from the southern heights where her family had found refuge after the American midday raid destroyed their home in the Bamberger Strasse. "Everything lay swathed in smoke, and there were fires still burning. But I saw the Frauenkirche. It still stood out, the dome. And I came back to my parents and I said, the Frauenkirche is still standing."

She must have been one of the last people to see it. The Frauenkirche disintegrated at around 10:45 A.M.

As a precaution against air raids, most of the host of glass windows in the early eighteenth-century structure had been bricked up from the outside, but for some reason a few windows on the north side—facing toward the Elbe—had not been sealed. During the night of February 13–14, the firestorm in its wrath had filled the streets of the Altstadt with burning brands, and it seems probable that some of these found their way into the Frauenkirche through these windows and started fires. But the actual collapse of the Frauenkirche was caused by the cooling that slowly followed when the fires finally ran out of fuel and died.

The intense heat inside had led to all manner of distortions, but the crucial factor was the bending of the girders that surrounded the pulpit. As they started to contract once more, due to the drop in temperature after the firestorm had subsided, they no longer properly fitted their supports. As a result, there was lateral pressure on the huge pillars that held up the cupola. One of these, to the southeastern side of the church, collapsed. The cupola momentarily settled, but then slowly began to subside to the southeast, where the pillar was missing. The weight was too much for the remaining supports. The comparatively soft local sandstone fabric of the walls, meanwhile, had been severely weakened by the high temperatures—sandstone loses its integrity at around 700 degrees Celsius.

The great dome began to collapse, pushing out some of the high walls, then crushing everything below itself. The shockwave as it slowly toppled onto the square of the Neumarkt caused the other, also fire-damaged, towers and stairways to crumble. The large, heavy cross, falling from three hundred feet up, crashed down on the disintegrating cupola. The vast weight of metal and stone of this high building, its towers and pillars fell to ground level, penetrating through the vaulted ceilings of the cellars and catacombs. All this took a matter of moments, but once it had happened, the dominating, iconic, seemingly eternal cathedral was no more than a gargantuan heap of rubble.

And as if to follow agony with insult, a short while later the Eighth U.S. Army Air Force returned once more to Dresden.

A LARGE FORMATION of Flying Fortresses of the First Air Division had set off that morning, February 15, 1945, from their bases in eastern England. Terrible weather—thick fog and ground mist—caused several bombers to be lost on takeoff and reduced the number of bombers that actually got air bound for the day's primary target—the Böhlen hydrogenation plant near Leipzig—from a planned 360 to 210. The cloud stayed thick beneath the bomber stream all the way across the continent. It was 10/10 cover at Böhlen, which made bombing impossible. This left the secondary target, where 7/10 to 10/10 was expected. Dresden.

This was the least successful of all the American attacks on Dresden. The squadron leader of one of the low units from the 401st Bombardment Group accidentally set off a set of marker flares four minutes early, resulting in almost two hundred high-explosive bombs descending in the Meissen area. A radar operator from the 401st's lead squadron admitted afterward that display problems on his system's screen may have led to bombs being dropped later than they should. Since there was damage in the southeastern suburbs and as far out as the town of Pirna—where forty-seven people died when bombs hit the Hermann Göring Housing Development, this seems likely.

The rest got to Dresden sometime before noon. During the ten minutes of the actual raid, no bombs fell on the designated target, the Dresden (that is, Friedrichstadt) "marshaling yards." The worst area of damage was actually the Südvorstadt—to be more precise the prison attached to the Justice Building in the Münchner Platz, whose structure had already been weakened by stray explosions on the fourteenth. Now, in this least effective of the Allied raids on the city, a big air mine scored a direct hit on the north wall of the prison building, killing thirty inmates but allowing many more to escape through the resultant blast hole. A number of political prisoners—mostly

Communists rounded up in Leipzig and many under sentence of death—were able to escape, as were quite a lot of captured Czech resistance workers who also, in some cases, faced an appointment with the notorious electrically operated guillotine in the bleak central courtyard of the prison—which was also destroyed by an American bomb that morning.

Thus was an extra platoon of Communist activists preserved for the postwar period. Most of the Czechs were less lucky. Still dressed in their conspicuous death row uniform of coarse black trousers and black short jacket, and identifiable as foreigners the moment they opened their mouths, most were picked up by police and army patrols. Among the few who did make it clean away were those who managed to rifle through abandoned luggage still strewn around the Hauptbahnhof underpass and thus provide themselves with German civilian clothes.

TWO DAYS AFTER THE GREAT RAIDS, despite the brief alarum caused by the latest American attack, people were starting to move around the city a little more. On February 15, eighteen-year-old Götz Bergander had already been out alone. The streets were mostly still blocked with rubble. The only way through was to clamber along the shattered, twisted railway tracks. In this way he got as far as the Hauptbahnhof. There he saw mounds of bodies piled up ready for collection. Already prisoners of war, watched by elderly Volkssturm (Home Guard) members, were starting to heave them onto carts. All the victims, mostly from the station's underground cellar complex, had suffocated. Bergander had been planning to look for a close school friend who lived in the vicinity, but felt too nauseated to go on. On young Götz's return home, his father gave him, for the first time in his life, a full glass of brandy.

The next day he and his younger brother ventured into what remained of the city and did a ten-mile circuit on foot. This time they ventured to the Elbe meadows. Less than three days after the British attacks, most of the corpses had been cleared from there, and work had already begun on recovering bodies from any streets that were accessible. The great shock was to realize that the Frauenkirche was no more.

It was as if St. Paul's Cathedral had vanished from London. Gone. I saw those two big pillars there and these mostly pale stones . . . piles of sandstone. After all I had seen, this was the last straw. Now I thought, there is nothing.

A woman sent a postcard to her absent daughter. It read simply: *"All three of us still alive, city gone."*

By then the writer, her mother, and her child had trekked out of Dresden and were living in a temporary camp. This was the fate of thousands who had no relatives in undamaged areas.

It was obvious there must be tens of thousands of dead in the city. Perhaps more. Rumors were rife. As Nora Lang and her little brother, having stayed in a reception center on the night of February 14–15, continued their trek to Winschdorf and their grandfather's smallholding, they passed the sign marking the city limits and then the huge municipal cemetery that lay in the heath land just beyond: the Heidefriedhof, or Heath Cemetery. They were to experience their own happy ending later that same day when they found their parents waiting for them at their destination, but first the two children had to witness the first of the city's dead being evacuated to their mass graves. "We stood by the road and watched as transporters came carrying bodies. Piles of bodies on them. We just stared."

PART THREE

AFTER THE FALL

25
City of the Dead

AS EARLY AS the morning of February 14, 1945, the chief of the Order Police (Ordnungspolizei) in Dresden sent a desperate message from his bunker on the outskirts of the city to his superiors in Berlin. He received an near-instant response from the Reichsführer-SS and head of the German police, Heinrich Himmler:

> The attacks were obviously very severe, but each initial air attack always conveys the impression that the city has been wholly destroyed. Take all necessary measures immediately. I am sending you immediately an especially competent SS leader, who could be useful to you in the present difficult situation. All good wishes.

That morning, aid teams were already on their way from the outside world. Troops from the large barracks on the edge of the Neustadt, which had been largely undamaged by the air raids, were officially allowed to cross onto the left bank only in the afternoon— under Wehrmacht regulations, anywhere east of the Elbe (including the main barracks) counted as part of the "rear frontline area." To leave such an area, even simply to cross the river from one part of Dresden to another, Berlin's explicit permission had to be sought.

The next basic service was rescue and clearance. Up to two thousand Wehrmacht troops and a thousand prisoners of war were immediately put to work, plus repair teams from Leipzig, Chemnitz, Zwickau, and Halle, and units of tunneling specialists from the mining towns of Freiberg and Sadisdorf. SS units were also seen in the city.

Exactly who was in charge remained at first unclear. All central administrative buildings had been destroyed or disabled. Himmler had sent down senior police people—the only major administrative organ still functioning was the Higher SS and Police Authority, in its bunker on the Weisser Hirsch—but the main organizer of practical help seems to have been Theodor Ellgering, chief executive of the Inter-Ministerial Air War Damage Committee based in Berlin. Ellgering took his orders from Goebbels, who had been given special powers in this area of the war by Hitler at the end of 1942. Ellgering visited Dresden on February 14, viewed the immediate aftermath, and then reported back to Berlin that evening. On the fifteenth he returned to Dresden armed with full powers from Goebbels, and quickly got to work, setting up a command post in a fruit fermentation plant on the edge of the city, briefing action teams and establishing his own communications system.

Despite the undoubted energy of the outsiders, it was at least two days before major improvements started to show on the ground for the population of Dresden. There was always a bowl of warm soup, right from the start, but within seventy-two hours of Dresden's destruction, the authorities were distributing six hundred thousand hot meals a day. On February 17, the party newspaper boasted that no one was going hungry in Dresden. At the same time, martial law was declared, which included the death penalty for looting and "rumor-mongering." Seventy-nine people would be found guilty of looting. The great majority were immediately executed.

And there was other necessary work, of the most terrible kind

Slowly, blocked streets and were being opened up and the grim work of emptying the cellars and shelters of their dead was beginning.

An important part in this work was played by Allied prisoners of war. There were several thousand in and around the city. The British were in various subcamps of Stalag IVB, the Americans in various places including—as we know from Kurt Vonnegut's famous semi-autobiographical novel, *Slaughterhouse-Five*—under the old abattoir area in the Ostragehege, not far from the Berganders' home.

"Corpse mining," Vonnegut called it—a darkly jocular description for a horrifying activity that would become one of Dresden's largest and most labor-intensive industries over the following days and weeks. In *Slaughterhouse-Five*, Vonnegut described his hero, Billy Pilgrim, being put to work. The time portrayed is two days after the raids:

Prisoners of war from many lands came together that morning at such and such a place in Dresden. It had been decreed that here was where the digging for bodies was to begin. So the digging began.

Billy found himself paired as a digger with a Maori, who had been captured at Tobruk . . . Billy and the Maori dug into the inert, unpromising gravel of the moon. The materials were loose, so there were constant little avalanches.

Many holes were dug at once. Nobody knew yet what there was to find. Most holes came to nothing—to pavement, or to boulders so huge that they would not move. There was no machinery. Not even horses or mules or oxen could cross the moonscape.

And Billy and the Maori and others helping them with their particular hole came at last to a membrane of timbers laced over rocks which had wedged together to form an accidental dome. They made a hole in the membrane. There was darkness and space under there.

A German soldier with a flashlight went down into the darkness, was gone a long time. When he finally came back, he told a superior on the rim of the hole that there were dozens of bodies down there. They were sitting on benches. They were unmarked.

So it goes.

The superior said that the opening in the membrane should be enlarged, and that a ladder should be put in the hole, so that the bodies could be carried out. Thus began the first corpse mine in Dresden.

In Vonnegut's novel, his Maori comrade died "of the dry heaves, after having been ordered to go down into that stink and work." In real life, others decided that they wouldn't take that risk. A British prisoner and his comrade actually went on the run rather than face more days digging in the corpse-filled cellars of Dresden or fishing body parts out of trees.

One British prisoner of war, Alec White, held in a camp outside the city, described being marched fifteen miles every day into the center, starting at 5 A.M., then laboring in burned-out factories or in the streets, where bodies were often piled up for collection after basements had been emptied of their dead.

The POWs, like the few remaining Jews, had been detailed to fetch bodies and clear rubble after previous raids, but the scale and the horror of the work to be done after the firestorm of February 13–14, 1945, was, like the experiences of those who survived the destruction of central Dresden, almost impossible to describe with any hope of authenticity.

All the city's fleet of mortuary wagons had been destroyed in the bombing. Nevertheless, within ten days, according to Ellgering, ten thousand bodies had been recovered, registered and where possible identified, and buried. Most had been carried several miles to the Dresden heath on the edge of the city and placed in mass graves (these were the transports Nora Lang and her little brother had stopped to gaze at on February 15 as they tramped to Winschdorf). But even this rate of disposal was not fast enough, as Ellgering admitted:

> It became imperative to further speed up the rate of work, for in consequence of the mild weather the corpses were beginning to decay . . . To avoid the outbreak of infectious disease, the Altstadt was declared a prohibited area . . . there remained no choice but to give permission for the corpses to be burned.

After the idea of burying them instead in the city parks was abandoned for public health reasons, a drastic but effective solution was found. Instead of carting and trucking corpses out to the cemetery, the dead from the streets and cellars of the Altstadt were transported to the great expanse of the Altmarkt, where flower markets had once been a famous feature, and less than five years earlier bands had played and vast crowds had cheered Dresden's Fourth Infantry Regiment as it returned from the war against the French, apparently victorious. A more terrible contrast than the scenes that commenced on February 21 could not be imagined.

The great water tank built the previous winter in the Altmarkt to supply the fire service had itself filled with the drowned and boiled bodies of those who had mistakenly sought refuge there. Once those were cleared, and rubble swept, the square was sealed off. Then began the work. Corpses were shipped in and laid out ready for registration and, if possible, identification. Searching for ways of keeping them off the ground—and allowing a draft under the planned mass funeral pyres—workers found a solution in the wreck of a nearby department store, where massive window shutters had survived the bombing.

They carried them from the ruins and set them down on the ground, making, as a contemporary grimly expressed it, "huge grill racks."

Large amounts of gasoline were trucked into the sealed city center. Teams poured petrol over the bodies as they lay piled on the shutters. Then the dead were burned at the rate of one pyre per day, with around five hundred corpses per pyre. The task was efficiently done—to reduce that number of human remains to fine ash without access to a purpose-built crematorium is a technically problematic process—under the supervision of outside SS experts. They were said to be former staff from the notorious extermination camp at Treblinka.

Between February 21 and March 5, when the last pyre was lit, 6,865 bodies were burned on the Altmarkt . Afterward, when the fire cooled down, it was estimated that between eight and ten cubic meters of ash covered the cobbled surface of the medieval square. The SS-Brigadeführer in charge of the burning had intended to transport the ashes out to the Heath Cemetery in boxes and sacks and bury them containers and all, but municipal parsimony triumphed. In the end the ashes were simply emptied out of their containers and into the prepared pits, thus enabling the valuable boxes and sacks to be reused.

That same week the police chief's office started to put together a lengthy, meticulous (if inevitably temporary) report on the air attack on Dresden and the damage and the death it had caused. The secret "Final Report of the Higher Police and SS-Führer for the Upper Elbe" originated within in the first two weeks of March and was submitted to the Reich commander of Order Police in Berlin on March 15, 1945. Parts of it were absorbed into other documents of the time circulating at the Order Police headquarters, including the nationwide "situation report" regularly sent out to regional police chiefs.

The "Final Report" contained the first official estimate of the death toll from the Anglo-American, but predominantly the British, raids on Dresden four weeks previously. Under "Injury to Persons" it stated:

> Assessment up to the morning of 10 March: 18,375 fallen, 2,212 seriously injured, 350,000 homeless and long-term rehoused ...The total number of dead, including foreigners, is estimated—on the basis of previous experience and assessments at the time of the bodies' recovery—at approximately 25,000. Beneath the masses of rub-

ble, especially in the inner city, may lie several thousand more fallen, who may for the moment remain totally irrecoverable.

The death statistics were not based on rough estimates. The process of registering and counting the dead—and their possessions—was meticulous in the extreme. Page after page of lists, organized street by street, record an extraordinary amount of detail about those who perished. A female Silesian refugee is found dead in the ruins of a hotel in the Neustadt. The large sum of money she has on her is carefully counted, her husband is sought for confirmation (he had been elsewhere when the raid hit, and survived). Many entries contain a name and the description "refugee." In remarkably few cases is there no indication of identity whatsoever. Cause of death is also generally stated. Many are classified as dying from "asphyxiation." Others are burned, often unrecognizably so. Then there are those crushed or struck by falling masonry or collapsing buildings. Here "struck dead" (*erschlagen*) is noted. Among these death rolls, there is usually at least one per page of the smallest but in a way saddest category of death: suicide. Perhaps burn wounds were too agonizing to bear, slow suffocation too awful a prospect, or life simply seemed insupportable with all loved ones lost. One story tells of a busy restaurant that suffered a direct hit. In the restaurant's shelter, days later, were found a large group of well-dressed diners, all of whom had been buried alive by the building's collapse. There were indications that life had gone on for some time after they were trapped. When the rescuers broke into the airless cellar, however, they found all the diners dead from neat gunshot wounds. A number of uniformed Wehrmacht officers lay nearby, each with his service revolver pressed to his shattered temples. Perhaps as the result of a pact, the soldiers had dispatched the civilians one by one before turning their pistols on themselves.

Nor did the state of the body when finally recovered necessarily express the cause of death, though it explained certain apparent anomalies. After the disappearance of his mother, sister, and brother during the firestorm night, Günther Kannegiesser had never been able to trace their bodies, despite persistent inquiries and searches of cemetery lists. Then, in 1997, after fifty years, he was interviewed by a Dresden newspaper about his continuing search, which had been given further stimulus by the opening up of public records in eastern

Germany after the fall of communism. A short while later, as a result of the article, he received a touching note from a man, now also in his late sixties, who as a boy had been a neighbor of the young Günther.

The long-lost childhood friend enclosed a letter written in the spring of 1945 by his own father to a brother-in-law, who at the time was a prisoner of war. On March 2, 1945, it transpired, the father had managed, using one of the "breakthroughs," to enter the basement area of No. 9 Zöllnerstrasse (the building where young Günther's family lived). He tells the story of his discoveries there with gentle precision:

> The heat was like in an oven. We were bathed in sweat. The walls could scarcely be touched, there wasn't a single piece of wood that the flames had not consumed. We counted 29 bodies spread through the basement rooms. They were burned and therefore no longer identifiable. I nevertheless ascertained clearly, and this is my innermost conviction, that all these human beings had experienced a quite gentle and peaceful death. They all lay there quite calmly and in a relaxed way, as if they had simply laid themselves down to sleep. The children were also huddled together, lying close to each other. This was how they must have been overtaken by death from asphyxiation. As so often in the circumstances, probably no one dared venture outside into the sea of flame. They felt safe in the basement, but the trouble was, smoke slowly penetrated there. We went through the same thing ourselves. You become increasingly tired and exhausted, then you want to lie down, because the air is better on the ground. Down there, however, you are subjected to inhalation of the poisonous but odorless carbon monoxide gas, and you sink into a sleep from which you never awake . . . Later, after they were all dead, the fire found its way into the basement of Number 9, and turned the place into something like a crematorium.

DRESDEN, it was already clear some weeks after the Anglo-American raids, represented, in absolute terms, the most catastrophic air assault on a German city since "operation Gomorrah" had devastated Hamburg in July 1943. Six-figure numbers for the dead at Dresden would be encouraged by the Nazi propagandists and are still quoted more than half a century later—though mostly by right-wing extremists attempting to gain converts to their cause by promoting the idea of a "German holocaust" worse than Auschwitz.* However, the accepted death toll both then and now remains between twenty-five thousand and forty thousand.

Of course, terrible as these numbers of dead were, they seemed, naturally enough, inadequate to those who had undergone the firestorm and seen the thousands of corpses littering the streets and parks. It is not unusual for figures to be drastically reduced from the initial estimate after a big air raid. A death toll of one hundred thousand or even two hundred thousand was widely believed after the British bombing of Hamburg in July 1943. The figure of twenty-five thousand victims of the Berlin raid of February 3, 1945—eight times what now appears to be the real number—still finds currency more than half a century later. Nor do bombed-out populations, awed by what seem like scenes of apocalyptic destruction and appalled at what has happened to their familiar districts, their friends and families, necessarily believe official figures. Almost always, they seem to think them too low and often continue to do so in the face of documentary evidence. In the case of Coventry, for instance, even thirty years after the German raid many still believed that "thousands" had actually been killed in the raid, but the government had "hushed up" the true number of victims to avoid damage to the morale of the British public.

The historic center of Dresden itself had come to resemble an uninhabited wasteland, empty—so it seemed—except for the rescue teams, the "corpse miners," and—while the mass burnings of the dead continued in the Altmarkt—the workers operating the grisly, improvised outdoor crematorium there.

Some weeks later, the establishment of a proper ration cards system allowed a reasonable estimate of the population of an area where, before the firestorm, many tens of thousands of people had been resident. It turned out that, astonishingly, there were living beings still to be found dwelling in the ruins of the Altstadt. They numbered around four thousand, mostly leading a troglodyte existence in cellars and other holes in the ground. Where once there had been shops, theaters, churches, and elegant apartment blocks, there were only blackened facades and rubble.

* For a more detailed discussion of the argument over casualty figures see Appendix B.

Entire areas had been cordoned off, especially where blocked streets and half-collapsed buildings created danger to civilians. The list of streets read like a tourist guide to old Dresden.

In the jargon of the clearing squads, they were known as "dead areas." This would later lead to rumors that in these districts the cellars had been sealed forever, with their many thousands of corpses still inside. The thirty-five thousand registered as "missing" would also figure large in the legend building.

THE OFFICIAL REACTION to the raid, as elsewhere in bomb-shattered Germany, was not concerned just with providing food and shelter for the survivors, and recovering and burying the dead. This was war—total war—and Berlin was concerned to restore as much of Dresden's usefulness to the war effort as soon and as completely as was humanly possible.

Among the first officials to reach Dresden from outside was a hard-headed military man whose orders had nothing to do with recovering the dead or saving the living. General Erich Hampe was the "Plenipotentiary for the Restoration of Railway Connections," head of a huge technical organization whose dedicated specialty was the repair of railway lines and facilities damaged by air raids. It was his pride that he could get the trains moving again more quickly than the enemy could ever have imagined.

Hampe and an aide appeared from Berlin within hours of the British raid. The general was appalled by the slaughter at the Hauptbahnhof and the chaos at the railway directorate. Even to him, who had seen so much destruction, this was a "special case." Hampe fetched a senior German railways executive from Berlin to restore order. Meanwhile, labor squads began to extract the hundreds of dead from the station area and the underground cellars—these were the corpses Götz Bergander saw piled up outside the Hauptbahnhof on the afternoon of February 15, ready to be piled into carts and trucks.

That same day, Hampe's repair gangs—mainly Allied prisoners of war, augmented with some forced laborers—began working round the clock on replacing track. The general had to bring in his own equipment and communications, since the administrative apparatus of the railways in Dresden was almost completely destroyed. The vital north-south line connecting the Neustadt station and Hauptbahnhof via the railway bridge was completely wrecked. The Neustadt goods yards were all but destroyed and the Friedrichstadt Marshaling yards were very badly hit. Eight hundred coaches and wagons had burned out as a result of the RAF raids alone. The American raid on February 14 knocked out the Friedrichstadt passenger station completely, and demolished another forty-five tracks in the Friedrichstadt Marshaling yards. To an inexpert eye, the damage must have seemed irreparable.

It was nevertheless true that the restoration of railway tracks counted among the least difficult repair tasks facing the authorities after major air raids. Once craters have been filled in and the surface made reasonably level, tracks can be relaid very quickly—especially if, as in Dresden, the main railway bridge had suffered relatively limited damage. Within days a slow, single-track connection between the Hauptbahnhof, and the Neustadt station across the river, and the goods stations between, had been established. Two weeks later the service was back to a reasonable level, though in the case of Hauptbahnhof-Neustadt route only two out of four tracks were usable. For trains traveling southeast along the Elbe, access was available only from a platform in an outlying part of the Hauptbahnhof.

This was not quite "back to normal." Moreover, for weeks after the raid, the Dresden railway directorate still had no functioning telephone system. This made control of work rosters, direction of traffic, and of coach and wagon utilization, problematical. All these were vital to the efficient operation of a complex rail system, and that system continued to suffer. The availability of qualified staff was much reduced. Many employees had been killed, seriously injured, or made homeless. Some remained simply "missing."

POINTING OUT that the entire Altstadt of Dresden, plus the inner eastern suburbs, had been "a single area of fire," the police declared almost twelve thousand dwellings, including residential barracks, had been totally destroyed. The list continued:

> 24 banks, 26 insurance buildings, 31 stores and retail houses, 647 shops, 64 storage and warehousing facilities, 2 market halls, 31 large hotels, 26 large public houses, 63 administrative buildings, 3 theaters, 18 film theaters, 11 churches, 6 chapels, 5 cultural-historical build-

ings, 19 hospitals including auxiliary and overflow hospitals and private clinics, 39 schools, 5 consulates, 1 zoological garden, 1 waterworks, 1 railway facility, 19 postal facilities, 4 tram facilities, 19 ships and barges.

The military targets noted as damaged were relatively unimportant except one: the Wehrmacht's main command post in the Tauschenberg Palace in the old part of Dresden. This was totally annihilated during the firestorm, and all the officers and men perished. Otherwise, military targets destroyed included the Wehrmacht library and the veterinary testing center for Military District IV, and many military hospitals. The barracks in which most of the sizable Dresden garrison lived lay a little less than two miles north of the Elbe, around the old arsenal area, and this district remained all but untouched by bombs. Since most of the soldiers—technically serving in a war zone now that the Russians had advanced west of Breslau—were at the time confined to barracks, there were remarkably few out and about in the city on the night of the British raid. This helps account for the low death toll among the military (around a hundred).

The barracks' survival caused disappointment among those who had hoped that at least some good would come from the raid. Pastor Hoch recalls that, although only fifteen, he had been ordered to report for military service later that month. As they huddled in the shelter during the British raids, his mother had expressed the hope that his call-up—and with it the prospect of a premature "hero's death"—would now be postponed. But the bombers spared the barracks in the Neustadt. Two weeks later, young Karl-Ludwig Hoch and others from the fifteen-sixteen age groups duly reported for their medical examinations.

> I saw that all the barrack buildings had their window panes intact . . . and the fat Nazis sat there under a picture of Hitler and I had to get undressed with the others. And it says in my diary, I wrote it at the time, "Most of them had no pubic hair." They were so young they hadn't yet entered puberty. And it was their job to give Hitler a few more nights with Fräulein Braun . . .

Industry was more seriously affected. Almost two hundred factories in Dresden suffered damage between February 13 and February

15. In 136 cases the damage was reckoned "serious," in 28 "medium-serious," in 35 "light."

Forty-one damaged or destroyed factories were mentioned by name as important for military production, with descriptions of level of damage and the probability of a resumption in production. Street addresses were provided, which allows their locations to be fixed. Twenty were in the eastern suburbs of Johannstadt and Striesen (including a few that trickled down toward the outer suburbs of Tollkewitz and Leuben); twelve were in the southern suburbs (Südvorstadt to Plauen); nine were in the Neustadt/Leipziger Vorstadt industrial area, across on the right bank of the Elbe. In the case of twenty-one factories, hundred percent stoppages of work are said to have resulted from the bombing. Estimates of when work could be resumed range from "not in foreseeable time" to a matter of weeks (often partial). In some cases it is indicated that the work will be resumed on alternative premises.

The industry worst affected by the bombing was the optical/precision engineering sector, in which the Zeiss-Ikon factories dominated. Zeiss-Ikon was by far Dresden's largest and most well known company. Of its almost fourteen thousand employees in Dresden, considerably more than half were, on average, absent throughout February 1945. Many had died in the bombing; other absentees would have been busy salvaging and clearing their homes.

Among Zeiss-Ikon's plants, the worst affected were the Delta-Works (100 percent destroyed), the Ica-Works (home of the company's research and development department), and the Mü-Works, all of which lay in the Johannstadt/Striesen area, plus the Petzold & Aulhorn plant in the southern suburb of Dresden-Plauen, which included Zeiss-Ikon's Alfa-Works (also 100 percent destroyed). Both the Delta-Works and Petzold & Arnhold were reported a hundred percent destroyed.

The large, modern Goehle-Werke factory in the Grossenhainer Strasse, where the Hellerberg Jews had been used as forced labor, was constructed in the 1930s to be bombproof and lay in any case away from the areas mainly affected by British bombing. However, even that factory, which before February 13, 1945, employed over four thousand workers, still reported only about half that number reporting for work two weeks later. In the case of the Ica-Works, a total workforce of around twenty-eight hundred workers was still reduced to less than five hundred at the end of February. For Zeiss-Ernemann, a total of two and a half thousand had become five hundred. The effect on armaments production must have been enormous. Disruption did not arise from damage to the factory buildings alone. Deaths of workers, damage to their homes and to essential infrastructure, including transport links, also played a vital part in drastic reductions in productivity.

Dresden companies were famous before the war for their cameras, which were exported all over the world. By the last years of the Second World War, the output of all these companies was devoted to war work. Balda, the next largest company, employed nine hundred workers, having before the war produced a highly successful cheap "box" camera. By 1945 it made mostly gauges for Luftwaffe aircraft. Their factory "suffered seriously" in the bombing. The Ihagee camera factory in the Schandauer Strasse, which employed (1943) more than 550 workers, also exclusively producing equipment for the Wehrmacht, was completely destroyed.

Things did not necessarily improve with time. H. Grossmann, a manufacturer of specialist machines and devices, and also a supplier to the Wehrmacht, had likewise been badly affected by the bombing at its factory in Chemnitzer Strasse, south of the city center. It claimed in April to have spent almost thirty-four thousand marks on clearance works between February 14 and February 28, with no income to balance this expenditure. Works were still continuing. Two months after the British raids, "a part" of the factory was now said to be suitable once more for war production.

The degree of destruction and disruption of industry in Dresden was major, but less than would have been the case if the British had systematically bombed the industrial suburbs.

Instead they bombed mostly the heart of the city, and that is what the world heard about over the following weeks. Josef Goebbels made sure of it.

26
Propaganda

ON FEBRUARY 14, during the regular afternoon press briefing held at SHAEF headquarters in Paris, a British wing commander of the public relations division had given an upbeat summary of recent air assaults:

> The Bomber Command effort last night, in which 16 aircraft were lost, which included a big double attack on Dresden as well as the attack on the Oil Plant at Böhlen, has been followed up this morning by a big Eighth Air Force attack in the same area, attacking transportation and industrial targets in Dresden, Chemnitz—which is a little further on into Germany—practically out of Germany and into Czechoslovakia ...

He expanded a little on the situation on the eastern front, especially in Silesia, where, as he breezily pointed out, "it appears that the main line of supplies to that front is almost bound to go through Dresden."

The next day the London press printed stories that emphasized the power of the blow delivered against Dresden. The mass-circulation *Daily Sketch* printed a still photograph taken from the RAF film of the raid, headed "Dresden Ablaze in First RAF Raid." The authoritative *Manchester Guardian* announced, "Blows by Over 3,600 RAF and US 'Planes Ahead of the Red Army."

So far, so good. The message about Dresden twenty-four to forty-eight hours later seemed clear and positive. A highly successful raid, one of a series designed to help the Russians. In Germany, public announcements were quite guarded. Goebbels didn't quite know how to approach this one, at least for internal consumption. To admit to the German people huge casualties in a city as symbolic as Dresden would be to risk seriously undermining what remained of the nation's morale. Instead, therefore, the Propaganda Ministry concentrated on the foreign press. Discrediting the Allies' bombing policy in the neutral countries might be relatively futile at this late stage in the battle, but it was at least harmless.

What Goebbels did not expect was that the Allied powers' own propaganda machine would come to his assistance, but that was what happened. The first inkling of a problem for the Allies came with a meeting in Paris on February 16, where Air Commodore Grierson of the RAF's press office gave reporters a briefing on developments in Allied air strategy. He was asked to discuss the reasoning behind the Dresden raid and similar attacks and said:

> First of all they are the centres to which evacuees are being moved. They are centres of communications through which traffic is moving across to the Russian Front, and from the Western Front to the East, and they are sufficiently close to the Russian Front for the Russians to continue the successful prosecution of their battle. I think these three reasons probably cover the bombing.

Another journalist then asked Grierson, as a follow-up, whether "the principal aim of such bombing of Dresden would be to cause confusion among the refugees or to blast communications carrying military supplies."

"Primarily communications," Grierson affirmed. "To prevent them moving military supplies. To stop movement in all directions if possible—movement is everything." He then added a fairly offhand remark about also trying to destroy "what was left of German morale."

The next afternoon, at around 5:30 P.M., an Associated Press correspondent, Howard Cowan, submitted a report for the approval of the censors at their headquarters in the Hotel Scribe, near the Paris Opera. The draft cable read:

> Allied air bosses have made long awaited decision to adopt deliberate terror bombing of great German population centres as ruthless expedient to hasten Hitler's doom.

More raids such as British and American heavy bombers carried out recently on residential sections Berlin Dresden Chemnitz Cottbus are in store for Reich with avowed purpose heaping more confusion on Nazi traffic tangle, and sapping German morale.

All out air war on Germany became obvious with unprecedented daylight assault on refugee crowded capital two weeks ago and subsequent attacks on other cities jammed with civilians fleeing Russian tide in east.

At first Lieutenant Colonel Merrick, the senior censor responsible, decided to stop the piece. Then Cowan returned, and he wouldn't take no for an answer. Merrick explained later:

After considerable discussion with him and checking of the guidance covering the conference by Air Commodore Grierson of 16 February, the story was passed with the one cut . . . The Cowan story moved at 18:28 hours.

It was one of the great propaganda mistakes of the war. The journalist Cowan's report on the post-Dresden press conference was immediately taken up in the American press and broadcast on Radio Paris, though not in Britain.

Nonetheless, it was received in the offices of the London newspaper offices, and caused alarm. How could a newspaperman, subject to official censorship, have been allowed to use the taboo phrase "terror bombing" to refer to Anglo-American air raids? Did this really represent a change in government policy?

Cecil King, a senior executive at the *London Daily Mirror*, added: "This is entirely horrifying . . . it gives official proof for everything Goebbels ever said on the subject . . ."

King was right about the propaganda gift that Howard Cowan's dispatch represented for the Germans. He was, however, wrong to see it as accurately reflecting SHAEF's official view. No one at the press conference had used the word "terror" or anything remotely like it.

Cowan's dispatch essentially interpreted Air Commodore Grierson's slightly woolly remarks at the press briefing in such a way as to draw radical and therefore newsworthy implications. The astonishing thing was not that he wrote the article, but that the censors allowed it. The Allied authorities quickly realized their mistake. Within hours officials had contacted AP to query the story. Back in Paris, a Reuters correspondent was primed to write a denial on SHAEF's behalf, and a wire went out to London just before midnight on February 17.

"The Dresden raid was . . . designed to cripple communications and prevent shuttling troops from eastern to western front and vice versa," the new dispatch declared. "The fact that the city was crowded with refugees at the time of the attack was coincidental and took the form of a bonus."

By February 19, Eisenhower's chief of staff and hatchet man, General Walter Bedell-Smith, had become involved regarding "the misinformation on bombing policy which appeared in yesterday's press." Plans were under way to reorganize the entire press department. The implication that "air chiefs"—as Cowan put it—decided policy also carried the unwelcome implication that they determined such matters independent of their political masters.

On March 6, in the British House of Commons, a Labor back-bencher launched a strong attack on the coalition government of which his party had been a pillar for almost five years. Richard Rapier Stokes, MP for Ipswich, was a devout Catholic and a decorated First World War hero. No pacifist, since 1942 he had nevertheless belonged to a small group of public figures who opposed what they saw as the indiscriminate bombing of German cities. The occasion was the first debate on the air war since February 1944, when the likewise antibombing bishop of Chichester had spoken in the House of Lords on the subject. Stokes got to his feet at 2:43 P.M. As he did so, the air minister, Sinclair—the man responsible for the policy under debate—pointedly left the chamber.

Stokes asked why Britain's Russian allies had never felt the need to indulge in "blanket bombing," while enjoying great military successes (he did not suggest that this might be because the Anglo-Americans were doing it for them). He read a report from the *Manchester Guardian*, based—he said—on a German telegraphic dispatch, which spoke of tens of thousands of people buried in the Dresden ruins, so badly burned as to make identification impossible.

"What happened on that evening of 13 February?" the newspaper asked. "There were a million people in Dresden, including six hundred thousand bombed out evacuees and refugees from the east. The raging fires which spread irresistibly in the narrow streets killed a great many for lack of oxygen."

Stokes then also read the entire text of Cowan's dispatch, thus placing it on the public record in Britain, and asked if its claims represented official policy. If so, he said, then why had the journalists' words, widely reproduced in America, broadcast on the radio, and even reported in Germany, been suppressed in Britain? Could it be that the British people were "the only ones who may not know what is done in their names?"

Later in the debate—five hours later, to be precise—an undersecretary from the Air Ministry replied on behalf of the government. He smoothly pointed out that Cowan's interpretation of the SHAEF briefing was technically incorrect and had been denied almost immediately. He also reasserted that indiscriminate bombing was never government policy.

> We are not wasting bombers or time on purely terror tactics. It does not do the Hon. Member justice to come here to this House and suggest that there are a lot of Air Marshals or pilots or anyone else sitting in a room trying to think how many German women and children they can kill.

Stokes continued to insist on the accuracy of his information, but the debate ended in anticlimax. There would be no further formal discussion of the bombing question before the war ended.

This was not to say that the leveling of Dresden, and the attention that its catastrophic fate attracted, passed everyone by. For the British elite, "Florence on the Elbe" was significant in a way that plain German cities such as Dortmund or Chemnitz or Wuppertal could never be. Many had visited Dresden as tourists, or even lived there, perhaps as students. It was said that after hearing of the raid, Violet Bonham-Carter, daughter of the First World War Liberal prime minister Asquith, marched to 10 Downing Street and demanded to speak to Churchill, who had served as a cabinet minister under her father. This formidable grande dame of British politics then soundly berated the most powerful man in Britain for bombing Dresden. She had attended a finishing school there in the city's golden days before the First World War.

Even the knights of the British shires could find themselves stirred. Sir Cuthbert Headlam, a former minister and stalwart of the Conservative Party in the north of England, wrote morosely in his diary on February 16, 1945:

> Dresden also is being smashed to pieces—it is an abominable business—but it cannot be helped in these enlightened days and no one now seems to have any compunction in killing crowds of civilians, so long as they are Germans or Japanese. It is not surprising in view of all that has been done by these two nations—it is nonetheless hateful to me—and one's only consolation (a faint one) that it will sicken people of war.

The disquiet rumbled on. Perhaps because of geographical distance, perhaps because most of the damage in Dresden had been inflicted by the RAF, the raid as such attracted less attention in the United States. Nonetheless, with the U.S. Army Air Force's strict adherence to "precision bombing" still an article of faith, among informed circles the unease was widespread—even at the highest level—and the raid's potential for undermining the Allies' crucial claims to moral superiority clearly understood.

On the same day as Richard Stokes's speech in the House of Commons, the Chairman of the Joint Chiefs of Staff in Washington, General Marshall, was forced to give a personal reassurance to Henry L. Stimson, secretary of war in Roosevelt's cabinet, that the Dresden attack accorded with previous practice and had been requested by the Russians.

Stimson seemed satisfied, but perhaps this was deceptive. A few weeks later he heard that a list had been drawn up of Japanese cities being considered as targets for the atomic bomb, if and when it was finally used. General Groves, the army man in charge of the Manhattan Project, had the list brought from his own office on the other side of Washington for Stimson's perusal. At the head of the list, as Groves's strongly preferred first choice, was the ancient imperial city of Kyoto, which, like Dresden, was a cultural center and also, like Dresden, contained a large industrial area. Groves described Secretary Stimson's reaction:

He immediately said, "I don't want Kyoto bombed." And he went on to tell me about its long history as a cultural center of Japan, the former ancient capital, and a great many reasons why he did not want to see it bombed . . . his mind was made up . . .

Stimson went even further. To Groves's embarrassment, the secretary of war called in General Marshall, whose office was next door, and without reference to Groves repeated his insistence to the chairman of the chiefs of staff: "I don't like it. I don't like the use of Kyoto."

Counterarguments were futile. Japan's most famous cultural center was taken off the target list for the A-bomb, and stayed off.

THE HISTORIC HEART of one of Europe's finest cities had been obliterated, along with most of the human beings who lived there. It represented to most Germans, and many other neutrals, an outrage, the apogee of terror.

On the day following the destruction of his home city, Günter Jäckel and the other wounded were evacuated from Dresden and traveled by truck and train into the rural fastnesses of the Erzgebirge. For the past two months he had been a wounded soldier, institutionalized and under medical supervision, sheltered (however unwillingly) from the reality of the wider war. He was shocked by a new, heightened atmosphere of fear and despair among the refugees: "Suddenly I was out there, right in the midst of the stream of refugees. Desperate people, ruthless people, women with baggage and children . . . loud and noisy and ruthless . . . there was already a feeling of collapse . . ."

The reaction of the Reich's elite was more complicated. To them the bombing of Dresden represented not just the destruction of a beautiful city but a final humiliation. Once Göring had promised that Allied bombers would never be permitted to do to German cities what the Luftwaffe had done to Warsaw, Rotterdam, and London. The Allies had proved him a liar, time after time. Just ten days earlier, on February 3, 1945, the U.S. Eighth Air Force had devastated the entire administrative center of Berlin in a daylight raid against which the Luftwaffe could offer no effective resistance. The Reich Chancellery and Goebbels's Propaganda Ministry had suffered severe damage. Allied power to inflict harm was now to all intents unlimited. Nowhere and no one in Germany was safe. For a regime that based itself on brute strength—and viewed failure as proof of inferiority—this was an absolutely intolerable situation.

Hitler heard the news of the destruction of Dresden's center with a face of stone, his fists clenched with suppressed emotion. Goebbels, reportedly shaking with rage, suggested to the Führer that they order the execution of tens of thousands of Allied prisoners of war, one for each civilian killed in Dresden.

On trial at Nuremburg, Albert Speer suggested that the Nazi leadership, especially Goebbels and the head of the Labor Front, Reichsleiter Robert Ley, wished to seize this opportunity of abandoning the Geneva Convention, in part so that Germans in the west would fight with the same bitter determination as they did in the east, where they already knew they could expect little mercy from the vengeful Soviets.

Whether Speer told the truth or not, there can be no doubt that Goebbels and Ley belonged to the *ultra* group among the Nazi leadership, who pressed for victory whatever the cost—moral, architectural, military, human. A short while after the bombing of Dresden, Ley wrote a ranting article under the title "Without Baggage," in which he bizarrely celebrated the destruction of Dresden as liberating Germany from the "burden" of its architectural heritage, and by implication its freethinking, humanist past. This was duly noted by the anti-Nazi Dresden-born author Erich Kästner, now living uneasily in Berlin and daily expecting a knock on his apartment door from the Gestapo. Appalled, Kästner copied Ley's words into his diary:

> After the destruction of beautiful Dresden, we almost breathe a sigh of relief. It is over now. In focusing on our struggle and victory we are no longer distracted by concerns for the monuments of German culture. Onward! . . . Now we march toward the German victory without any superfluous ballast and without the heavy spiritual and material bourgeois baggage.

In these convictions Ley was not alone. Several Gauleiters wanted to rebuild their historic cities in a way that would reflect Nazi ideology and attitudes rather than the "weak" Christian-humanist past; such men saw the destruction of those cities as just such an opportunity.

Bomb-damaged historic buildings considered reflective of the wrong kind of history had already been quietly neglected until they fell down or became irreparable.

In the meantime, however, Goebbels knew how to use what he had from the Anglo-American attacks on Dresden. Within a short while, wild and terrifying estimates of the numbers of dead were in circulation. Neutral newspapers, especially in Switzerland and Sweden, were fed by German diplomats with details of the raid, including a photograph of a child from Dresden bearing terrible "phosphor" burns.

The Germany Foreign Ministry's propaganda operation in the later phase of the war had constantly emphasized the "terror" aspects of the Anglo-American raids, listing architectural and cultural treasures destroyed—and giving energetic publicity to statements by public figures in the Allied camp who were opposed to the bombing of German cities (especially the tiny but vocal London-based "Bombing Restrictions Committee"). The day before the British attack on Dresden, another Foreign Ministry document, "Against the Arch-Enemy of Europe" had arrived at the Berne Embassy. It devoted five closely typed pages entirely to a polemic against Sir Arthur Harris as the chief engineer of cultural destruction, not just in Germany but in Europe:

> The curses of a continent and the hate of generations will be visited upon him, and his benighted name will be remembered only as that of one of the most evil and cruel violent criminals ever to have raged against human life, property and happiness.

When the firestorm raid on Dresden occurred, the German propaganda machine in the neutral countries was primed for attack, and probably all the more effective for the fact that its servants' shock and outrage must have been quite genuine.

On February 16, the Goebbels-controlled German News Bureau (DNB) issued a press release covering the Allied raids on Dresden and Chemnitz. No casualty figures were given, but the case was quickly set out that would form the basis of the German propaganda offensive: Dresden was proclaimed as a city without war industries, a peaceful center of culture, clinics, and hospitals, and scorn was poured on the

Allied claim to be attacking Dresden as a transport center, since the goods stations were "on the periphery of the city." The press release pointed out that before the war the British educated classes had lived, studied, and undergone cures in Dresden, although this had not stayed the RAF's hand:

> It was ... wholly possible for the residential areas of Dresden and its cultural-historical center to have been spared. The use of incendiary bombs shows that the cultural and residential areas were deliberately attacked and destroyed ... The explanation can only be that they desire to obliterate and annihilate the German people and all its remaining possessions.

The accusation was all the more telling for having a core of truth. For years the German propaganda machine had characterized all British and American raids on German towns or cities as "terror attacks," until the phrase had all but lost its meaning. This time, however, the use of the word "terror" had real resonance. It suddenly hit home. Most educated Europeans had heard of Dresden, many had visited the city, and most thought of it exactly as the Goebbels-inspired press release described. By February 19, instructions had also been issued for German missions abroad to produce illustrated leaflets, to be entitled "Dresden, Victim of the Air War," using plates and text to be supplied by the Foreign Ministry.

It was almost three weeks before all the material arrived, including two photographs of horribly injured children. A secret telegram to Berne suggested a new title for a leaflet—"Dresden—Massacre of Refugees"—and also indicated an official interview with the Swedish newspaper *Svenska Dagbladet* (February 25) where the headline said: "Rather 200,000 than 100,000 Victims."

The numbers game had begun.

These figures being quoted were not based on official estimates. These had been published neither in Germany nor abroad, though by the beginning of March—as was clear from the Dresden Police and SS "Final Report"—the authorities were beginning to get an idea of the probable actual casualties: around twenty-five thousand, though no estimates had yet been made public.

All the figures bandied about in the foreign press were therefore a mixture of feverish guesswork, hearsay, and probable manipulation by Goebbels's ministry. On February 17 the Stockholm *Svenska Morgonbladet*, for instance, cited highly inflated figures "privately from Berlin." It said "2.5 million people," had been in Dresden at the time of the raids—four times the city's actual population, implying the presence of almost two million refugees. The deaths and injuries "must run into the hundreds of thousands . . . since all the cinemas, inns and churches were overflowing with refugees, who had not been able to gain access to air raid shelters." The article's fantastic claims reek of manipulation by the Ministry of Propaganda's "spin doctors." Columns of fleeing vehicles had been strafed with such fiendish accuracy that the wrecks of farm carts and military vehicles "blocked the streets," another neutral journalist claimed, making it impossible for aid from other towns and cities to reach Dresden. This dramatic account, it turned out, was based on the testimony of one eyewitness who had reached Berlin some days after the raid.

The initially fantastic estimates of refugee numbers dropped to a more realistic level over the next days, as clearer news from Dresden filtered through to the Berlin, where the journalists were based. Nevertheless, the figure of one hundred thousand to two hundred thousand dead stubbornly continued to appear in the Swedish press. This was noted by the Foreign Ministry and duly used for propaganda purposes. Clearly, such rumors were encouraged—and may even have originated in briefings from the Propaganda Ministry officials who liased with the neutral press and steered the foreign journalists in approved directions.

There is also good reason to believe that later in March, copies of— or at least extracts from—a supplementary Dresden Police and SS "Order of the Day 47" (*Tagesbefehl Nr 47*) which had been circulated a week after the "Final Report"—were leaked to the neutral press by Goebbels's Propaganda Ministry. In this document, however, the revised estimate of dead already recovered in Dresden during the past four weeks (just over 20,204) was doctored with an extra zero to make 202,040. The police's continuing prediction of a total death toll of around twenty-five thousand was also multiplied by ten to give a quarter of a million anticipated dead. Even the number of corpses incinerated in the Altmarkt—6,856—was given an extra zero, taking it up to 68,650.

The German people remained in ignorance. Naturally, rumors

abounded and may have been spread, like those that informed the foreign press, by official sources. The endangered Nazi elite was eager to draw a picture of the evil fate that would await Germany if enemies as cruel as those who had bombed Dresden were to emerge victorious. If the German people could no longer be persuaded to fight for final victory, let them fight out of despair and terror.

"Enjoy the war," went the double-edged graffiti found on ruined walls all over the Reich, "for the peace will be terrible."

In Dresden it must have seemed as if the dead were legion, and legion upon legion. Who would not believe that, between the piles of corpses in the streets and the parks, and the unknown thousands surely burned to ash or melted to viscous puddles in the city's cellars, the Allied raids must have cost the lives of not just tens, but hundreds of thousands? Fritz Kuhn, then a prisoner of war in Russia, remembers his father, who survived the bombs, writing from Dresden of "150,000 dead."

The slot into which Goebbels chose to insert his first nationally published, official statement about the bombing of Dresden was *Das Reich*, the weekly newspaper that he had personally founded, and in which he made many of his most important (or ominous) pronouncements. On March 4, an unusually lengthy article appeared entitled "The Death of Dresden, a Beacon of Resistance." It was powerfully crafted and, for an official regime statement, remarkable in its frank portrayal of what the enemy air forces could do to a German city:

> The three air attacks on Dresden . . . have occasioned the most radical annihilation of a large, continuous urban area and, in relation to the number of inhabitants and of attacks, by far the most serious losses of human life. A city skyline of perfected harmony has been wiped from the European heavens. Tens of thousands who worked and lived beneath its towers have been buried in mass graves, without any attempt at identification being possible . . .

There was no corresponding talk of factories and slave laborers and garrisons and troop trains, or of the secretly declared "defensive area" of Dresden—only of cultural treasures and innocent artistic pleasures, all now lost forever. The article ended with an appeal that exploited the city's dead with a kind of passionate shamelessness:

This is not a campaign for sympathy; we are simply dragging the enemy's methods of war into the light of a fire that he has lit himself. His aim is to force us by mass-murder into surrender, so that he can then carry out a death-sentence on—as the other side expresses it— the surviving remnants. In response to such a threat, there is only the way of fighting resistance. Only the blind cannot see this, and only the weak—who have already given up the struggle—can shrink back from following this way to the end. The blind were led by the sighted out of the burning city, and no one who escaped the flames with the prize of life can seriously consider casting it into the Elbe.

According to testimony at Nuremberg by the chief of the Propaganda Ministry's press division, Hans Fritzsche, Goebbels had already made his own private estimate of about forty thousand dead at Dresden. He informed his underlings of this during the same ministerial conference at which he also announced the soon-abandoned plan to execute equal numbers of Allied prisoners of war in revenge. Goebbels would have known, after almost three years' experience as guiding hand of the Reich Committee for Air Raid Damage, that final casualty figures usually came out at a fraction of the initial estimates. Such wild calculations were often made—and committed to paper— under the direct influence of the shock and horror of the attack, not to mention the observers' understandable dismay at the scale of damage to buildings.

Nevertheless, it was in the interests of the regime to have apocalyptic estimates of casualties at Dresden in circulation. These would modify the attitude of the neutral press in Germany's favor, and perhaps even at this late stage could mold neutral public opinion. As Richard Stokes's speech in the House of Commons showed—partly based on the material put out by Goebbels's own German Press Agency—the British Parliament could also be influenced and even the British prime minister.

The destruction of Dresden was bound to exercise an independent power of its own, one that could not but affect the Allies' claims to absolute moral superiority. However, the extent of the wide, long-lasting ripple of international outrage that followed the Dresden bombing represents, at least in part, Goebbels's final, dark masterpiece.

27
The Final Fury

DRESDEN WAS by no means the last of the major attacks launched by the RAF against German towns and cities in the first months of 1945. In the course of March 1945 Bomber Command dropped more than sixty-seven thousand tons of bombs on Germany. This represented not just the greatest tonnage of any single month between 1939 and 1945, but was only slightly less than *the entire tonnage dropped during the first three years of the war.*

Not all these attacks in 1945 were on urban areas. The transport and oil campaigns continued unabated—and it is now clear from Sir Arthur Harris's dispatches that, reluctant though he may have been, he devoted more effort in these directions during the last months of the war than either his often tactless remarks indicated or public opinion believed. It is, however, undeniable that during this period many of the city raids, like that against Dresden, were especially devastating. The RAF might still experience "cock-ups" and failures, but these were—unfortunately for the Reich's urban population—fewer now, especially as the weather improved and Luftwaffe fighters appeared only sporadically in the skies over Germany. Not that the campaign had become entirely risk-free from the aircrews' point of view—more than four hundred British bombers were lost between February 13, 1945 and the end of the war.

Ten days after Dresden, Bomber Command achieved another "perfect raid" against the town of Pforzheim in southwestern Germany. On that night of February 23–24 1945, with no flak over the town and the aircraft able to mark and bomb from only eight thousand feet, it is estimated that 83 percent of Pforzheim's built-up area was destroyed and 17,600 of the town's inhabitants died—one in four of its population (compared with roughly one in twenty in Dresden). In terms of percentage of the population killed, it was by far the most lethal attack of the war. The town, known as the "Gateway to the Black Forest," made precision instruments for the Wehrmacht, and had some importance as a transport center for the western front.

The same could not be said for Würzburg, which was bombed by the RAF exactly one week later. More than eleven hundred tons of bombs, mostly incendiaries, fell on the compact old city within seventeen minutes. The ancient cathedral and university city on the river Main in northwestern Bavaria seems genuinely to have harbored very little industry. What it did boast, however, was a great many beautiful medieval and rococo buildings, and some unique libraries. Ninety percent of its city area was destroyed, and some 4,000 of its 107,000 inhabitants died. As in Dresden, the public and the emergency services were inexperienced in matters of air warfare, and proper air raid shelters were nonexistent. Victims died in gas-filled cellars or were burned as they tried to reach the river Main.

Würzburg fell to the American Seventh Army less than three weeks later, following six days of fierce fighting amid the city's ruins, which—like the jagged crater landscapes of Stalingrad and Aachen—favored the desperate defenders.

It was not all great slaughter for little gain. Bomber Command completed its unfinished business with Chemnitz and demolished the factories there as planned, wrecking a third of the built-up area in the process. The Ruhr, Germany's last remaining great industrial area, continued its torment, with boys from Dresden manning their flak guns there to the end. Of the Ruhr cities, Essen was devastated for the last time in the second week of March. Dortmund also suffered terribly. Eleven hundred Lancasters, Halifaxes, and Mosquitoes visited the city on March 12, delivering almost five thousand tons of bombs in this single raid. Bomber Command inflicted damage at Dortmund that by postwar British estimates "stopped production so effectively that it would have been many months before any substantial recovery could have occurred."

The air offensive continued to its terrible climax, but in the five or six weeks since the bombing of Dresden, the overall political and strategic situation had changed dramatically.

In mid-February the western Allies were still to some extent recovering from the shock of the Ardennes offensive. They also remained firmly stuck west of the Rhine, as they had been since the previous November. The Russians had begun their offensive into eastern Germany early at the West's request, to reduce the pressure on the Anglo-American forces. In response to this "favor," the western Allies' use of their air power to shatter the German cities behind the Russian front—including Dresden—had been intended to provide demonstrable practical recompense.

Crossings over the Rhine had begun on March 7, and on the twenty-fourth a massed amphibious assault, combined with a parachute drop to secure the key town of Wesel (which had been all but obliterated by Allied air attacks), meant that Allied troops were soon present on the east bank in irresistible numbers.

On March 27 the Germans' final great hope failed. On the evening of that day, at 7:21 P.M., the last V-2 rocket fell on London. Launched from a mobile platform near The Hague, the missile hit a block of flats, Hughes Mansions, in Whitechapel, East London, and killed 134 of the residents. All the victims, give or take a few servicemen on leave, were civilians—and the great majority, as it happened, Jewish. To avoid being overrun by Allied troops, the German rocket launch unit, *Gruppe Nord*, then began withdrawing into Germany with its remaining "miracle weapons." These were never used again, or at least not by representatives of the Third Reich.

The next day, March 28, 1945, Sir Winston Churchill unleashed a thunderbolt in the corridors of Whitehall that, unlike the crossing of the Rhine, remained secret until many years after the end of the war, but was no less telling for that. It came in the form of a memorandum to General Ismay, his chief of staff (Churchill also held the post of minister of defense), which read:

> *28 March 1945. Prime Minister to General Ismay (for Chiefs of Staff Committee) and the Chief of the Air Staff*
> It seems to me that the moment has come when the question of bombing of German cities simply for the sake of increasing the terror, though under other pretexts, should be reviewed. Otherwise we shall come into control of an utterly ruined land. We shall not, for instance, be able to get housing materials out of Germany for our

own needs because some temporary provision would have to be made for the Germans themselves. The destruction of Dresden remains a serious query against the conduct of Allied bombing. I am of the opinion that military objectives must henceforward be more strictly studied in our own interests rather than that of the enemy.

The Foreign Secretary has spoken to me on this subject, and I feel the need for more precise concentration upon military objectives, such as oil and communications behind the immediate battle-zone, rather than on mere acts of terror and wanton destruction, however impressive.

The document caused consternation and confusion among the General Staff—especially in view of Churchill's four years of support for the strategic bombing campaign and, just two months earlier, his personal involvement with, indeed insistence on, the bombing of the eastern German cities including Dresden.

"This was, perhaps," as the official historians of Bomber Command commented with exemplary restraint, "among the least felicitous of the Prime Minister's long series of war-time minutes."

The mention of Dresden shows how, even just a few weeks later, the raid was no longer seen just as an uncommonly effective operation, but had instead come to signify something uncommon of an entirely different and less desirable sort. Perhaps Stokes's barbs in the House of Commons three weeks earlier had penetrated and festered beneath the prime ministerial skin. The mention of the foreign secretary's intervention seems to indicate that the German propaganda campaign on the Dresden issue had met with an uncomfortable degree of success in the neutral countries.

Whether Churchill's concern arose from the horrifying consequences of the Dresden raid alone seems doubtful, but Dresden had already come to symbolize something. As he went over the air war in his mind, it would not have been unnatural for Churchill to have questioned the current bombing policy. With the Allies' situation dramatically improved in the past month, and the defeat of Germany now clearly only weeks away—what was the point of continuing to area-bomb cities? It frankly no longer really mattered how much or how little the Germans could produce in their factories, or how more or less well their telephones worked. They could not, in any case, move their newly produced armaments and equipment to where they needed to be. And as the British and American land forces moved to encircle the Ruhr, a huge chunk of the enemy's productive facilities would soon come under the control of the occupiers—who would also then be obliged to house and feed the newly conquered population.

The logic of the prime minister's intervention was quite clear. It was his way of expressing himself, and the terms of reference he used, that were shocking.

To speak, even in a secret memorandum, of "bombing . . . simply for the sake of increasing the terror, though under other pretexts," when the public stance of the government had always been to deny any such policy, broke all the rules of discretion. It was like a confidential version of the disastrous dispatch by the journalist Howard Cowan, which had caused such dismay back in February. The chiefs of staff, to whom the memorandum was circulated that night, reacted accordingly.

At the Chiefs of Staff Committee the next morning, the counterattack began. This being Whitehall, the first step was a holding action. A "detailed study" would be necessary, it was declared, before the issues raised by Churchill's memorandum could be properly considered. And Sir Arthur Harris would have to be consulted.

Having been shown the memorandum even before the morning meeting, Deputy Chief of Air Staff Bottomley had written to Harris, requesting his opinion. Bottomley did not simply hand over the text of the memorandum. He tactfully paraphrased its contents rather than quoting them directly. The response he received was swift and characteristically aggressive.

Harris said that he considered the allegations of terror bombing an insult both to the Air Ministry's policy and to Bomber Command's methods of pursuing this policy. He insisted that the destruction of the Germans' industrial cities had fatally weakened the enemy's war effort and opened the way for the rapid advances that the Allied land forces were now making. Why should such attacks be abandoned now unless it could be demonstrated that they would neither shorten the war nor preserve the lives of Allied soldiers? Harris consciously echoed the famous words of the nineteenth-century German chancellor Bismarck—who had said he did not consider the Balkans "worth the bones of one Pomeranian grenadier"—saying that he did not consider the whole of the remaining cities of Germany as "worth the bones of one British grenadier."

Heedless of diplomacy, the AOC Bomber Command crashed on. Harris and Churchill resembled lone rhinoceroses rampaging around neighboring parts of the Whitehall jungle, while others fled for cover: Harris continued scathingly:

> The feeling, such as there is, over Dresden could be easily explained by a psychiatrist. It is connected with German bands and Dresden shepherdesses. Actually Dresden was a mass of munitions works, an intact government centre, and a key transportation centre. It is now none of those things.

There was some justice in Harris's observations about Dresden's contribution to the German war effort. In its origins the attack on Dresden was in any case not a pure "terror" raid but an amalgam of military horse trading between the eastern and western Allies and intelligence-led wishful thinking. All the same, the tone of this raw, hasty reply to Bottomley's query was ill chosen—at least if Harris was looking for friends in the government or among the chiefs of staff. Nor did it answer the Churchill memorandum's reasonable questions regarding the final point and consequences of such raids at this stage of the war, or whether attacking in such force and with such a lethal lack of discrimination was any longer a supportable policy. Perhaps Harris's reply could not, given its author's continuing attachment to that policy.

Meanwhile there followed a flurry of meetings of the chiefs of staff—four in twenty-four hours. One of these, at 5:15 P.M. on March 29, was attended by Churchill, who took the chair, a fairly rare event. The topic was coordination of military strategy with the Soviets during the coming offensive, but it is not out of the question that discussions were also had about the bombing memorandum. When the chiefs of staff met again at 11 A.M. the next morning (with chief of Imperial General Staff General Sir Alan Brooke once more in the chair), they noted that "the Prime Minister has instructed that his minute on bombing policy should be withdrawn . . ."

The crisis had been averted. The mercurial Churchill had been coaxed back into line. Perhaps the original memorandum was hypo-critical. It was certainly outrageously revealing of the moral conflicts at the heart of British bombing policy. But then what could be expected from a statesman notoriously prone to self-contradiction; who would at one moment call for more bombers, more attacks on German cities, and then—as happened in the summer of 1943 when he was shown film of the appalling destruction in the Ruhr—weep and plaintively ask his colleagues: "Are we beasts that we should do such things? Are we taking this too far?"

Churchill's memorandum also, perhaps, showed a certain instinc-tual politician's feeling for subliminal changes in the country's mood. Possibly he sensed that, from being war-weary and vengeful, the British people had become—though still eager to see an end to the war—concerned about the things that were being done in their name to gain the final victory. And perhaps he also foresaw that Dresden, whatever the true circumstances surrounding its destruction, would in the future symbolize that change of heart. A "raid too far," as the say-ing would go.

On April 1 the chiefs of staff formally accepted a more anodyne for-mulation of the memorandum. It no longer mentioned terror, or Dresden, though it made essentially the same points. "We must see to it," the note now said, "that our attacks do not do more harm to ourselves in the long run than they do to the enemy's immediate war effort."

On April 4, the chiefs of staff produced a reply, which agreed "at this advanced stage of the war no great or immediate additional advan-tage can be expected from the attack of the remaining industrial cen-ters of Germany." Nevertheless, it did not agree to abandon the princi-ple of area bombing altogether. It might be necessary against towns or troop concentrations behind the enemy front, if resistance stiffened once more. If the Nazi leaders established new centers of power, or there was evidence of concentration of U-boats and their facilities in, say, Kiel, area attacks might be resorted to.

As a communications and transport center behind the enemy front, Dresden would probably have qualified as a target for area bombing, even under the new conditions.

The order was duly issued to Harris on April 6 and permission given for it to be published (which showed a certain keenness to assuage public opinion). Harris, however, insisted that to issue such a declaration would make the enemy's task easier, and might enable him to concentrate his air defenses in areas where tactical attacks were still likely, thus unnecessarily endangering the lives of bomber crews. He may have been right in this, but he probably didn't want to publish it anyway.

On the night of April 14–15, 1945, Bomber Command carried out its last city raid—against Potsdam, the still largely undamaged seat of the Prussian kings. It was actually the first time that Bomber Command's heavy bombers (as opposed to Mosquito raiders) had entered the Berlin Defense Zone since March 1944. The aiming point was the center of Potsdam, specifically the Prussian Guards barracks and the railway facilities, so it was not technically an area raid, though it could be thought to look like one. The bombing destroyed a great deal of the old city, as in Dresden, and, also as in Dresden, caused disproportionate casualties. Five thousand citizens of Potsdam—probably not all civilians, given the nature of the town—died just days before the Russians arrived. Like the inhabitants of the Saxon Residence, the citizens of Frederick the Great's chosen royal seat had grown so used to many alarms for nearby Berlin—but never any bombs—that they failed to take proper precautions.

Two days later American aircraft launched a final, devastating raid on Dresden. By mid-April the line through the city, no matter how imperfect in its operation, was the sole surviving north-south connection in the rapidly shrinking German Reich. The Eighth Air Force was given orders to break that link, thus effectively dividing the rump of Nazi power into two.

The April 17 raid was a big one, the largest of the American attacks and the largest single assault on the city if the two British attacks on the night of the firestorm are reckoned separately. Almost six hundred aircraft took part. Flying Fortresses of the First and Third Air Divisions dropped around fifteen hundred tons of bombs on Dresden. The force took its time; the attack was spread over a period of almost an hour and a half. This was a true precision operation, aiming specifically at railway targets. There was some natural drift, and corresponding damage throughout the largely ruined city, but the rail yards were mostly where the bombs fell—to devastating effect.

Clearly aimed at transport targets used by the German military, this was perhaps, the most straightforwardly justifiable of all the Allied air raids on the city. Something between four hundred and five hundred civilian victims (including foreigners) also paid with their lives, more than died in any other attack except that of February 13–14, 1945. The American aircraft were crammed with up to three tons of bombs; with no need for diversions and spoofs at this late stage in the war, space that had once been needed for extra fuel could now store extra destructive power.

From that day until the end of the war and beyond, Dresden remained completely out of action as a major railway junction. At last, just days before the end, the Eighth Air Force had succeeded in its long-held aim.

The ruins of Dresden fell to the Russians on VE-Day, May 8. For some time now, Saxony's capital had been the only major German city still under Nazi control.

As the last day of the war faded into the first morning of peace, Dresden slipped seamlessly from the hands of one set of totalitarians into the grasp of another.

28

The War Is Over.
Long Live the War.

THE RUMBLE of Russian guns had been audible in the city for days.

The Red Army had advanced rapidly to within a few hours' drive of Dresden as early as February. It had then halted, perhaps concerned about overextended supply lines. In March it became clear that the coming final Soviet offensive across the Oder would be directed principally at Berlin. It was left to the Americans and their British allies to push into Thuringia and Saxony, even though these areas had been confirmed at Yalta as part of the Soviet Zone in the great partition exercise that would follow the German surrender. Add to this the fact that one of the few successful German counteroffensivess of the war's final phase drove the Soviets back near Bautzen, east of Dresden—involving, among other Dresden-based units, four machine-gun battalions motorized in converted buses from the city's transport pool—and you have reason enough why Gauleiter Mutschmann remained in control and his devastated capital under National Socialist rule to the very end.

On April 16 Dresden had finally been officially promoted from "defensive area" to "fortress"—a propaganda device rather than a practical difference. It was announced that anyone hanging white flags from windows or preaching defeatism would be punished by death.

Two days later, the Americans entered Saxony's second city, Leipzig.

Many Dresdeners hoped in vain that the U.S. Army would also take their city before the end of the war. Schörner, the fanatical Nazi general now in charge of the remnants of Army Group Center, had meanwhile abandoned plans to defend Dresden to the death like

Breslau or Berlin. He withdrew his troops south through the city and
into the Bohemian redoubt. Nevertheless, there was some fighting
around the city as the Soviets finally approached. As late as the afternoon of May 7, a unit led by a certain Major Köhler claimed seven
Soviet tanks destroyed at Wilsdruff, just west of the city. That same
day eight Wehrmacht soldiers were executed by a battlefield court-
martial for downing their weapons and declaring their unwillingness
to fight on.

In Dresden itself, the situation had become grotesque—almost
comical, except for the judicial murders. When the news of Hitler's
death reached the city a week earlier, Mutschmann had issued a typically defiant declaration and ordered an indefinite period of full public
mourning for the late Führer. Pastor Hoch, then fifteen, remembers
the bizarre atmosphere of those days:

> Dresden was the only city that experienced eight days of National
> Socialism without Hitler. Here in Dresden the Nazis were still in
> power from 1 May and there was an order that all public buildings . . .
> there were hardly any because they had all been destroyed . . . should
> be draped in black because the Führer had died "at the head of his
> troops." And the boats that still sailed on the Elbe also carried the
> flag—the Nazi flag—at half-mast for eight days. That was unique in
> Germany . . .

On May 8 things got trickier still. The language of flags was both
dangerous and easily expressible:

> The general population was supposed to fly flags at half-mast too.
> Here there were of course all these people who used to hang out
> their swastika flags. Victory over Paris. Victory wherever. But of
> course now they suddenly didn't have any, because they had burned
> them. The Russians were so near that they could have turned up at
> any time.

The day before the Russians arrived, an SS unit was assigned to
blow up the Loschwitz Bridge over the Elbe (along with the main
bridges in the city), but two courageous citizens of Dresden decided
to prevent it. They sneaked out during the night to cut the detonator

wires on the bridge. In the morning when the sappers tried to blow the charges, nothing happened. So the striking structure from the 1890s, nicknamed by the locals the "Blue Wonder" because of its color, was saved. That same day, May 8, 1945, the Red Army finally marched across it into the devastated Saxon capital.

The Soviets met opposition only from a few diehards. In the heart of Dresden, Professor Rainer Fetscher, a distinguished eugenicist, physician, expert in sexual hygiene, and self-proclaimed "bourgeois democrat," hoped to negotiate a peaceful handover of the city. Accompanied by a representative of the Communist underground, he advanced under a white flag toward the Soviet lines but was shot dead while still in no-man's-land. A well-known and respected figure in Dresden whose private practice (after the Nazis dismissed him from the Technical University) had become a meeting point for anti-Nazis of all stripes, Fetscher became a postwar martyr. There is still controversy over whether he was murdered by fanatical SS-men or a trigger-happy Red Army soldier.

Dresden was now at the Soviets' mercy.

On the day after the city's surrender, one woman recalled:

> The Russians went into the houses, began to search them and took whatever they wanted with them, without those affected being able to do anything about it . . . They looted and violated—in the latter case old or young, it mostly didn't matter . . . all these abuses were forbidden by the Russian military authorities, though outrages kept occurring, usually because of drunkenness . . .

Punishment for such crimes seems to have been severe, but enforcement highly sporadic. Where an officer chose to enforce military law, soldiers would be shot on the spot, but all too often he did not, or there was no one in authority around to intervene. Soviet soldiers would operate in bands, seizing women—often in the street—and taking turns raping them, while armed comrades kept watch. These incidents went on for some months, and they were not few in number.

A letter from a Herr G. to the Soviet-installed high burgomaster of the city at the end of May 1945 detailed how his wife and daughter had both been raped by drunken Russian soldiers in their own home—the daughter so many times that she needed hospital treatment.

To judge by previous events here, it is clear that Saturday and Sunday are the most dangerous days, during the time between 10 P.M. and 4 A.M. How is anyone supposed to work hard during the daytime, when he gets no sleep all night? I ask you most politely to send us help here as soon as possible. If nothing can be done, then perhaps the wives and the better daughters of the Nazis should be put at their disposal. They are, after all, clearly identifiable . . .

Statistics are impossible to confirm. The Red Army seems to have kept no records, at least ones that are accessible, and—as Anthony Beevor discovered when he wrote out about the mass rapes in Berlin—many Russians vehemently deny that such things ever happened. One postwar figure put the number of women raped in the entire Soviet Zone of Occupation at half a million—around one woman in thirteen of the population. On a local level, the estimates are usually based on anecdote, though in Pirna near Dresden the municipal medical officer examined just over eight hundred women who had been sexually assaulted by Soviet soldiers, mostly in May 1945. More than one third had been infected with gonorrhea.

The Russian authorities showed some genuine concern for the welfare of the German population. On May 16, 1945, the Red Army released thirty thousand tons of potatoes, ninety-five hundred tons of wheat, and eleven hundred tons of meat and other provisions to cover the Dresdeners' emergency needs. By May 20, hundreds of food stores and bakeries had reopened for business, and a rationing system was initiated to avoid outright starvation. Food and housing problems were, initially at least, no worse for the average German civilian in the east than in the west.

What the inhabitants of Saxony, including Dresden, did *not* get from their Soviet "liberators" was political freedom. The German Communists to whom the Soviets quickly handed day-to-day control in the towns and cities—and soon in the provincial governments of the Soviet Zone—also refused to acknowledge that the comrades of the Red Army could have indulged in any atrocities.

Max Seydewitz was Communist premier of Saxony in the immediate postwar period, later director of Dresden's art collections, and author of a book on the city's destruction. His attitude was typical of the German apparatchiks who ruled what became Russia's westernmost and most rigidly loyal satellite, the "German Democratic Republic." His version of the Soviet conquest of the city tells the story that he and his colleagues wanted the people of Germany to believe, in denial of their own experiences:

> The stunned and silent population, mostly passively confined into its cellars, was saved from hunger by the Soviets, who energetically got to work restoring the shattered supply routes. Through their example they woke Dresden's population from its daze, and at the same time they carried out the first, decisive acts that roused the dead city to a new life . . .

This fairy tale of the Red Army as Dresden's wholly benign liberator and savior was one small but vital part of a rapidly growing tissue of myths, obfuscations, and suppressions, which was soon to make east Germany very different from the west. In the Western-occupied zones, which rapidly progressed to real self-government, there were still restrictions, but at least problems could be talked about, complaints could be made, and grievances could be expressed. In the east, including Dresden, there were soon a host of things that could not be acknowledged or discussed—unless the Communists specifically permitted them—and that situation was to continue until 1989. By that year the people of Dresden had been accustomed to holding their tongues for fifty-six years.

To talk about the atrocities committed by the Red Army during those early months became taboo. To talk about the western Allies' bombing of Dresden was, however, soon be permitted—though when and how would, like everything else, be decided by the Communist authorities.

ON VE-DAY mutual congratulations were exchanged between the various victorious British commanders, including Air Marshal Harris and his superiors. On May 10, in a Special Order of the Day, Harris also sent a heartfelt message to the men of Bomber Command. He allowed himself a few Churchillian flourishes and even a rare touch of sentimentality:

To all of you I would say how proud I am to have served in Bomber Command for 4 ½ years and to have been your Commander-in-Chief through more than three years of your Saga. Your task in the German war is now completed. Famously have you fought. Well have you deserved of your country and your Allies.

This was nobly thought and said, but distinctly ignoble thoughts were rapidly forming in Harris's mind regarding the status of himself and his command in the new postwar world. On May 13 Churchill's VE-Day speech was broadcast on the radio. Harris, who listened to it that afternoon in the company of the American air force commander, General Ira Eaker, was astonished and appalled that in the prime minister's long litany of courage and victory, and his listing of the major campaigns—the Battle of Britain, the Blitz, the work of the Royal and Merchant Navy in the Battle of the Atlantic, the Desert and Mediterranean Wars, and so on—Bomber Command's great, four-year-long battle over the skies of Germany was not directly mentioned, bar an oblique reference to the damage done to Berlin.

This amounted to the prime minister's first public expression of the distancing process that had begun with his memorandum on bombing of March 28, in which Dresden initially featured so prominently. Churchill was already deeply distrustful of Stalin's motives and anxious at the possibility of a continued Soviet advance into western Europe. He may also have been reluctant to say anything to offend the newly conquered German population. It was on its support, and possibly on a rearming of the German military, that any hope of halting such an advance would in part rest.

Following the same train of logic, however, the other major hope for beating back Soviet aggression would have been the power of Bomber Command. If the devastating British raids toward the end of the war—and most especially that against Dresden—were part of just such a strategy of preemptive intimidation against Russia, why such a downbeat attitude toward Bomber Command in victory? With such a thing in mind, would the prime minister and his advisers not instead *play up* the bombers' achievements?

For some months it had also been clear that, while Bomber Command aircrews would be entitled to the service medals of the theaters in which they had participated, the command as a whole was unlikely to get its own campaign medal, despite having clearly played such a major part in the defeat of Germany. Since the ground crew, administrative staff, and even senior staff officers had carried out their duties in the United Kingdom, it was argued in official circles that they were entitled only to the Defense Medal, awarded to all those who had served on the home front. At the end of May Harris wrote a letter to Sinclair, the outgoing minister for air, copying it to his own immediate superior, Portal, and (for information) to Lord Trenchard, the veteran interwar RAF commander with whom Harris regularly corresponded:

> I must tell you as dispassionately as possible that if my Command are to have the Defence Medal and no "campaign" medal in the France-German-Italy-Naval War then I too will have the Defence Medal and no other—*nothing else whatever*, neither decoration, award, rank, preferment or appointment, if any such is contemplated or intended.

To his letter to Trenchard, Harris added a personal postscript: "I started this war as an Air Vice-Marshal. That is my substantive rank now. With that and the 'Defence' medal I shall now leave the Service as soon as I can and return to my country—South Africa. I'm off."

In the royal Birthday Honors, Harris had been coopted as a Grand Commander of the Bath, a more prestigious award than the plain knighthood he had been awarded some years earlier. Harris could not reasonably refuse the honor—it being a gift of the king rather than the government—but wrote to Portal to express his embarrassment and ask that he be spared any repetition. Even the high honor he had been awarded was "dust and ashes to me" because of the perceived slur on the men he had commanded. It now seems likely that Harris's personal feelings were also associated with the often-quoted "mystery" of why—unlike most other senior British commanders—he was never awarded a peerage.

Within weeks Churchill was out of power and a Labor government had been elected. The first honors awards to come up were those in the 1946 New Year's list, where Harris's name was tellingly absent. This has been put down to postwar revulsion against Bomber

Command. The public had undoubtedly been affected by the shocking newsreels that had come out of occupied Germany, showing the devastation and suffering inflicted on the wartime enemy. Even airmen who had now gone in at ground level were forced to admit a certain involuntary frisson of horror

Typically, when taxed by Churchill about Harris's omission from the honors list, the dour Labor prime minister, Clement Attlee, retorted tersely that such lists had to be limited and that Harris's promotion to full marshal of the Royal Air Force on January 1 was "not an inadequate recognition of his service."

Sir Arthur Harris spent his remaining time in the RAF mainly touring Allied and friendly countries as they celebrated the end of the war—including trips as an honored guest in America, Norway (where he was granted a twelve-Spitfire Norwegian air force escort), Sweden, and even Brazil. He was not invited to the grand Allied victory parade in Berlin. Toward the end of August, his impending retirement from the RAF was announced. His successor at Bomber Command would be Vice Marshal Bottomley. Harris was fifty-three years old.

Harris would write his war memoirs with what many considered indecent haste, drawing on the dispatches that he had delivered to the Air Ministry in the autumn of 1945. The latter were, however, not to be published until the 1990s. The official statistics contained in the dispatches were not allowed to be mentioned in his memoirs, which were published in 1947, earning Harris around 10,000 pounds—a very considerable sum at that time. But then the man who had led Bomber Command at its destructive height was not, unlike many of his critics, a rich man.

This all-but-retirement coincided with the end of the war with Japan and the end of the brief supremacy of saturation bombing. Harris had noted earlier that same year that the age of the bomber was almost certainly over. He confided to Churchill just a few days after the devastating attack on Dresden that he saw the bomber's reign as a brief, passing phase, just as that of the battleship had been at the turn of the twentieth century. Rockets, he told the prime minister, were the weapons of the future. Well, there was certainly no need to train rockets and feed them, or keep up their morale, or prevent them from asking questions about what they were doing. Then, in August, came the atom bomb. The American air forces' massive firebombing raids on Tokyo and other centers of population and industry had brought Japan to the brink of defeat. Atom bombs on Hiroshima and Nagasaki tipped the single remaining Axis power into surrender. Where just months earlier it had taken hundreds of bombers to reduce a town or city to rubble, now it took just one.

The now-retired AOC of Bomber Command left for South Africa on February 14, 1946, the day after the anniversary of the bombing of Dresden.

The air marshal was to return frequently, and was an honored guest at the coronation of Queen Elizabeth II. He then worked in America for a while, but for the last quarter-century of his life he settled back in England, in a pleasant riverside residence, the Ferry House at Goring-on-Thames, in Oxfordshire. Fate gave him a postal address easily confused with the late commander of the Luftwaffe.

From this comfortable base, Sir Arthur Harris (Bart) attended Bomber Command reunions, was feted by his former aircrew (the "old lags"), worked on an authorized biography with a former staff officer at High Wycombe, Dudley Saward, and died at the ripe age of ninety-one in 1984.

Sir Arthur was an especially keen and generous supporter of the Boy Scout movement. He mellowed a little, some say, and even conceded that with hindsight the "oil plan" might not have been such a crazy idea, but he never apologized for anything.

ON WEDNESDAY, FEBRUARY 13, 1946, the day before Sir Arthur Harris left for South Africa, the first anniversary ceremonies were held at Dresden.

The twenty memorial ceremonies of various kinds that took place were under the strict guidance of Major Broder of the Soviet Military Administration. A note by one of the major's assistants to Walter Weidauer, the acting Communist mayor of Dresden, imperiously explained Broder's stipulations:

> Anything that makes 13 February appear as a day of mourning is to be avoided. He will conduct discussions about political events for 13 February on 29 January. It is the major's opinion that if a false note is struck when 13 February is commemorated, this could very

easily lead to expressions of anti-Allied opinion. This is to be avoided under all circumstances.

There was to be no critical mention of the role of the western Allies in the bombing. The official line was that the Nazis, and more specifically Mutschmann, were responsible for everything terrible that had happened to Germany and to Dresden. An article in the Soviet-produced German-language "Daily Newspaper for the German Population" entitled "The Scum of Humanity," singled out the Saxon gauleiter in the most damning terms:

> It was he [Mutschmann] who, together with Hitler, turned Dresden into one of Germany's armories, into a powder keg, i.e., a source of reinforcements, which supplied the material for the annihilation of peace-loving peoples . . . He conjured up the unwholesome powers by whose exercise Dresden was destroyed. His playing with fire rebounded—though not directly on Mutschmann, who possessed a personal bunker made of ferro-concrete . . .

There were no official commemorative gatherings during the two years that followed, although when each February 13 came around the Communist-controlled press issued similar statements. The emphasis was on looking forward, on rebuilding, rather than brooding on the catastrophic results of the Hitler regime. One newspaper declared in 1947: "Today it is our sole vow to do all we can to rebuild our Dresden within a peaceful, truly democratic and united Germany."

Two years later the same newspaper named a figure of around thirty-two thousand dead—thirteen thousand buried, five thousand incinerated on the Altmarkt, and fourteen thousand still buried under the ruins. The last projection was to prove false. As for the commission that was supposed to have investigated the matter in 1946 and decided on thirty-five thousand dead, there is—as officials of the present-day City Museum assert—no record of any such commission in the city archives. All the same, there seems little question that no serious political advantage was taken of Dresden's terrible fate for the first four years after the end of the war.

Then in 1948–49 came the Berlin airlift and the definitive splitting of Germany, the Communist coup in neighboring Czechoslovakia (its border just thirty miles from Dresden), and the final descent of the Iron Curtain across the heart of Europe. Soon the Soviet Union had tested its first atomic bomb. The cold war was truly under way, and within a very short time Dresden became one of the most well-placed pawns on its virtual propaganda chessboard. The tone in the commentary of the Communist-controlled press changed dramatically:

> The days of horror that were the 13 and 14 February are at the same time an indictment of the Anglo-American conduct of the war, which through this deed covered itself not with glory but with dishonor . . . the horrific annihilation of Dresden could not be justified by reference to the final defeat of the fascist army, for this army had totally ceased to exist as a serious opponent . . .

That in mid-February 1945 the German army had ceased to exist as a serious opponent for the Allies might have surprised the fanatical German General Schörner, who at that time still had a million men under his command in Army Group Center, with Dresden at their backs and antitank ditches and traps under construction. Equally entitled to claim misrepresentation would have been the 79,000 Russian soldiers who were to die and the more than a quarter of a million who would be wounded in the taking of Berlin ten weeks later—not forgetting the 125,000 Berliners who perished in the street-fighting; or the troops of the western Allies who took three weeks to cross the Rhine in serious numbers in March; or the crews of the four hundred British bombers and scarcely fewer American bombers lost between then and the end of the war.

The distortions increased as the years went on. In 1950 an official committee was formed in East Berlin. Its task was to organize meetings all over the recently founded German Democratic Republic (GDR) to commemorate the 1945 raid on Dresden. The aim now was to oppose "the American war-mongers."

> The National Front of democratic Germany struggles against the destroyers of Dresden, the warmongers of today!
> American bombers destroyed Dresden—with the help of the Soviet Union we are rebuilding it!
> Because we love peace, we hate the American makers of war!

Wild statements and even wilder figures started to be bandied about—and this time the Communist authorities issued no corrections. In Freiberg, west of Dresden, it was declared:

320,000 human beings including 150,000 re-settlers were murdered with bombs, phosphor and sulphur. This was a crime against humanity, committed against defenseless women and children, and we shall never forget that the Americans were responsible . . .

It was variously reported in the tame press that forty-five thousand and one hundred thousand Dresdeners had attended the demonstrations in their home city, at which a leading official of the East German Information Ministry declared that the bombing of Dresden had occurred because "Wall Street wanted to make it impossible for the Soviet Union, its supposed ally, to help the German people after the end of the war."

In June 1950 the Korean War broke out. On February 13, 1951, massive meetings called to commemorate the firestorm anniversary throughout the GDR were told that the current U.S. leader, Harry Truman, had ordered the bombing of Dresden (he had been vice president at the time). Two years later in 1953, when Eisenhower had succeeded Truman in the White House, they were told (by Walter Weidauer) that it was Eisenhower who had been personally responsible. No distortion was too extreme. It was claimed that the attack had been launched "without the knowledge and against the will of the Soviet military leadership." The phrase "Anglo-American Air Gangsters"—Goebbels's invention—started to be used by high East German officials, as did "terror attack." In 1954 the death toll was officially quoted as "hundreds of thousands."

The stories that emerged from the catastrophe were persistent, accepted by many if not all Dresdeners. Some, like the tale that Gauleiter Mutschmann had been informed in advance of the air raid, and that his underlings had been glimpsed removing priceless carpets and furniture from his villa the day before, could be found in almost every bombed German city, with only the name of the Gauleiter changed. In Dresden's case, though, there were extra twists, which lent a frisson of privilege, of specialness, to the city's justified tale of suffering, including accusations that a German-American camera manufacturer, Charles A. Noble, had guided the bombers to their targets in Dresden, and a story that the Americans had come within an ace of testing the atomic bomb on Dresden.*

The general line by the mid-1950s was clear: The British and the Americans had bombed Dresden partly out of brutal spite but also as part of a campaign to destroy the areas of Germany that were designated as part of a future Soviet Zone. This line would remain roughly the same, while casualty figures seesawed up and down, along with the temperature of the cold war, until the 1980s.

* See Appendix C for details of these tales.

29

The Socialist City

BEFORE 1914, even before 1939, the world had flocked to Dresden.

During the brief, violent flowering of Hitler's Greater German Reich, Dresden found itself at the geographic heart of that dream of a forcibly Germanized Mitteleuropa. Enthusiastic plans for the city's expansion were made on that basis. Then a fine city, which had thought itself safe, was devastated beyond imagination. A cultural icon found itself under Russian occupation, no longer at the heart of the great central European cultural and economic community but reduced to the status of a provincial city in a remote corner of a small, defeated satellite country.

The main preoccupation just after war—as elsewhere in Europe but especially in Germany's shattered cities—was bare survival. A roof over one's head. Enough to eat. Dresden's population in August 1945 was down to 360,000, roughly half the peacetime total. A job, any sort of job, that brought in some kind of salary was precious. Things were all the more urgent and difficult if menfolk remained prisoners of war for many years. This last was especially likely for those taken captive by the Russians, who used them as labor for postwar reconstruction in the Soviet Union. But what of reconstruction in Dresden? The first task was to clear ways through the rubble, and then clear the rubble itself. Most main thoroughfares were re-opened in the second half of 1945. Now came the gigantic task of clearing a vast, devastated wilderness that covered eight square miles.

Dresden, like other German cities, created in those grim first postwar years a new breed of hero—or more accurately, because of the man shortage, heroine—the *Trümmerfrau* (rubble woman). These were the women who cleared the worst of the bombed-out shells, dug out the rubble, separated it, and dressed the stones if possible for reuse. The rubble was sent out of the city, first on carts, reminiscent of the same process after the near destruction of Dresden by the Prussians in 1760, and then on a system of narrow-gauge railways, the famous *Trümmerbahnen*, whose tracks began to be laid in 1946.

The first of a number of great mechanized sorting centers was established in the ruins of the Johannstadt, on what had formerly been the leafy expanses of the Dürerplatz. There the rubble would be graded into fine or coarse rubble, or reusable sandstone, and sorted by size. Reusable stone and reusable bricks would be kept back, and the unusable material transported, again by narrow-gauge railway, to various tipping points on the outskirts of the city, where it would be buried. Similar sorting centers were later established elsewhere.

A great long pit, between a mile and quarter and a mile and a half long, and a hundred yards across, curved around between the riverside drive of the Käthe-Kollwitz-Ufer (formerly Hindenburgufer) and the Elbe. It took most of the rubble from the central and eastern ruins of the Altstadt, while that from the west went to make up an artificial mountain of rubble in the Ostragehege, north of Friedrichstadt and close to the original RAF aiming point. Another artificial hill of debris at Dobritz received all that remained and could not be reused from the suburbs of Johannstadt and Striesen. It was hard work, and rations in the early years were meager. A high accident rate in the workings was attributed to widespread physical weakness.

Postwar estimates reckoned 32–40 cubic meters of rubble for each of the 627,000 inhabitants of prewar Dresden (1939)—compared with 16 cubic meters per Berliner, and 21.65 cubic meters for residents of Frankfurt-on-Main. Out of a total of 220,000 dwellings before the war, 90,000 lay in ruins. Only 21 percent of all the dwellings within the city limits remained wholly undamaged. Herbert Conert, one of the prewar planning directors who had been kept on by the new rulers, thought that rebuilding Dresden would take "at least seventy years."

The priority for the new Communist authorities was, Where should the people of Dresden live? As well as the historic Altstadt, entire inner suburbs had been destroyed: Johannstadt, Striesen, a great deal of the Südvorstadt, most of Strehlen, the Seevorstadt, and a great deal of the Neustadt. These were not, as it happened, the strongly working-class areas of Dresden. Those areas lay to the west, north, and northwest of the city—districts such as the western part of Friedrichstadt, Pieschen, Mickten, Trachau, and Cotta. What some have seen as unfairness on the part of the Allies—should not these areas of industrial development and workers' housing have been the very ones to be bombed?—may be true in one way, but the geography of the firestorm does contain one grim coincidence. The areas most thoroughly destroyed in the Allied air raids were also the areas where in the last free elections before the Nazis took power, Hitler's party had received its highest proportions of Dresden's vote: Johannstadt, Seevorstadt, Pirnaische Vorstadt, inner Neustadt, Südvorstadt, west Striesen, and Strehlen (the inner Altstadt hovered on the brink of the highest tercile of Nazi-voting districts).

Among architects and planners, there was much talk, in the immediate postwar period, of finding a "balance" between tradition and modernity in the rebuilding of Dresden. The new Communist rulers had other ideas.

Admirable as the great rubble-clearing exercise might have appeared—and to a great extent was—it served the new regime in more ways than the most obvious. Just as this massive state-run effort got under way, the authorities announced that the preservation and possible rebuilding of privately owned property (including many of the buildings around the Altmarkt that had been declared salvageable and scores of partly damaged Baroque town houses) and of religious buildings (including the eminently restorable seven-hundred-year-old Sophienkirche whose twin spires rose proudly between the bustling Postplatz and the Zwinger pleasure gardens) would have to be paid for from the resources of private owners and the churches respectively.

As for the rubble-clearing exercise itself, this was pushed ahead without much regard to whether the areas concerned could possibly be rebuilt. The excuse was that there would be shortages of new building materials for the foreseeable future, and so recycling was the only option. It was also a fact that the Russians were demanding large quantities of building materials and recovered metals to be sent back to the Soviet Union, including even the melted-down bronze from the statues that had ornamented Dresden's many parks and street fountains.

The shells and facades of medium and seriously damaged buildings began to be dynamited soon after the end of the war, for safety reasons. This was often enough legitimate—many Dresdeners who were in the city at the time will tell stories of pedestrians being crushed by ruins that suddenly tumbled into the street, or of near escapes from such a fate—but the rate of such preemptive destruction soon reached levels that led to protests from the city's architectural and historical advisers. There was not, however, much that they could do. Dresdeners trying to save what could be saved of their beloved city were forced to stand by helplessly as potentially restorable historic buildings were either blown up (without even the chance to capture them in photographs) or made unstable (therefore dangerous, and therefore liable to be demolished) by the careless use of explosives on neighboring ruins.

This by no means benign neglect on the part of the Communist authorities was only to some extent the result of postwar lack of resources. The new regime's policy soon became clear. A small number of high-profile architectural gems—the royal castle, the Hofkirche, the opera house, the Albertinum, and others close to the Elbe, were to be rebuilt in the long term. The rest of the center of Dresden was to be given over to the "democratic home building" program beloved of Walter Weidauer and the rest of the comrades.

The result was, finally, the destruction through neglect of what was left of bourgeois Dresden. The big royal palaces, ironically, were preserved—the centers of "parasitism"—but what was left of the many square miles of fine housing and public architecture that had been, in many ways, the city's true glory, was allowed to disappear. Dresden would become, as the official party line expressed it, "a socialist city." From the center to the outskirts, Dresden became a city of broad highways, flanked by row after row of rectangular, uniform cement-block apartment houses (Plattenbauten—slab buildings—as the East German slang word expresses it). This was the "socialist city."

By final official Ukase of 1962, the Frauenkirche was left as a ruin, to remind everyone of the evils of war (though also perhaps because the rebuilding would have been ruinously expensive). Otherwise, by the 1980s, the Elbe skyline had largely been restored. The State Theater, the Brühl Terrace with the Academy of Arts, the opera house, the Albertinum, the New Town Hall, the Hofkirche had been rebuilt and—last of all, for some in the party had opposed its restoration on ideological grounds—the massive work of re-creating the Saxon royal *Schloss* had begun. It is still not quite finished.

As for the dreary rest of the inner city and its near suburbs, nothing much had changed since Kurt Vonnegut revisited Dresden in 1965. Here was a man who feasted his eyes on old Dresden in 1945, during the last weeks before its destruction. "Oz," he had observed simply—like the fairy-tale residence of the wizard in the L. Frank Baum fable and the Judy Garland film. Twenty years later, the author of *Slaughterhouse-Five* described what had once been the renowned "German Florence" as looking like Dayton, Ohio.

The Elbe was still unchangingly beautiful; the outer reaches of the city, with their nineteenth century villas, were more or less undamaged, though subjected to long-term neglect. The wider physical setting of Dresden remained unmatched in its attractions.

All the same, little survived of what had once made this a pearl among cities, except its people. The easy harmony of scale and proportion—humanism realized in warm, vulnerable sandstone—was mostly gone, replaced by Stalinist gigantism and post-Stalinist communal living.

Part of the improvement that Weidauer and the other apparatchiks claimed was, to an extent, justified. Their assertion that this "modern" kind of housing was, for most ordinary Dresdeners, an improvement on the picturesque rookeries that had once seethed damply behind the elegant facades of the Altstadt, contained truth, but it ignored much of what makes life worth living. Compare the postwar planners in London, who tore down the neighborly slums of the East End and replaced them with tower blocks that brought hygienic living and efficient use of space, but in the wake of these benefits also isolation, petty crime, and social breakdown.

Despite the rhapsodizing in Weidauer's and Seydewitz's books about its lost beauty, the Communist planners after 1945 distrusted old Dresden—for much the same reason that radical Nazis such as Ley declared their crazed, defiant "good riddance" when it was destroyed. To the Nazis, old Dresden represented a tolerance and decency they despised; to the Communists, a "bourgeois" individualism and attachment to style that had no part in the productive proletarian future.

Squeezed into the bottom southeast corner of the walled-off

GDR, the lie of the land around Dresden also meant that the city was one of the few areas of East Germany that could not receive Western television. Most East Germans were loyal citizens of the "Workers' and Peasants' State" by day, but by night avid watchers of West German entertainment shows, dramas, and—ominously for the regime—news programs. Not so in the case of Dresden, sunk unreachably in the deep bowl of the Elbe valley, where signals from Western stations could not penetrate. The area was mockingly dubbed the "Valley of the Clueless" (Tal der Ahnungslosen) because its people always seemed the last to know what was happening in the outside world. Dresdeners could, of course, receive Western radio—but the joke had some point. If every picture tells a story, no pictures meant serious gaps in the city's common narrative.

During the forty-five years between the end of the war and the fall of the Berlin Wall, Dresden became isolated. Already terribly traumatized by the events of February 1945, the people of the city threw themselves into building homes to replace the tens of thousands lost, and into creating some semblance of a revived culture. But beneath the surface of the happy, shining new socialist city, unexpressed pain festered; rumors and fantasies bred in the cramped darkness of Dresden's collective memory.

30

The Sleep of Reason

"TO BURN AND DESTROY an enemy industrial center," read 5 Group's orders on February 13, 1945.

The brutally utilitarian description of Dresden willfully ignored the complicated nature of the target as well as the true aim of the raid, but at the same time was a horribly accurate summary of what happened on the ground. Dresden was indeed burned and destroyed, in the space of a night. Some of the city's industry was consigned to the flames with it. As were between twenty-five thousand and forty thousand human beings, an architectural heritage created over centuries, and a treasured, enviable way of life.

In 1942, almost exactly three years before the destruction of Dresden, the first area bombing directive had allowed bomber formations to attack housing and morale as well as obvious military and industrial targets. This had been a sign not of strength but of weakness. It had been an admission that accuracy by night was so poor that precision targets simply could not realistically be attacked. The choice had been between bombing urban areas in Germany by the square mile or not bombing them at all; and for a British military elite under huge pressure to "do something" against the Nazis, it was no choice at all. Lübeck and Rostock had been the immediate result. Then came the Ruhr, Hamburg, Berlin . . . And the German casualties mounted.

This British determination to strike back ruthlessly reflected the mentality of an island under siege, conducting forays against the enemy from within its sea moat. The Germans might have seized control of the continent, but they would not be safe in their dominance. Moreover, the average German, it was thought, would not—and should not—be safe either. In 1870–71 and in 1914–18 the Reich's civilians (with the exception of those in East Prussia during the first weeks of the First World War) had looked on as their armies fought over other peoples' lands. As late as 1943 Goebbels had claimed with satisfaction that the extent of German conquests put the Reich in the position of being able to wage war—permanently as he then saw it— thousands of miles from its own borders. As far as land operations were concerned, this was true, but it did not reckon with Anglo-American air power. The RAF's bombing raids provided encouragement to the beleaguered British population and reminded the millions in Germany who had voted for and supported Hitler, and the millions more who tolerated him as long as he remained successful, that war always came at a cost.

As a leaflet dropped in many thousands over the Reich reminded the Germans: "Europe is a fortress. But it is a fortress without a roof."

The British had discovered the firestorm at Hamburg, and had decided it was worth trying for that impressive phenomenon every time they attacked a city. Why not?

None of this is to say that area-bombing raids were all that Bomber Command could do. Their work against specialist German targets— above all the artificial oil plants—was efficient. The same went for attacks on rail and other transportation targets. Harris himself was amazed at how accurate his aircrew proved when assigned to daylight tactical bombing after D-Day (though it must remembered that the clear, long summer days were the best time for such attacks). By the autumn of 1944, when he was able to return Bomber Command to the task of bombing Germany, an Allied task force was firmly established on the continent, and Britain was no longer an island power at bay. Technology had improved. H2S and Gee, and all the other technical miracles had slowly increased the accuracy of the RAF's bombing, even at night and even in poor weather.

However, ineffective raids such as the near-fiasco at Chemnitz, less than twenty-four hours after the destruction of Dresden, also showed how easily things could go wrong. Even a short while before the end of the war, the Americans, though bombing in the daytime, were also still having problems with living up to their promise of "precision bombing" unless conditions were absolutely clear. Between September 1944 and April 1945, only 30 percent of bombs dropped by the

Eighth Air Force were aimed visually, while 70 percent were dropped blind using H2X. Of that 70 percent dropped by radar-aided means, it was estimated that only 2 *percent* fell within one thousand feet of their aiming points. However, what were the western powers supposed to do? Ground their entire air forces except when conditions (in northwest Europe, in winter) guaranteed an accurate visual fix on the target? They could have. But they did not.

Respected historians of the air war, most prominently Anthony Verrier and Max Hastings, believe that after the summer of 1944 the RAF could and should have stopped bombing cities and concentrated on precision attacks. By not doing so, in fact by stepping up the bombing of Germany's urban populations, the British, and to some extent the Americans, lost the moral high ground. This may be true, and since the end of the war the priorities of Bomber Command have been subjected to intense examination and widespread criticism.

All the same, there are a number of problems with this belief, some political and some practical. It is hard to believe that in the winter of 1944–45 a voluntary Anglo-American withdrawal from city bombing would have been acceptable to Allied public opinion—especially after German resistance stiffened in the autumn, to be followed in December by Hitler's counterattack in the Ardennes, which cost tens of thousands of Allied (mostly American) casualties. And then there were the V-1 and V-2 raids on Antwerp, Paris, London and southern England, which cost thousands of civilian lives and were terrifyingly indiscriminate by their very nature. Not every English voter would have approved of the MPs who were constantly leaping to their feet in Parliament to demand that this or that undamaged German city be bombed with all urgency, but firm opposition to city bombing was confined to a minority, even late in the war. It was generally perceived that the Germans had "asked for it," that they had "sown the wind and must reap the whirlwind."

As for the proposition that such devastation came to be inflicted when the war was "almost over," it must be said that no one knew when the war was going to end. The fact that Germany insisted on prolonging the battle long after defeat was inevitable (and therefore continued to bring retribution upon itself) if anything hardened the hearts of a war-weary, embittered Allied public.

Attitudes would, as we know, change, but not yet. "Bombing," as

Professor Richard Overy has commented, "provoked a real heart-searching only once the conflict was over."

On a practical level, aircrew had been trained, at enormous expense, to carry out the big "city-busting" raids. The huge bureaucracy of planners and administrators had been trained to support and guide them in their task. Area bombing had become a habit, even an addiction. The RAF knew how to do it. And it guaranteed a result. As the Anglo-German writer W. G. Sebald put it in his recently printed lecture on the German experience of the air war:

> An enterprise of the material and organisational dimensions of the bombing offensive . . . had such a momentum of its own that short-term corrections of course and restrictions were more or less ruled out, especially at a time when, after three years of the intensive expansion of factories and production plants, that enterprise had reached the peak of its development, in other words its maximum destructive capacity. Once the material was manufactured, simply letting the aircraft and their valuable freight stand idle on the airfields of eastern England ran counter to any healthy economic instinct.

It would be nice to think that the deadly course of the bomber war was pursued as the result of continual moral struggle and reevaluation. Nice, but wrong. Sebald's explanation is frankly more plausible.

It was a matter of what worked (or was perceived at the time to work). The British Air Staff had originally learned lessons from its careful analysis of the German attack on Coventry in November 1940. The longest-lasting damage, and the most crucial, had been the harm done by the Luftwaffe to the city's sewage, power, and communications services. Factories can be moved, or put underground, or rebuilt fairly easy, but the complex and often convoluted systems for living that are developed in a city over decades can be far harder to put back together. As Goebbels observed in his diary after a meeting with Speer in April 1943, "Industrial damage can be dealt with far more easily than damage to private homes." Hence the logic of concentrating the bombing not on the outskirts of Dresden, and the *Autobahn* and all the bridges, but the heart of the city, where housing is dense and everything comes together. And that was exactly what 5 Group of RAF Bomber Command were good at. Burning and destroying "centers."

In that sense, the attack on Dresden was routine. Many, perhaps the majority of Dresdeners who lived through the firestorm, believe that the attack on their city was somehow special in its malevolence, cunning, and destructive intent. So how truly exceptional was the bombing of Dresden? In its effects, certainly. A fine city filled with beautiful buildings was destroyed in the course of one terrible night and a morning. Tens of thousands of noncombatants died or suffered horrific injuries. The attack was planned with malevolent skill, and carried out efficiently. But in its conception? Perhaps not. The author of one of the authoritative German histories of the bomber war, Olaf Groehler—himself from Leipzig—comes closest to the truth when summarizing the escalation of bombing during the final months of the war:

> With respect to the air war, the bombing of Dresden stands out from the sequence of continuous, heavy air raids because of its destructive scale. Nourished by rumors and legends, it spread like a shockwave all over Germany. The destructive effect of the attack on Dresden shattered the mold of what had become customary, of all previous experience. But if one analyzes the planning documents for the city attacks undertaken in early 1945, one must recognize that in many cases these resembled the style of attack used against Dresden, right down into the details. Often it was only the favorable or unfavorable weather situation, or the way the city was built (including its shelters), or how much experience the population had gained during the war years, that determined the final extent of obliteration, destruction, and death.

In other words, in practical terms Dresden was one heavy raid among a whole, deadly sequence of massive raids, but for various unpredictable reasons—wind, weather, lack of defenses and above all shocking deficiencies in air raid protection for the general population—it suffered the worst. Even such a statement may not be entirely correct. Darmstadt, Kassel, Pforzheim, and Würzburg were smaller, but arguably they suffered no less. Proportionately Pforzheim suffered much more, losing a sixth of its population—17,600 human beings—

and an estimated 83 percent of its built-up area on the night of February 23–24, 1945. The bomber force that did this to the town at the gates of the Black Forest was less than half the size of that involved in the "double-punch" attack on Dresden ten days previously.

Groehler's remarks about the routine quality of the Dresden operation also leads to thoughts about one of the other postwar quandaries about the raid: that under the guise of aiding the Red Army on the eastern front, it was deliberately planned to intimidate (or deter) the advancing Russians. There is indeed a puzzle here, for in his magisterial work *The Bomber War Against Germany*, where he makes the remarks quoted above, at the same time the distinguished East German historian devotes a sizable excursus to asserting exactly that. In other words, the operation was routine, and proved awesomely destructive largely by chance—and yet it was also carefully planned (as Groehler put it) to "demonstrate the strength of British-American weaponry to the Soviet Union."

The chief piece of evidence here is an internally circulated summary of the work of Air Intelligence from 1945, which was loaned to the author Max Hastings by a former officer at Bomber Command's headquarters in High Wycombe. It contains a copy of briefing notes sent out to squadrons and wings before the Dresden raid and reads as follows:

> Dresden, the seventh largest city in Germany and not much smaller than Manchester, is also [by] far the largest unbombed built-up area the enemy has got. In the midst of winter with refugees pouring westwards and troops to be rested, roofs are at a premium, not only to give shelter to workers, refugees and troops alike, but also to house the administrative services displaced from other areas. At one time well known for its china, Dresden has developed into an industrial city of first-class importance, and like any large city with its multiplicity of telephone and rail facilities, is of major value for controlling the defense of that part of the front now threatened by Marshal Koniev's breakthrough.
>
> The intentions of the attack are to hit the enemy where he will feel it most, behind an already partially collapsed front, to prevent the use of the city in the way of further advance, and incidentally to show the Russians when they arrive what Bomber Command can do.

Of itself, the little sting in the tail of this document does not necessarily imply that the official plan was to intimidate the Russians, and Hastings does not dwell on this. It could be interpreted as a kind of "live" version of the somewhat macabre "blue books," containing photographs of bomb damage in German cities, that Harris frequently sent to Moscow throughout the war to "show what Bomber Command can do," and which Stalin is said to have examined with satisfaction.

More serious are the remarks made by the British master bombers for the two attacks on Dresden, Peter de Wesselow and Maurice Smith, to the author Alexander McKee in the 1970s. Groehler reports that de Wesselow said that the attack should "impress the Soviet Army with the power of Bomber Command" and in Smith's case that "the destruction of an untouched city of this sort would make a significant effect on the Russians." Both these are accurate as far as they go, except that Groehler does not mention that Smith went on to add, slightly more enigmatically: "We had the impression that the Russians discounted what Bomber Command could do. Although we did not regard them like the other Allies, we respected the Russian army and wanted to help them against Hitler." Likewise, the de Wesselow quotation is diluted if one quotes the entire sentence, which includes important qualifying words. "I *think* we knew," de Wesselow recalled, "and were *probably* told that it was to help, and still more impress, the Russians with the power of Bomber Command." More than thirty years later, who can be sure that what they *think* they were told was colored by subsequent reading and the power of postwar controversy?

Perhaps the content of the more thorough briefings given to key figures such as Smith and de Wesselow was closer to that outlined in the document unearthed by Hastings. They were close enough to the top to receive tidbits of gossip (Smith does mention hearing that the "operation was of special interest to Churchill"). As far as ordinary aircrew are concerned, though some express concern about being sent in to bomb cities containing refugees, none recalled to this author any hint that the raid on Dresden might have the purpose of "impressing" the Russians in the sense of intimidating them (which is the only way in which Groehler's assertion makes sense). At this stage in the war, leftist feeling was widespread (1945 was to see the election of a Labor government with a huge majority). For many in the armed forces as well as the civilian population Soviet Russia was still a much-admired ally and Stalin viewed in the guise of avuncular, pipe-smoking "Uncle Joe." Any insertion even of an oblique anti-Russian element into a routine attack briefing might have created angry bewilderment rather than additional motivation.

Groehler also does not confine the accusation to the Dresden raid alone. Read carefully, what he says is that such a putative demonstration of power involved "the reduction to ashes of Dresden *and other cities*." Here he becomes vaguer and at the same time more plausible. Dresden was a big raid, but no bigger than a considerable number of others at that time directed against the urban areas of Germany. It is not impossible to believe that the crescendo of bombing over the period around Yalta might have had a dual purpose—hastening Germany's end while at the same time quietly hoping to deter the Soviets from continuing their advance to the west once the war was won. In general, this is not an impossible mix of motives.

It is, however, in this case a mystery why Churchill should get cold feet about the same "terror bombing" in his notorious memorandum at the end of March 1945, when one would have though that the potential Soviet threat (which it was allegedly intended to deter) had in the meantime increased rather than diminished. The material so far available is open to differing interpretations and is in any case circumstantial. Even a clear hint in an official document of the time would help, but none seems to have come to light.

Whatever the truth—and without that clear evidence an open verdict still seems the most plausible—Herr Groehler's assertions certainly fit the anti-Western line in the German Democratic Republic. It was during the regime's final years that Herr Groehler wrote his otherwise brilliant book (though it was published a few months after the fall of the Berlin Wall).

In any case, the escalation of bombing was not confined to the big cities. Other appalling massacres seem scarcely to be noticed in the general historical record (even Pforzheim merits mention by the official history of Bomber Command only as a footnote, and then merely as one of a list of targets). And then, for instance, there is the daylight attack mounted on March 12 by the American air force against the

German port of Swinemünde on the Baltic coast. This again occurred at the request of the Russians, who had advanced to within fifteen to twenty miles of the town. Conditions were poor, so H2X (radar bombing) was used. Many German naval craft and port installations were thought to have been destroyed, but the raid—if local German accounts are to be believed—also killed up to 23,000 refugees from the East who were waiting in mile-long lines for boats to take them across the bay to the western shore. The actual casualties were almost certainly lower, but according to a modern German writer still ran into five figures. "It brought," as the official American history comments, "an exciting variation from the normal routine." The official USAAF report classified Swinemünde as a "transportation target."

Ultimately, Dresden must be placed alongside the other terrible raids carried out on Germany in the last two years of the Second World War, and especially in the last months, and regarded in that light: as, largely accidentally, the most devastating and horrifying. When moral questions are asked, however, it seems hard to justify seeing Dresden as a special case that exists in a dimension of its own. Dresden had beautiful buildings. So did many other German towns and cities. Jörg Friedrich's book contains a solemn inventory of the churches, palaces, historic houses, libraries, and museums consigned to the flames by the Allied bombers, from Goethe's house in Frankfurt to the bones of Charlemagne from Aachen cathedral; from the irreplaceable contents of the four-hundred-year-old State Library in Munich to the rococo glories of the archiepiscopal seat of Würzburg, a city that was its own work of art.

Bombing did not (could not) distinguish between the two countries that were Germany—the proud, ancient, humanist Germany that had rooted itself and flourished over the centuries, despite all the worst efforts of its overambitious rulers and greedy neighbors; and the violent, aggressive Germany that the Nazi hijackers of those traditions had ruthlessly created after 1933.

Martin Mutschmann deserted his capital, Dresden, at some time during the twenty-four hours preceding VE-Day. He fled to the still-unoccupied Erzgebirge, to the house of an acquaintance, where he was arrested either by local Communists or by Soviet troops on May 10. The fallen Gauleiter was later subjected to intensive questioning:

INTERROGATOR: What do you have to say about the air attacks on Dresden?

MUTSCHMANN: It's terrible, the quantity of valuables that were destroyed in one night. Dresden was city infinitely rich in artistic treasures and many other things. Now almost all of that is *kaput*.

INTERROGATOR: So you're not at all concerned about the human victims? It seems you think only in terms of material valuables?

MUTSCHMANN: Of course, a very great number of human beings also died. But I just meant that artistic treasures can't be replaced.

INTERROGATOR: How could such great losses of human life come about?

MUTSCHMANN: Dresden was not sufficiently prepared for an air attack. It's true I tried to do something about building shelters, but I got nothing from the higher-ups, no labour and no materials, cement etc. People criticised me because I had shelters built at my house in the city and on my estate in Grillenberg. But these were purely private projects, whose construction I was able to finance from my private means. A shelter-building program for the entire city was not carried out. True, I had to reckon with a large-scale raid against Dresden, but nevertheless I kept hoping that nothing would happen to Dresden.

The mix of motives, the moral obtuseness, and the ragbag of excuses express perfectly the mentality of the philistine functionary, taking refuge in an artistic heritage he neither understood nor knew how to preserve. This is the authentic voice of the regime, taking responsibility for nothing, fleeing into a strident ignorance when taxed with his part in the destruction of a city and a country that deserved so much better than he and his like were willing to provide. Dresden would have been saved for all humanity in the centuries to come but for the brutal dreams of conquest, enslavement, and genocide that Mutschmann and his like harbored almost to the end.

Mutschmann's precise final fate is unknown. He is variously described as dying of maltreatment in Dresden in 1948 (when he would have been nearly seventy years old) or being shot in the Moscow Lubyanka by the Russian secret police sometime before 1950.

When the first Soviet troops pushed across the German border into East Prussia toward the end of 1944, looting and raping and burning as they advanced, they were, as Anthony Beevor records, "disgusted by the plenty" they found everywhere, in town and country alike. The neat houses, the evidence of comfort and material well-being at every turn, served only to enrage them further. A Red Army sapper said to his superior:

> How should one treat them, Comrade Captain? Just think of it. They were well off, well fed, and had livestock, vegetable gardens and apple trees. And they invaded us. They went as far as my *oblast* of Voronezh. For this, Comrade Captain, we should strangle them.

This is perhaps the great, still-unanswered question about Germany and the German people between 1933 and 1945. With the vast material and spiritual riches of places like Dresden at your disposal, why place all that at risk by launching a ruthless, in large part genocidal attack on the rest of Europe? Whatever the Nazi ideologues might say, Germany did not lack *Lebensraum*. Did anyone really expect the world to fight back while wearing kid gloves, in order not to damage Germany's artistic treasures or kill German civilians?

Those who concentrate on the war in western Europe miss the sheer massive scale of the mass slaughter inflicted on the civilian population in European Russia by the invading Germans. The vastness of such a mosaic of violence makes it hard to grasp: Because of this we seek instinctively for the personal, the particular, the apparently clear-cut case. It is rarely mentioned that almost exactly the same number of Soviet citizens died as a result of bombing during the Second World War as Germans: around half a million. Why are there no shelves of books emotively recalling the fate of the forty thousand human beings—many of them women and children and refugees—who died in the Luftwaffe's systematic bombing of Stalingrad, which began with a thousand-bomber raid and lasted over four days in August 1942, even before the siege had begun? Or in the bombing of Minsk, which included the central hospital? Was it morally right for eight hundred thousand Russians, again mostly civilians, to die by bombing, shelling, and starvation in the German siege of Leningrad? The conventions of war allow almost any tactic of destruction against a defended fortress town and the people within it once it has refused to surrender. But is such a thing, on such a scale, more or less *moral* compared with the bombing of Dresden?

By the time Dresden was destroyed, Hitler—who would order the blinds lowered in his railway car when passing through a bomb-devastated city—was blindingly, nihilistically clear in his judgment about the imminent fate of the Nazi project and what it would mean for the German nation. In mid-March 1945, the rapidly ailing Führer told Speer:

> If the war is lost, the people will be lost also. It is not necessary to worry about what the German people will need for elemental survival. On the contrary, it is best for us to destroy even those things. For the nation has proved to be the weaker, and the future belongs solely to the stronger eastern nation. In any case only those who are inferior will remain after this struggle, for the good will already have been killed.

Hitler spoke with crude, Darwinian contempt of a German nation whose majority had trusted him to restore order and prosperity, had later followed him into a world war, then finally given their all to him at the front, in the bombed cities and in the factories. Misguided, perhaps, but brave and industrious beyond question. Even the Allied bombers had not totally crushed the Germans' will. Weakened, exhausted, most of them perhaps secretly hoping for the end—any end—to the war, they had kept working and fighting almost to the final day.

Nonetheless, after Dresden the end came relatively quickly, sooner than the Joint Intelligence Committee's forecasts predicted. Günter Jäckel had felt a new sense of chaos, of panic, as he was being evacuated into the mountains after surviving the bombing of Dresden, and was clear that the raid on his hometown was responsible for this shift in mood. The horrors of the bombing, and Goebbels's propaganda, may have raised anti-Allied feeling to new heights, but at the same time there was an unignorable sense of disappearing possibility. Germany could not fight back, or defend itself anymore. The end of the Third Reich would certainly come. The only question was, would it come without too many more horrors, or only after the fate of Dresden had been visited on other cities again and again by the huge fleets of Allied aircraft that could now bomb Germany at will?

Götz Bergander is skeptical that the destruction of Dresden short-
ened the war by much, but nevertheless:

> It is true that most Germans no longer believed in victory, but nev-
> ertheless they could not imagine unconditional surrender. The
> shock of Dresden contributed in an fundamental way to a change of
> heart. This expressed itself at that time in the words: Better an end
> to terror than terror without end.
>   "Terror without end"—that was for most Germans the bomb-
> ing war.

It became fashionable among writers in the postwar period to dis-
miss city bombing, not only as immoral but also as essentially useless.
There seems, however, little doubt that the strategic bombing campaign played a major role in the defeat of Germany (if not perhaps the
"knockout" one that Sir Arthur Harris and his supporters dreamed
of), and growing evidence that it may even have proved decisive. Early
postwar surveys made the mistake of confining cost-benefit analysis to
a kind of simple accounting of notionally lost German production.
Especially when Speer took over the government's war industries
portfolio and introduced long-overdue efficiency measures (aided by
the growing political trend toward a "total war" ideology among more
radical Nazi leaders such as Goebbels and Ley), German armaments
production continued to increase. This trend continued until the end
of 1944, and it was therefore assumed that Allied bombing had been
almost entirely ineffective.

More recent studies, especially those of Professor Richard Overy,
have taken a broader view and also included the massive financial and
material costs involved for the Reich in creating a complex and sophisticated aircraft tracking and air defense system, in rebuilding and relo-
cating industrial and military installations, and in feeding, housing,
and caring for victims of the escalating Allied bombing. This not only
took weapons and equipment from the frontline land troops, but also
vastly reduced the number of offensive aircraft available on all fronts,
especially in Russia. Moreover, while the ever-aggressive Hitler
demanded more bombers, the constant need for night and day fighters
to keep the Anglo-American bomber fleets away from German cities
and factories meant that fighters were always given priority over a new

generation of long-distance bombers, which might have enabled the Luftwaffe to take the fight to the enemy. From 1943 Germany was always, as the sporting metaphor goes, "on the back foot" as a result of the strategic bombing campaign.

At the beginning of January 1945 Albert Speer and other leading officials met and summarized the effect of relentless Allied bombing on production during 1944. Germany, they calculated, had produced 35 percent fewer tanks, 31 percent fewer aircraft, and 42 percent fewer trucks than planned. All this was due to intensive Allied bombing of the Reich's industrial centers—which even in cases defined as "precision" would have caused "spillage" (the World War II American euphemism equivalent to the modern "collateral damage") and in others would have been a by-product of area bombing, where civilian casualties were ruthlessly factored in.

On the last day of January 1945 (coincidentally the twelfth anniversary of the Führer's accession to power), Speer sat down and wrote a memorandum to Hitler in which the armaments minister frankly admitted defeat in the struggle to continue supplying German armed forces. "Realistically," he wrote later, "I declared that the war was over in this area of heavy industry and armaments . . ."

Overy continues:

> The indirect effects were more important still, for the bombing offensive forced the German economy to switch very large resources away from equipment for the fighting fronts, using them instead to combat the bombing threat. By 1944 one-third of all German artillery production consisted of anti-aircraft guns; the anti-aircraft effort absorbed 20 percent of all ammunition produced, one-third of the output of the optical industry, and between half and two-thirds of the production of radar and signals equipment. As a result of this diversion, the German army and navy were desperately short of essential radar and signals equipment for other tasks. The bombing also ate into Germany's scarce manpower: by 1944 an estimated two million Germans were engaged in anti-aircraft defense, in repairing shattered factories and in generally cleaning up the destruction.

Significantly, Overy highlights the "high tech" aspect of production shortages, especially the optical and electrical/communications industries. This was precisely Dresden's specialty. During 1944 and early 1945 production was actually stepped up, and plants and labor forces (especially slave workers) moved to Saxony in general and Dresden in particular. According to figures supplied by armaments inspectorates all over the Reich, in the autumn of 1944 the Dresden Military District was the most popular destination for such industrial dispersals, presumably because of its distance from either front and its perceived relative safety from air attack. With Upper Silesia falling into Russian hands, and the Ruhr under constant and withering bombardment, at the beginning of February Saxony was one of the few areas where production remained relatively unhindered and even increased.

There is still an enormous amount of detailed research to be done on the more obscure aspects of the German war effort, which one suspects could be immensely revealing. An interesting footnote from around the time of the heavy Allied bombing of Dresden sheds a little light. Five days before the raid, on February 8, 1945, Radio-Mende of Dresden wrote to the local energy board. The letter requested priority in the provision of electricity for a new factory, since the company had transferred some of its production to a surprising secret location in the vicinity:

> We are maintaining a dispersed production facility in the premises of the State Porcelain Manufactory, with the address:
>
> State Porcelain Manufactory
> Section "Scharf"
> Meissen on the Elbe.
>
> This production serves provision for the frontline and falls within the stipulations of the emergency armaments program . . .

By this stage in the war, some far from romantic items were being produced under high security at the historic porcelain factory. "Dresden" china, later much cited as evidence of the peaceful nature of the city's industry even in wartime, actually came from Meissen, twelve miles downstream, though there were porcelain-painting studios in Dresden. But now even in Meissen, in the workshops where once shepherdess figurines had been crafted, they were making key items of modern communications equipment—telex terminals

(*Fernschreiber*) for the Wehrmacht. The letter from Radio-Mende, uncovered by a sharp-eyed local historian, shows not just how efficiently the dispersal of vital war industries was continuing, but how cunningly, or cynically, it was pursued. Had the Allies (legitimately) bombed the former Royal Manufactory, and inevitably along with it the picturesque and ancient town of Meissen, it is not hard to imagine the outcry that would have erupted—as the German industrial bureaucrats were perfectly well aware.

Allied air planners (and the reader) could be forgiven for thinking that, since German armament factories could be found anywhere, why should they not bomb everything?

THE FACT that one of Europe's finest cities was almost entirely destroyed—while much, though by no means all, of what made it a legitimate target for bombing survived—can be criticized and condemned. The extent and manner of the loss of human life, most of it by normal standards classifiable as innocent even if the city itself was not, still wrenches at the heart six decades later. And it is impossible, even as one shrinks at the crassness of a man such as Mutschmann weeping crocodile tears for the "valuables" lost, not to wish the beautiful buildings and the artistic treasures of Dresden back to life.

This does not mean that the Allied bombing of Dresden cannot be justified. Dresden was not an "open city," but a functioning enemy administrative, industrial, and communications center that by February 1945 lay close to the front line. RAF Bomber Command struck at Dresden in the way it had been attacking German cities for years, which sometimes caused great destruction and sometimes did not. In the case of Dresden—because of unseasonably good weather exactly over the city, an unexpected absence of opposition, a lack of the usual "cock-ups," the inexperience of the city's people, and the local Nazi leadership's appalling neglect of air raid protection—it wrought something terrible and apocalyptic. This was the "raid which went horribly right," the consequences of which have haunted successive postwar generations and will continue to do so.

It is true that much of what has been thought and said about Dresden since its destruction owes a great deal to the efforts of first Nazi and then Communist propagandists. Nevertheless, once the war was over and we started to look around for symbols to understand it by, the popular instinct rightly picked out, and continues to pick out, what happened on February 13–14, 1945 as a warning of excess. Dresden remains a terrible illustration of what apparently civilized human beings are capable of under extreme circumstances, when all the normal brakes on human behavior have been eroded by years of total war. The bombing of Dresden was not irrational, or pointless—or at least not to those who ordered and carried it out, who were immersed deep in a war that had already cost tens of millions of lives, might still cost millions more, and who could not read the future. Whether it was wrong—morally wrong—is another question. When we think of Dresden, we wrestle with the limits of what is permissible, even in the best of causes.

Götz Bergander, son of Dresden, eyewitness to its suffering, and the first objective historian of its destruction, summed up succinctly but tellingly:

> What began as routine led to an inferno and left behind a signal. What seemed capable of achievement only on paper—the coming together of favorable circumstances for the attacks—was suddenly an accomplished fact.
>
> But wasn't that what the supporters of area bombing had always wanted? Too late came the question of whether they had really wanted it.

Or, as the painter Goya—also no stranger to horror—expressed it with even more economy: "The sleep of reason brings forth monsters."

Commemoration

THE STALL SELLING FLOWERS and wreaths by the wrought-iron entrance gate is doing excellent business. In the parking lot of the huge Heidefriedhof cemetery, amid the heath land beyond the edge of Dresden, solid citizens step from their Mercedes and BMWs and begin a dignified procession down the long avenue to the memorial site.

Nearby, battered Volkswagens and East German Trabants cough to a halt and disgorge young men with shaved heads, cut-off jeans, and heavy boots. But even they are on their best behavior today. They carry wreaths in the old imperial colors of black-red-white, coded symbol of neo-Nazis in a state where public display of the swastika is illegal.

I get off at the cemetery stop, having taken a tram out to the Wilder Mann terminus of line no. 3 and traveled the last couple of miles or so by bus.

Access-privileged vehicles glide past us as we walk between the neatly cared-for graves. There are the Dresden city officials, including the high burgomaster. There are the buses carrying survivors of the firestorm, who are permitted to travel right as far as the monument—both as a due honor and a sign of respect for age, for most of them are in their late seventies and upward now. Finally the smallest but in some ways most conspicuous group arrives: the representative from the British embassy in Berlin, in his chauffeur-driven car, and the man from the American consulate in Leipzig, hidden behind the tinted windows of a spanking-new white Jeep until he too emerges into the February daylight.

The two emissaries of the western Allied powers greet each other, and then begin to work the small huddle of German VIPs. It is a mild day, slightly overcast but dry—not unlike February 13, 1945, except that more than five and a half decades have passed and this is the third February 13 of the twenty-first century. The skinheads disappear into the trees until it is time for the annual ceremony; though some of the older, less aggressive-seeming rightists lurk with vague distaste on the fringe of the respectable political crowd.

The 2002 wreath-laying ceremony takes place at 11 A.M. It lasts about fifteen minutes, and is carried out with dignity by mainstream mourners and neo-Nazis alike. The rightists' wreaths refer to "terror bombing." On the square stone memorial where the wreaths and flowers are laid is a plaque:

> How many died? Who knows the number?
> On your wounds we see the suffering
> Of the nameless ones who burned here
> In a fiery hell of human making.

I see the faces of several Dresdeners whom I have interviewed or met. Their faces are drawn, concentrated, their gazes inward. They go to the same funeral every year—for the ones they knew and loved, for their remembered city, and perhaps also for the bits of their selves that they also lost that night in February 1945.

Afterward, the Anglo-American representatives leave fairly quickly. For the rest of us, the walk back to the entrance is slow and quiet.

February 13, 1945, has become symbolic not just for Dresdeners, and as such ceremonies go, it is the largest and most-reported in Germany. There are marches, meetings, and concerts. All have a theme of mourning, reconciliation, and peace. Except in the case of the far right and (though in smaller numbers) the extreme left.

The rightists march in the evening of February 13 to the statue of the "rubble woman" by the New Town Hall, because they are not allowed to march right through the center of town. Some skinheads have started to wear the *kaffiyeh*, the Palestinian scarf, adding to the complexity of the signs they give out. Their overall message, however, is clear. Hundreds of thousands of Germans definitely died at Dresden, the "bomb holocaust" was directed against Germans, and in the Second World War the Allies, not the Germans, were the true war criminals. The numbers who attend these marches increase steadily, year by year. Those of their fellow countrymen who have tired of always being labeled the villains, or never thought they were in the first place, are drawn to such events.

Provocatively, or rather perversely, their long-haired leftist opponents claim to welcome the bombing of Dresden, to praise Bomber Command. Two shriek some incoherent adolescent version of this to the mainstream crowd assembled in the darkness of the Neumarkt by the Frauenkirche, which is finally being rebuilt by international subscription (a chief mover in this was Pastor Hoch, in the early years after the fall of East Germany). Most of the citizens look bemused or mildly outraged, which is presumably precisely the kind of attention the kids want. They retreat into the darkness, hurling a few parting insults in the direction of the "bourgeoisie."

It is a dignified affair in the center of town. The scene in front of the Frauenkirche—which later merges into a prayer vigil in the now-renovated crypt of the church—is middle-class, calm, but with a deep undertow of emotion. Candles have been placed in their hundreds by the perimeter of the building site that will one day soon be the city's old Protestant cathedral reborn. People appear at microphones and tell their stories. They are simple but unbearable. Unbearable, perhaps, precisely because of their simplicity. Films are shown, including one of Coventry, which has become Dresden's sister town in the intervening years. There is another memorial meeting for the victims of the Dresden raid around the remembrance stone in the Altmarkt, where almost seven thousand bodies were incinerated in the weeks following the slaughter. This is livelier, more political, with singers and speeches about the evils of Nazism as well as the horrors of the firestorm.

DRESDEN IS NOW part of the Federal Republic of Germany. In 1989 its citizens played their role in the fall of the Communist dictatorship by first besieging a train from Prague—filled with East Germans who had sought refuge in the West German embassy there and had been allowed to exit by a panicked East German government—and then, when the Communist authorities tried to disperse them, undertaking a massive peaceful protest that stretched all the way up to Prager Strasse from the rebuilt Hauptbahnhof. The decisive moment came when the beleaguered colonel commanding the "People's Police" unit, faced with a choice between allowing the demonstrators to stay or breaking up the crowd by force, chose to let them stay. The message went out to other parts of the moribund German Democratic Republic: No one was shot in Dresden—no Tiananmen Square in the GDR—and mass demonstrations began in Leipzig, Berlin, and elsewhere. Within days, old the post-Stalinist leadership was gone, and within a year Germany was reunited. The police colonel remains a popular man in modern Dresden, and is apparently big in real estate.

In 1990 came the first free expressions of feeling about the bombing of Dresden. On that anniversary David Irving came, trailing clouds of glory and a considerable entourage, to be applauded as the man who "told the world" about Dresden's destruction. Two years later, two events centered attention on the newly free Dresden. The first was the unveiling of a statue of Sir Arthur Harris outside St. Clement Danes Church in London (a church with strong RAF connections) by the queen mother. German politicians, including the high burgomaster of Dresden, registered their objections, questioning why such a provocative act was considered necessary so long after the war. During the ceremony itself protesters in the crowd shouted accusations of mass murder.

In October of that same year Queen Elizabeth II visited the "new" provinces of Germany (as the former GDR areas are known) and included Dresden in her itinerary. The feeling among many in the city was cool, even sullen. She was booed by elements of the crowd at the Kreuzkirche after she failed to apologize formally for the 1945 raid on Dresden. The matter did not die now that the naturally varied feelings about this could be discussed in the city. There are hopeful signs. In 2000 the foundation for the rebuilding of the Frauenkirche was presented with a duplicate golden cross to replace the one destroyed in 1945. It was funded by the British counterpart of the foundation, the Dresden Trust, led by the indefatigable British friend of the city, Alan Russell. It was sculpted by the son of a Bomber Command pilot, and formal handover was made by the duke of Kent in fluent German. There were no catcalls.

Dresden at the time of writing (2003) has financial difficulties as Germany struggles with industrial and financial problems. The supply of subsidies from West Germany has inevitably started to dry up, and new tax income in East Germany is still not enough to make good the deficit. The center looks beautiful, with the restaurants and cafés full and the opera house, theaters, and other venues prospering, it seems. Soon work on both the *Schloss* and the Frauenkirche will be completed, and Dresden will have back the main components of its prewar skyline.

Massive investment by the German government since 1989 has given the Altstadt and the Neustadt alike a fresh, attractive look. In the latter, areas that suffered lighter damage in the bombing are home to ethnic restaurants and funky clubs, many of which thrive on the spending power of Dresden's large student population. On the other hand, the permanent population continues to decline. It is under half a million now. Tourism—or at least international tourism—has not yet really recovered. Despite such showpieces as the gleaming new high-tech Volkswagen plant just north of the Grosser Garten, and a plethora of service and catering industries, unemployment is high. The old Communist-built apartment houses may have been improved by generous coats of paint, but in the neat open spaces between them you see young people sitting idly during the day, and some of those young men and women bear a close resemblance to the ones who flock to the far-right demonstrations.

More than one of my interviewees, grandmothers and grandfathers as they are now, drew my attention to the "youth" problem. Ironically, it is these children of communism who offer the greatest real threat to the new Germany, and with it to the new Europe my generation has inherited, where wars are unthinkable and national rivalries are a matter for the soccer pitch and the Eurovision Song Contest rather than the Panzer attack or the bombing raid.

With the new freedoms have come also the beginnings of proper and systematic research about Dresden's contentious history. The archives are open, and academics and private enthusiasts are hard at work laboring over a mountain of neglected material. It is now that something close to an objective study of the city's history under the Nazis and its final fate can start to be made—although no such comprehensive book has yet been written. Sadly, many of these new historians find themselves relatively unhonored in their own city. Such scholars will smile and tell you that people in Dresden do not believe there was industry here, or soldiers, or terrible things, none of the things you found elsewhere in Nazi Germany. People believe that Dresden was a city strictly, not to say exclusively, devoted to the arts and culture. An innocent city. This is why the British bombing was an outrage.

The politics of remembrance in Dresden are complex, as complex as memory itself. There is a strong pacifist element in the city, perhaps stronger than most places in Germany, which is itself the most antiwar of all the large European nations. It is a genuine conviction, stubbornly held and based on terrible experience, that Americans especially have found hard to understand in the turmoil that has accompanied the opening years of the twenty-first century. Now in his mid-seventies, Christoph Adam still will not do or say anything that might glorify war or militarism. Under the Communists he suffered because he would not be conscripted into the "People's Army." He is clear that this aversion to war arises from his experiences on the firestorm night. In varying degrees this is true of the overwhelming majority of those who survived the experience.

The far right, neo-Nazis of various stripes, of course do not oppose war—only wars against Germany. Being in some cases cunning, they use pacifist language to heighten the suffering of their own people during the Second World War, with Dresden at that suffering's apogee. Dresden was innocent. Germany was innocent. In practical terms, it is difficult to know what they want from this. A rematch? More likely, to the far rightists the bombing of Dresden is a useful tool for attracting wider support. After all, it is the one thing in World War II that most reasonable Germans agree was terrible and undeserved. The rightist slogans repeat themselves, building on the foundation of the Dresden firestorm. *Why do we always hear about the Jews?* the argument goes. *We Germans suffered just as much, if not more. Ours was the true holocaust. The Anglo-Americans set out to destroy German culture. That is why they made war on Germany, and that is why they destroyed Dresden.* For them there will always be hundreds of thousands of dead, boiling lakes of phosphor, deliberate, brutal Allied tactics of cultural annihilation and mass murder. There has to be. Their political success depends on it.

The danger is that the far right also says a lot with which plenty of solid Dresden citizens—for all their decency and peaceableness— would find it hard to disagree. The Anglo-American attack on Dresden, a city innocent and untouchable in the view of many, even in the era of Mutschmann, was *sinnlos* (senseless) and a crime. This was the line for more than four decades under the Communists, backed up by official propaganda, and it is still widespread dogma, not just among neo-Nazis. Other places deserved to be bombed, maybe, but not Dresden. There was nothing bad here. No industry, nothing to do with the war. *There was no industry in Dresden. There was no reason to bomb Dresden.* These and other claims are fervently maintained, even in the face of clear evidence to the contrary. They have become, for many, part of their communal identity.

The bombing, and the differences in its historical interpretation even among Dresdeners themselves, are extremely, even worryingly political in Dresden. Dr. Helmut Schnatz, who wrote about the problems presented by reports of the American strafing of civilians on February 14, 1945, was subjected to a hail of invective when he attended a panel discussion in Dresden in the spring of 2000. Rational discussion was all but impossible. Known far-rightists—including one later imprisoned for his activities—heckled him from carefully chosen positions around the floor, quoting Nazi propaganda, and were enthusiastically applauded by the respectable mass of the audience. There is an unofficial boycott of Dr. Schnatz's book. It is nowhere on display in the city's bookshops, though if you ask them most will order it for you.

Like the man who lived through the bombing of Guernica and passionately recounted the verifiably untrue story of his experiences, the survivors' stories in their own way are true, because that is *their* experience. To quote one final time from Götz Bergander, born Dresdener and witness and chronicler of the air raids:

> The difficulty in disturbing the Dresden legends is that they are built on a basis of truth, namely on personal impressions left behind after a few terrible hours when people's very lives and beings were threatened. Those who were able to save themselves, who had to go through the experience of the walls of flame, the firestorm, the countless previously unknown sights and sounds, are afterward understandably ready to defend their subjective perceptions. So they really believe they were machine-gunned from the air in the middle of the night, and that they saw phosphor pouring down in great fiery curtains to engulf streets and houses.

There was no reason to bomb Dresden.

The truth is far messier. In a genocidal war between great nations, the "German Florence" fell to the most destructive new weapon of that war, the bomber. Slaughter from the skies—in a three-year-long air battle over Germany, where almost every day there was fighting—inevitably became bureaucratic, routine, inflexible, locked onto squares in a grid and often fueled by political rage. The British, especially, went to war reluctantly and had moral force on their side, but that did not make their leaders or their generals kind or humane, and perhaps it could not be expected to do so. The Germans, on the other hand, despite their high cultural achievements—exemplified by Dresden—were felt to have excluded themselves from the community of civilized nations by their behavior in the East and in the occupied countries.

On both sides, the only thing that mattered was winning the war. And once that overriding aim was decided, so was the fate of a city such as Dresden. Many of those in authority there during the war years realized this far more acutely than the general population.

Dresden contained plenty of things worth bombing, but it also contained even more things that in a better world would have been somehow preserved, even if that meant sparing the entire city, armaments factories, marshaling yards, and all. Think of the German commander who ignored Hitler's order to burn Paris, or the American artillery colonel who refused to shell the ancient university town of Heidelberg. But by 1945, in the air war, individual choice—except perhaps for the occasional bomb aimer who deliberately over- or undershot the target—was all but irrelevant. All that mattered was the things worth bombing, and everything other was hardly considered.

Then the war ended, the fighting ceased, and the world awoke from its terrible dream. The defeated had only survival to think of, but among the victorious nations this was when people started to turn to one another in shamed amazement and ask: *Did we really do that?*

A FEW MINUTES BEFORE 10 o'clock, I leave the crowd waiting in front of the Frauenkirche. Soon the time will come when the first Lancasters swept down along the Elbe toward the Altstadt on the night of February 13, 1945. I walk down the Münzgasse, once again filled with bustling cafés and restaurants, and up the steps to the Brühl Terrace.

The river is the color of antique pewter in the gray half-light. There is almost no one about.

I look to the northwest, the bombers' approach route. Just then a clock begins to strike the hour. After it finally falls silent, there is a moment's pause. Then the church bells start to peal solemnly in churches all over Dresden, the same way they have done every year since the firestorm, marking the day and the hour when Bomber Command's fleet swept down toward its helpless target.

And as it first learned to do on this night almost sixty years ago, a whole city holds its breath.

APPENDIX A

The "Massacre on the Elbe Meadows"

UNDER THE RUBRIC "Enemy Air Opposition," the Eighth Air Force's Intops summary for February 14, 1945 says:

> The GAF [German Air Force] reaction on this day was strikingly weak and almost entirely ineffective . . . Only two of the escorting fighter groups were in significant combat. At 12:15–12:20, one P-51 group encountered approximately 12 FW-190s at 18,000–20,000 feet on a southerly course south of Chemnitz and an additional 15–20 FW-190s and Me-109s in the same area at 30,000 feet. This group apparently also encountered the three e/a [enemy aircraft] reported attacking a squadron of bombers in the Dresden area . . .

Alden Rigby was not part of any of the groups that encountered German aircraft. He comments:

> We didn't really leave altitude all over the target area . . . I didn't see much of the city at all as I recall. It was just poor weather . . . of course the British had been in there the night before, it could have been smoke . . . we would have gotten them [the Flying Fortresses] over the target and headed for home. I'm sorry I can't tell you more, but when you're in that kind of a situation you're not observing a lot of things. What you're doing is keeping up in the air and getting them off the target.

By contrast, here is a postwar German report, describing the fighter escort's alleged activities at Dresden at around the time Al Rigby described:

The city itself was not worth bombing. But on the outskirts, toward which hundreds of thousands had fled, it was more worthwhile. And for the fighters and fighter bombers there was plenty to do: namely to hunt the Germans in packs along the country roads . . .

While they [the Flying Fortresses] destroyed the houses in the suburbs, fighters and fighter-bombers chased at low-level along the country roads, attacked the farmsteads of the surrounding villages with cannon-fire and bombs.

The above extract is from an alleged "factual report in eyewitness accounts," first originated in 1952 by a writer named Axel Rodenberger for a series in a popular German magazine, *Das Grüne Blatt*. It was printed in book form the next year to considerable success, the most recent edition being issued in 1995 by a well-known German publisher (though with an explanatory introduction admitting its limitations as an accurate historical account). Rodenberger's book is, however, still accepted in some surprisingly serious quarters as "history."

Every alleged fact in the extract is wrong. The American raid was not directed at the suburbs (though due to considerable disorganization, bombs did indeed fall in the outlying areas). Most of the Flying Fortresses attempted to bomb somewhere in the region of the Friedrichstadt or Altstadt marshaling yards, quite legitimate targets close to the heart of the city and still largely undamaged. There were also no fighter-bombers on the February 14 Dresden raid. These misleading statements could arguably have been due to the ignorance of the author and have not influenced later historical analysis. The shocking part of the quoted piece, however, is the vivid and apparently detailed description of American planes chasing innocent civilians down country roads, of escort fighters—piloted by men like Al Rigby—engaged in the systematic, frenzied slaughter of German civilians escaping the burning city.

Parts of Rodenberger's book were translated and published in America, though they aroused no great attention. Also in 1952, General Hans Rumpf—the former head of the firefighting forces—published a book (*Der Hochrote Hahn*) in German about the air war. In this, he claimed that both during the night and the next day at noon, the British and the Americans had descended to low level and strafed and machine-gunned civilians even as they were also being blasted by bombs. To claim that British aircraft did so is to claim an impossibility given the circumstances, the discipline of the British units, the temperatures at ground level, and the types of aircraft involved. No one now takes it seriously. The accusation against the Americans is, however, more plausible, and this is the one that stuck.

A few years later, a work published by a powerful apparatchik of Communist East Germany, Max Seydewitz—the first Communist prime minister of Saxony in the years after the war and a member of the party's Central Committee for many years—further stoked the fires. Seydewitz described the alleged atrocity:

> Then the pilots flew over the Elbe meadows, which were black with human beings who had escaped the burning city, and there they flew low and fired at the people on the ground in broad daylight.

Seydewitz mysteriously transferred Rodenberger's country roads to the city, but the accusation was the same. Both the British and the Americans were again accused of strafing the city, with the USAAF concentrating on machine-gunning civilians in the Elbe meadows and the Grosser Garten as well as in other parts of the city, including the Johannstadt hospital. Seydewitz's book was half a serious attempt at history, half an anti-Western rant. Its aim was to prove that the Anglo-Americans were as bad as the Nazis and the bombing of Dresden a deliberate war crime. Seydewitz's account is tailored to that Cold War priority.

But the big "breakthrough" of the strafing story came in 1963 with the publication of David Irving's international best-seller, *The Destruction of Dresden*. Irving did not give credence to the stories of the British machine-gunning civilians from low altitude during the Dresden firestorm. However, he not only supported the American strafing story; he filled in all manner of extra details unmentioned by Seydewitz:

> ... the Mustang fighters suddenly appeared low over the streets, firing on everything that moved, and machine-gunning the columns of trucks heading for the city. "Strafing by machine guns was observed during all the raids," states the Dresden police chief's official report. One section of the Mustangs concentrated on the river-banks, where masses of people had gathered. Another section took

on the targets in the *Grosser Garten* area. These strafing attacks were apparently designed to perfect the task outlined in the air commanders' directives as "causing confusion in the civilian evacuation from the east." Civilian reaction to them was immediate and universal. "We were in the Lennéstrasse just by the *Grosser Garten*," related one woman, evacuated with her ministry from Berlin to Dresden. "I and one or two others were able to save ourselves beneath some wooden benches. The fighter aircraft came right down and a woman near us suddenly screamed out, shot in the stomach . . ."

. . . the American fighter planes also strafed the Tiergartenstrasse, the road bordering the *Grosser Garten* on the southern side. Here the remnants of the famous *Kreuzkirche* children's choir had taken refuge. Casualties on record here including the choir inspector, seriously wounded, and one of the choirboys killed . . .

Well, there was indeed there a choirboy killed by Allied air power, and a wounded inspector found in the Tiergartenstrasse, but they had not been killed by American machine guns. Their injuries had been inflicted by British bombs as they fled the school the previous night. This is perfectly clear if one reads the original, acknowledged source of Irving's information: Seydewitz's book. This was based in turn on the account by the cantor of the Kreuzschule choir, Professor Mauersberger. Mauersberger found the injured inspector and the dead choirboy after the second British raid, in the small hours of the morning, before a single American P-51 had warmed up its engines for the long trip ahead.

As for the Dresden police chief's assertion that strafing by machine guns was observed during *all* the raids, since no one takes seriously the accusation that British planes carried out such attacks, quoting such a statement is not, in itself, worth much. The police chief was simply reporting what he had been told. Had large numbers of innocent civilians been massacred by the river, as was later claimed, it would have been recognized as a major infraction of the laws of war. It is very hard to see that the cautious, neutral phrase "strafing was observed" would apply.

Even in the most recently revised edition of his book Mr. Irving continues to describe these alleged attacks as being carried out by the Twentieth Fighter Group. "It will suffice," he says, "for the purposes of this narrative to describe the 20th Fighter Group's role in the operation." Accordingly, he goes on recount how "simultaneously with the end of the American attack at 12:23 P.M. the thirty-seven P-51s of the 20th Fighter Group's 'A' Group hurtled low across the city." The problem with this characteristically vivid description is that the Twentieth Fighter Group was, as we know, at the time more than eighty miles away escorting the attack not against Dresden but against Prague. The latter did indeed end at 12:23. The Dresden raid ended at 12:31.

Götz Bergander pointed out this inconsistency in 1974, having checked the records in the National Archives in Washington, D.C. Mr. Irving's corroborating evidence also suffers from comparison with the official record. "Others," he maintains, "attacked vehicles on the roads leading out of the city, crowded with refugees. According to one account, a P-51 of the 55th Fighter Squadron [a squadron belonging to Twentieth Group] flew so low that it crashed into a wagon and exploded." It is quite true that a Mustang of Twentieth Group was lost in this way. We even know that the aircraft in question was flown by Lieutenant Jack D. Leon. So why doesn't this decisively prove that strafing occurred at Dresden? Because, again, the problem is that the place where the pilot met his fate was near Donauwörth in western Bavaria—170 miles west of Prague, whence he was returning (and approximately 210 miles from Dresden).

The occasion of the low-level attack that went wrong was not some massacre of the choirboys in the Tiergartenstrasse in Dresden, but an attack on a vehicle carrying a Wehrmacht colonel, commander of a local repair facility. He was killed in his exploding truck. Leon either hit the vehicle or was perhaps knocked out of the sky by the explosion. No other pilot of the sixty-seven aircraft that made up the Twentieth Group's contingent that day was lost during operations. No pilot from any of the escort groups died or was lost over Dresden itself.

Irving does not refer to any other documentary evidence for the activities by fighter aircraft from groups that were (unlike the Twentieth) actually at Dresden. He simply includes them in his account of the alleged massacre, saying that along with the (in fact absent) Twentieth, "the 'A' groups of the other three fighter groups operating over Dresden" also "hurtled low across the city."

This is a difficult subject. Many survivors of the raid on Dresden vehemently and with patent sincerity insist that they (or people they knew) were indeed attacked by low-flying aircraft after the noon American raid, and will provide vivid descriptions of the horrors that ensued. These are people who suffered terribly, and who are tormented by their experiences to this day, and it is hard to deny them their vision of what happened. There are, however, also many survivors who say they saw and heard nothing of the kind—despite being likewise in the affected area, or nearby, at the time or soon after.

Günther Kannegiesser, for instance, walked past the alleged massacre scene a short time after it was supposed to have occurred. He had delivered the women and their children to the relatives in Laubegast, and had been sitting down to a midday snack when the American bombers arrived. On his return later that afternoon, his route took him along the Elbe meadows. There were horribly wounded bodies and signs of bomb damage from both the night and midday raids, but no heaps of corpses belonging to machine-gunned victims of fighter attacks. A friend of his from the neighborhood, Fritz Gieseler, was actually in the Grosser Garten (where survivors were also alleged to have suffered strafing attacks) during the American raid:

> At the time of the attack at midday on February 14 1945, he had just come back from Löbtau to the Grosser Garten. He was just south of the gate refreshments room . . . and he saw no low-flying fighters there. And we both knew what a strafing attack looked like . . .

A policeman, Werner Ehrlich, was also in the same area at the time, and told Götz Bergander forty years after the event:

> I was in the Grosser Garten during the midday raid, lying beneath a tree in the main avenue; it is true that there were bundles of stick incendiaries toppling down on us, but no actual shots. As if by a miracle, I wasn't hit by an incendiary stick, which opened up like magicians' bouquets when they struck and spread themselves around . . . I also heard nothing in my official capacity about all those low-flying fighters that are supposed to have done for all those people—allegedly heaps of them.

The tale of the strafing attacks against civilians rests totally on often contradictory personal statements recorded years after the event, which seem to become wilder as the distance between the raid and the recollection grows.

There is no documentary confirmation whatsoever. In neither the American Army Air Force documents available, nor—perhaps more significantly—in German accounts originating at the time are such daylight strafing attacks mentioned. It is claimed that perhaps the American pilots, ashamed in retrospect of having slaughtered civilians, "covered up" their actions, and hence the lack of reports. In the case of German records, however, this cannot be the case. On the contrary, accusations of Allied massacres of civilians in such low-level swoops were a favorite theme of Goebbels's propaganda machine, and significant strafing by fighters, where it occurred, was invariably noted in official accounts of air raids, usually described by some such formula as "terrorizing the civilian population."

One would expect a major Allied atrocity against helpless, bombed-out civilians in broad daylight to have been given great prominence in official reports. The actual bombing of Dresden is reported promptly, in grim detail, the day after it occurred. But neither on February 15 nor on February 16 (the days on which any such strafing attacks on Dresden would also have been mentioned) does the Wehrmacht High Command's report see fit to mention any strafing or low-level fighter activity. On the seventeenth there are accounts of strafing. However, they refer to attacks in the southern and western parts of the Reich. As for the reaction of Goebbels's ministry, that is another matter. One thing can be affirmed: Not a word was said by the propaganda maestro or any of his underlings about daylight strafing of civilians at Dresden between the day of the raid and the end of the war.

David Irving also states (again referring to Twentieth Fighter Group):

The "B" group fighters had to maintain visual contact with the bomber boxes; the "A" group pilots were briefed that as soon as the bombers' attack on Dresden was over they were to dive to rooftop level and strafe "targets of opportunity." Columns of soldiers being marched into or out of the wrecked city were to be machine-gunned, trucks strafed by cannon-fire, and locomotive and other transportation targets destroyed by rockets. Both "A" and "B" groups would withdraw from the bomber formations at 2:25 P.M. at a point near Frankfurt, where escort duty would be taken over by P-47 Thunderbolts.

No documentary reference is given for this "briefing," which clearly amounted to an order to strafe ground targets (inevitably involving civilian casualties) immediately after the raid. As it happens, the written orders for the First Air Division's attack on Dresden that day are available at the National Archives in Washington. They contradict everything in Irving's description—except, perhaps, the mundane final point regarding replacement of the escort detail by P-47s at Frankfurt.

The folder is titled:

8th A.F. FIGHTER FIELD ORDER NO. 1622A.

SUPPORT B-17'S AND B-24'S.

14 FEB. 1945

MISSION NO. 830.

It stipulates fighter escorts for all thirteen hundred or so American bombers in daylight action that day and issues special instructions for their modus operandi.

Point X in the order for February 14 contains two important provisions:

(3) EVERY ATTEMPT WILL BE MADE TO CONSERVE GASOLINE.

(4) ANY STRAFING WILL BE DONE ON WITHDRAWAL AT GROUP LEADERS DISCRETION IF NO E/A HAVE BEEN ENCOUNTERED OR ARE EXPECTED. ONLY "A" GROUPS WILL STRAFE. AIRDROMES WILL NOT REPEAT WILL NOT BE STRAFED.

This is no command for hell-for-leather attacks on ground targets but a call for extreme caution on particularly long, potentially difficult trips during which efficient, close fighter escort could be crucial. The specific orders issued to the Twentieth are also available, as is the text of the briefing. They include the above instructions and also add that ground targets, specifically to be sought on the return trip, should especially include "transport facilities."

As Götz Bergander sums up:

> This then is the order in which instructions were supposedly issued that would increase the confusion on the escape routes out of Dresden to panic levels—the order that was carried out with merciless perfection. It contains not a word of any of this. A plan so extraordinary in the history of strategic air warfare as one involving low-level attacks, heading practically into the last of the falling bombs, to hunt human beings in the target area, would have been mentioned in some form or another during this briefing . . .

It is sometimes hard to interpret contemporary documents. Working through the folder in which the field order is contained, the reader comes across a teletype short summary of the fighters' involvement in the day's raids. In this document, there is to be found the following paragraph:

> 356 GRP had one squadron in combat over Dresden. No enemy aircraft seen by other groups. "A" groups went to deck for strafing on withdrawal.

This looked, on the face of, as if it could describe exactly what the proponents of the "massacre" claim—the "A" groups flying low and strafing after the bombers had left Dresden. Except that, on careful reading, it is clear that the document is referring to *all* of the fighter groups that went out that day, not just those that accompanied the First Air Division. It is simply stating that they obeyed orders, in that "A" groups went down to low level and strafed "on withdrawal." And in the context of the report, "on withdrawal" does not mean "after the bombers had finished their bomb runs" but *once the fighters had withdrawn from their escort duties*. As Alden Rigby said, the escort fighters' job was "keeping up in the air and getting them off the target." Withdrawal (often shorted to "w/d") might therefore happen a half-hour later or two hours later (this is usually indicated in the records), depending on the perceived level of continuing risk. Until then, unless encountering enemy aircraft, fighters were under orders to stay with the bombers.

In all the records in the National Archives in Washington, there is not a single reference to low-level attacks or strafing runs on targets in or near Dresden. But there are detailed accounts, including real-time "flash reports" received during the actual raid, by all the fighter groups (to some extent dovetailing with reports by the bomber groups), which account for the exact movements—times and locations given—of all the groups involved in the admittedly somewhat confused operations of the First Air Division in the middle hours of February 14, 1945. Among these were frank accounts of strafing activities on the return trip (as instructed), including trains, trucks, and other communications targets. A mistake was admitted after a pilot fired on a hospital train, failing to make out the red cross on the roofs of the cars. All incidents occurred more than one hundred miles west of Dresden.

Plus there *was* at least one hostile encounter over Dresden between American and German fighters. Small change for all but those involved, but perhaps more important than it seems.

As the First Air Division's Headquarters Report said, enemy air opposition to the February 14 raid was "strikingly weak and almost entirely ineffective." The 352nd Fighter Group, responsible for covering the rear groups of the bomber stream, reported: "Uneventful escort to target, Dresden," and this is fairly typical.

The 356th Fighter Group had an encounter near Chemnitz (in which the 364th may also have been involved), during which P-51s claimed one enemy aircraft shot down from a pack of twelve FW-190s. The rest of the Germans quickly disappeared into high cloud and did not reemerge. According to reports immediately after the raid, however, the 356th once more spotted enemy fighters:

> At 12:35 3 FW-190s were observed to make a pass at a bomb group to which 356 was not assigned. One flight attacked the enemy aircraft . . . These e/a may also be the ones reported by a bomb group as attacking them in the target area.

There was obviously a brief, but possibly intense dogfight around Dresden a few minutes after the last bombers had dropped their bombs and turned for home. The teletype report to HQ First Division by the commander of the 306th Bombardment Division (who had at first overshot the target and were forced to make another run) confirms that some bombers on a second run at the target were attacked by three German fighters (FW-190s) and a brief battle ensued in the air. As the fighters maneuvered for a second attack, the report says that "they were immediately bounced by P-51 escort," which pursued them.

This incident is important because, alongside the reports of strafing in Dresden that day, there were also several accounts of a battle between American and German fighters taking place over the city. The Americans were chasing the enemy, and there were times when the planes went low (quite common when one fighter was trying to shake off another) along the valley of the Elbe and then the city.

Dr. Helmut Schnatz is a retired high school history teacher from the Rhineland. As a boy, he witnessed the devastating bombing of his native city of Koblenz and also observed strafing and aerial combat over his home. He went on to become official chronicler of the air war in his part of Germany and to write numerous authoritative articles on the subject, including one on British low-level attacks on targets around the western German city of Trier. He had become fascinated, along with other historians, by the often quite glaring contradictions to be found in accounts of air attacks (and, by association, other extremely traumatic events). In particular, he was and remains interested in the differences between various individuals' eyewitness versions of the same event, the descriptions found in official records (which may also be various and contradictory), and clearly established facts (such as the weather, topography, the capabilities of different machines).

Having read Götz Bergander's work, Dr. Schnatz decided in the mid-1980s to undertake a thorough investigation of the Dresden "strafing" phenomenon.

In his book, *Tiefflieger über Dresden? Legenden und Wirklichkeit* (*Low-Flying Aircraft over Dresden? Legends and Reality*), Dr. Schnatz examines all aspects of the events with the sharp eye of a man who has been studying every element of the air war—military, political, and technical—since he was a boy. He examines reports of British low-level raids during the night, confirming that they were physically impossible—possibly arising from observations of the relatively low-flying (but unarmed) marker planes, and the ease with which showers of rapidly falling incendiary sticks might be mistaken for cannon or tracer shells. But his main interest is the alleged American attacks, since these are the most plausible and, if they happened, the most horrifying.

Dr. Schnatz comes to unmistakable and convincing conclusions. For his research—which took ten years—he consulted former fighter pilots, bomber pilots, other air war experts, archives, and even psychologists, and collected dozens of accounts by and interviews with survivors of the destruction of Dresden. The last include many convincing and level-headed accounts, but also some that are clearly exaggerated and inaccurate (though nevertheless published in German and other newspapers and magazines without challenge).

There are, for instance the two tales published in the same special edition of a German weekly periodical. In one, a woman claims to have survived on an *ice floe* in the frozen river Elbe for some hours at a temperature of "minus 20° Celsius" and while thus engaged to have witnessed repeated daytime strafing attacks. In the same edition, another woman says that "the waters of the Elbe were ablaze with phosphor." The newspaper makes no comment on this apparent contradiction, or on the fact (as many witnesses confirm) that even before the firestorm it was quite a mild winter's night—Günter Jäckel, for instance, spent the night outside in a meadow in his bedclothes without suffering ill effects.

Dr. Schnatz's conclusions are:

- There were no orders for the fighters to launch low-level strafing of Dresden.
- Even had this been so, documents regarding the precise course and duration of the operation indicate that no time could have been taken out for strafing over the target.
- The swift declaration of the all-clear by the air raid authorities in Dresden (less than fifteen minutes after the last bomb was dropped at 12:31 P.M.) indicates that there could have been no additional air attack once the bombers had gone.

- The P-51s at Dresden were operating close to the limit of their fuel capacity and could not have undertaken extensive low-level strafing so far from base without endangering their own safety.

- They did strafe, but only far to the west of Dresden, mainly against railway targets, when their escort duties were over and running out of fuel was no longer a problem.

- Given the agreement between German as well as American accounts of the raid on Dresden, there can be no question of American records' being falsified or sanitized after the event.

As to what actually happened, Götz Bergander originally suggested that 356th Fighter Group's dogfight near and over Dresden spilled down into the Elbe valley and, briefly, over the city itself. The timing is right. Schnatz shares this view, and has done extra work with eyewitnesses and also with technical experts, which to his mind confirm this. The suggestion is that at least three fighter aircraft—possibly more—were involved in a howling, low-level chase over the city in which wing-mounted guns may have been fired. (And certainly civilians might have been hit, injured, and killed.)

The traumatized, bewildered Dresdeners, still stunned from the new bombing attack a matter of minutes earlier, and inexperienced in air raid matters, thought it was a murderous, low-level strafing attack of the kind they had been warned to expect, and about which they had heard such extravagantly violent stories.

Both Götz Bergander and Helmut Schnatz believe that this would be sufficient to account for many of the legends that have arisen since about the machine-gunning of (in some accounts "thousands of") women and children in the Grosser Garten and the Elbe meadows that noontime of Ash Wednesday, 1945. Many did indeed die there, but from bombs not bullets.

There is not a single expert German historian who will pronounce the incident as fact in any work written now. Tellingly, even the 1982 updated edition of Max Seydewitz's book about Dresden (celebrating his ninetieth birthday), which on its first publication in 1955 had played a key role in spreading and legitimizing the strafing story, suddenly contained not a word about the alleged American machine-gun massacre by the river, or the less probable British equivalents during the night. All references to these (as well as to American strafing attacks on the Johannstadt hospital) had been quietly expunged. Recent research among the records of Seydewitz's East German publisher indicate that this was due chiefly to Götz Bergander's conclusions in his book, which had been published in West Germany a few years previously (though it had remained banned on the eastern side of the Wall).

None of this means, however, that no one believes in the low-flying fighters anymore. On the contrary. Historians who deny it are accused by many survivors of "mocking" the dead and the maimed, and even of falling for an Anglo-American "conspiracy" to hide the truth.

APPENDIX B
Counting the Dead

THE YEAR 1952 had seen the appearance in German of Axel Rodenberger's memoir *Der Tod von Dresden* (*The Death of Dresden*). The book gained best-seller status in West Germany, where the author now lived after fleeing Communist rule in East Germany. It is a feverish collection of rumor, hearsay, and personal observation—occasional sensational stories, such as those of crippled soldiers leading blind comrades from a blazing military hospital, are taken straight from the notorious article in Goebbels's *Das Reich* of March 4, 1945. Rodenberg's passion was understandable; he had witnessed the horrors of the firestorm. The book is nevertheless almost entirely without value as history, although this doesn't seem to have stopped its being cited even by would-be serious writers. The book also includes overly inflated estimates of the death toll at Dresden. Rodenberger names a figure of 350,000 to 400,000 dead in Dresden allegedly recorded by a local official of the Propaganda Ministry in a report to Berlin. Although Rodenberger describes this man as sitting down and dictating to his secretary, and claims to quote this report "word for word" (*wörtlich*), no such document has never been discovered, and it is unclear how Rodenberger could have gained access to it in the first place.

In 1955 the Communist politician Max Seydewitz's *The Unconquerable City* (*Die unbesiegbare Stadt*) was published, with the East German government's full weight behind it. Seydewitz quoted the official figure of 35,000 dead, while allowing that it might have been a few thousand more.

Soon, however, figures on the level of Rodenberger's were once more in circulation. In 1963 came the publication of the internationally best-known work on the bombing, David Irving's *The Destruction of Dresden*. A university dropout in his mid-twenties who had worked in the German steel industry, learning fluent German, Irving said he became interested in the bombing of Dresden as a result of conversations with a coworker who came from there. Irving gained access to the limited original sources available at the time, conducted invaluable interviews with many surviving senior RAF officers, including Harris, Saundby, Bottomley and others, and managed to gain some access to documents and lower-echelon figures in Dresden itself. Strikingly, he produced a figure of at least 135,000 dead for the raid.

This figure came from a man named Hanns Voigt, by then resident in West Germany, who had been in charge of the missing persons center in Dresden in the weeks after the raid. Although the "street books" kept by the searchers clearing the cellars, shelters and other places where the dead were to be found only went up to between thirty thousand and forty thousand, Voigt insisted that this was a minimum figure. Irving accepted Voigt's higher personal estimate—an informed guess based also on separate accountings of personal belongings, garments, jewelry and so on, plus the numbers of missing unaccounted for by the end of the war—that at least 135,000 must have died. Vividly told, well written, and based on what seemed like exhaustive, thoroughly checked research, Irving's book became an international best-seller and launched him on a brilliant, controversial career.

In 1963, with East Germany all but closed to outside researchers, there was as yet no firm evidence that contradicted Herr Voigt's estimate. The "Final Report"—referred to in the text—is a convincingly thorough summary, but twenty years after the bombing of Dresden it was still not available. In the postwar confusion, it seems, all copies of this key report went missing. They were still unavailable in 1963, when the appearance of David Irving's book on Dresden, with its dramatic casualty estimates, conditioned the English-speaking world's view of the city's destruction for an entire generation.

In fact, the only new estimate to come to light in the Sixties seemed to drive the death figures even higher. Shortly after the first publication of his book on the destruction of Dresden, Irving was given a supposed copy of *Tagesbefehl No. 47* (TB47—see chapter 26) that cited figures of 202,400 dead already registered and 250,000 anticipated. These figures had been in circulation for many years—Irving had dubbed them fake in his book's first edition—but on seeing apparent authentication, he publicized the new, even more horrifying figure widely.

In 1965, Walter Weidauer, Communist high burgomaster of Dresden and would-be historian, published his book on the bombing—*Inferno Dresden*—in East Germany. It also did not refer to the "Final Report," but a short while later, after he gave a lecture in the town of Bad Schandau, Weidauer was approached by a woman named Frau Jurk. She handed over a closely typed, eleven-page document, which came from among the papers left by her late father-in-law, a former colonel in the Order Police in Dresden. It was the "Final Report." Colonel Jurk had actually written the report at the behest of the Dresden commander of the Order Police and signed it on his behalf. Weidauer reproduced the entire report in an appendix to the second edition of his book, causing a considerable stir.

Confirmation (and some extra information regarding damage to private homes) was provided by the almost simultaneous discovery in the West German Federal Archive in Koblenz of a long-lost copy of the situation report for air attacks on Germany issued by the Reich commander of the Order Police (nr 1404) of March 22, 1945. This presented Berlin's summary of the information coming in from Dresden a little over a month after the firestorm raid—in condensed but essentially unaltered form.

The definite figures in those documents (the Final Report and Situation Report) were between 18,000 and 22,000, estimates of final numbers around 25,000, so how could numbers of (exactly) ten times that level be cited in a document circulating at around the same time? Moreover, the text of T847 also included the explanation that the figures were being issued as an exceptional measure in order to scotch rumors of gigantic casualty rates: "Since the rumors far exceed the reality, open use can be made of the actual figures."

If the "actual figures" were up to 250,000, as the supporters of the fake TB47 had insisted, of what could the "rumors" possibly have whispered? *Millions* of dead? Clearly exceptional permission had been granted, in view of the widespread panic and fear aroused by wildly inflated casualty rates in Dresden, for the real (if temporary) numbers of dead to be cited by those in authority. And the real numbers were 20,240 and 25,000 respectively,

There was indeed only one explanation. A zero had been added to these figures for propaganda purposes. Irving duly wrote a letter to *The Times* of London in which he admitted that the two six-figure estimates quoted in the so-called "TB47" were "probably" a Nazi fake.

The macabre argument over the death toll at Dresden still continues. Evidence comes and goes, but there is a basic divide between those who agree that the figures were between twenty-five thousand and forty thousand, and those—still including Irving—who insist in the face of the documentary evidence that the deaths went into six figures, in some cases into several hundreds of thousands. Ironically, it is the British writers on the subject of the Dresden bombing—David Irving and Alexander McKee—who set the numbers higher (though not as high as do the neo-Nazi websites and pamphlets that seek to equate the destruction of the city with the Holocaust).

David Irving has never let go of the Voigt figures altogether, however, and his estimates have crept up again since the 1980s, finally settling in the latest edition of his book (self-published/online edition of 1995) at "up to 100,000." This is based, it seems, on taking the twenty-five thousand established dead, adding the thirty-five thousand reported missing as of March 1945, and then some.

Alexander McKee, writing almost twenty years after Irving's book was first published, at one point cites without comment a diary entry claiming that the authorities had counted "256,000" dead. He ends up taking the thirty-five thousand figure and then remarking that— given the poor air raid provision and the numbers of anonymous refugees in the city (which like most writers of the time he vastly overestimates)—it "might easily be doubled to 70,000 without much fear of exaggeration. But no one will ever know for certain." In other words, McKee simply decided to double the number because he didn't think it sounded like enough. A lower figure, it must be said, would not have enabled his book to be published under the impressive title *The German Hiroshima* (*Das Deutsche Hiroshima*) in Germany.

Götz Bergander's cool and objective account, *Der Luftkrieg in Dresden* (*The Air War in Dresden*), first published in 1977 and reprinted and revised in 1994, arrived at a death toll of around forty thousand. Bergander accepted the rough viability of twenty-five thousand to thirty-five thousand dead, but then added some thousands because of his own conviction (based not on a whim but on his own observations at the time) that a somewhat higher toll was likely. Bergander also performed a service by finally finding an authentic example of the TB47 directive in the keeping of a former reservist, Werner Ehrlich, who had made just such a copy as part of his duties. (Irving's source, citing the figures with faked extra zeroes, had been a carbon of a typed copy of a copy of unknown provenance). However, the most important single discovery came after the end of Communism and the fall of the East German regime.

Earlier writers, including Irving, had perforce accepted figures provided by Herr Zeppenfeld, the head gardener of the Heidefriedhof, the huge cemetery on the heath outside the city limits. There, by general agreement, the vast majority of the dead from the air raid were buried in mass graves. Zeppenfeld, who had commanded one of eight teams charged with the recovery and burial of bodies, was quoted by Seydewitz in his 1955 book as saying that they had counted all the bodies buried there, plus the ashes of the "nine thousand" incinerated on the Altmarkt between the third week of February and the second week of March 1945 (though, as we know, this number was almost certainly too high if we accept the total of 6,865 corpses noted in the genuine version of TB47). The total had come to 28,746. Adding the numbers of dead buried elsewhere in graveyards gave a rough reckoning of thirty-five thousand dead altogether from the raid, though Seydewitz also thought that "the number of dead is certainly greater than the 35,000 established at the time of burial."

In 1993, however, four years after the collapse of communism, documents from the municipal cemetery office were found in the Dresden City Archive. These provided, for the first time, a detailed official breakdown of how many bodies had been buried by the Dresden authorities after the raid, and where. The burials had been undertaken with great thoroughness (like almost everything else associated with the aftermath of the Dresden raid), and totals reported regularly to the city authorities. The total buried in the Heidefriedhof between February and the end of April 1945 turned out to have been 17,295, including the ashes of the incinerated victims from the Altmarkt. An additional 3,462 were buried in the Johannisfriedhof, 514 in the Neue Annenfriedhof. The total of registered burials was 21,271—more than 7,000 fewer than head gardener Zeppenfeld's deceptively exact estimate for the Heidefriedhof alone, which was undocumented and confessedly based on ad hoc counting methods.

Herr Friedrich Reichert of the Dresden City Museum thinks that around two thousand were buried in various other cemeteries around the city.

Reichert estimates that around twenty-five thousand people died in the bombing of Dresden. It is a figure that tallies with the total cited by the Soviet-supported authorities in the city shortly after the war, but is lower even than that suggested by knowledgeable and level-headed observers such as Götz Bergander. The fairest estimate seems, therefore, to lie between twenty-five thousand and forty thousand. This makes the loss of life in the city less than the total for Hamburg (although Hamburg possessed at least twice Dresden's population), and as a proportion of the total population, less than that for towns such as Pforzheim or Darmstadt.

As for the assertion that, even after VE-Day, in Dresden "thousands of victims were still being recovered each week from the ruins," this must be an overstatement. The official figures are quite clear. According to Walter Weidauer, high burgomaster of Dresden in the postwar period, between May 8, 1945 (when Irving claims that "thousands" of bodies were still being found each week), and *1966*—a period of more than twenty years—a total of only *1,858* bodies were recovered from the ruins of the city. Weidauer also states that there is no substance to the reports that tens of thousands of victims were so thoroughly incinerated that no individual traces could be found. Not all were identified, but—especially as most victims died of asphyxiation or physical injuries—the overwhelming majority of individuals' bodies could at least be distinguished as such. Since 1989—even with the extensive excavation and rebuilding that followed the fall of communism in Dresden—no bodies have been recovered at all, even though careful archaeological investigations have accompanied the redevelopment.

None of this is to minimize the appalling reality of such a vast number of dead, so horribly snatched from this life within the space of a few hours, or to forget that most of them were women, children, and the elderly. Wild guesstimates—especially those exploited for political gain—neither dignify nor do justice to what must count, by any standards, as one of the most terrible single actions of the Second World War.

APPENDIX C
Legends of the Fall

AFTER THE FIRST WORLD WAR, two enterprising German brothers, Heinz and Karl Spanknöbel, immigrated first to Switzerland and from there, in 1922, to the United States. Heinz was political, an early Nazi enthusiast who for a while was "Land Leader of the NSDAP in the USA." Later he returned to Germany, where he established a factory. Karl Spanknöbel stayed in America, took U.S. citizenship, and changed his name to Charles Adolf Noble. A former Adventist minister and health-food retailer, he ended up owning a photographic processing business in Detroit that became one of the largest in the country. His son John Noble later claimed that constant exposure to the harmful chemicals used in the film-developing process at that time gradually affected his father's health. By 1937 Charles Noble was looking to change businesses—and countries.

In 1938 Noble acquired a modest but profitable camera factory in Dresden, employing around one hundred workers, from a Herr Thorsch. Thorsch had Swiss nationality but was half Jewish, so did not feel confident that he would be left unmolested by the regime. In fact, the deal agreed with Charles Noble a.k.a. Spanknöbel was not so much a sale as an exchange—Thorsch got Noble's unwanted business in the United States, and Noble got a viable manufacturing business in the country he hadn't seen for nearly twenty years. There are some mysteries: In 1943 Noble was accused of taking a disproportionate profit from the "de-judification" transaction, though if we are to believe the Noble family's story, they were already being persecuted by the authorities for having assumed American nationality. Moreover, this accusation comes from Max Seydewitz, who had a variety of other urgent political motives for blackening Noble's name. According to

John Noble, it was a fair, even generous agreement where each side got what he wanted.

Anyway, the "Camera Workshops Charles A. Noble," as it was soon renamed, prospered under his management. He developed a small-format single lens reflex camera. This camera, the Praktica, and its successor, the Praktiflex, were very successful. The Nobles took up residence in a spectacular mansion called the Villa San Remo, which had been built in the 1890s for Crown Princess Luise of Saxony. Set in its own grounds on the Bergbahnstrasse, in the exclusive spa suburb of Weisser Hirsch high above the city, the palatial residence boasted a superb view of Dresden and the Elbe valley, especially from its mock-Renaissance tower.

Even after the United States entered the war on the Allied side, the Noble factory was permitted to continue in business under Charles and his son John, although soon the making and selling of mass-market cameras gave way, as in the rest of Dresden's consumer manufacturing sector, to production for the war effort. The Noble factory produced filters for aircraft engines and parts for aiming devices. Herbert Blumtritt, historian of the Dresden camera industry, expresses suspicion: "that in the Second World War a company which remained in the family possession of a U.S. citizen was not placed under German administration . . . or its owner interned—in fact continued to produce for the armaments industry—must remain forever beyond comprehension, or has deeper causes."

Noble was, of course, German-born, and his brother Heinz at one point a prominent American Nazi. He also seemed to be managing his factory perfectly to the government's satisfaction. However, Max Seydewitz supplied a story that combines political conspiracy with romantic betrayal. According to Seydewitz, Noble acted as a mediator between the Nazi regime and the Americans, a two-way channel supplying intelligence to Washington and also allowing the German leadership to pursue a hope of a separate peace or at the very least favorable treatment from the western Allies. Seydewitz continued:

> Noble's tasks naturally also included intriguing against and spying upon the Soviet Union, in which the gravediggers of Germany all too willingly supported him. In exchange for the information about the Soviet Union, which he received by radio from representatives

of his masters in Wall Street and passed on to the Nazi leaders, he received from them confidential information about Germany, which he communicated from his transmitting station at the Villa San Remo to those same American taskmasters.

The plot thickens further. According to Seydewitz, all the information that Sir Arthur Harris and the commander of the Eighth Air Force, General Spaatz, had about Dresden came from Noble. From Noble the western Allies knew that the city was overflowing with refugees and the wounded, and more:

> They knew precisely the position of the densely populated districts of the inner city, the position of the Zwinger and the Frauenkirche, the location of the other cultural monuments and churches. They knew perfectly well that in February 1945 Dresden was unprotected by flak or by fighter aircraft. They were also fully informed of where in Dresden the military targets were situated—and no bombs were dropped on these on February 13th. The fliers were expressly forbidden to drop any bombs on the Weisser Hirsch area, for the Anglo-American High Command wanted, at all costs, to avoid endangering the invaluable agent who lived in that district. For this reason, Weisser Hirsch was one of the very few areas of Dresden that was protected from the air attacks of the 13th and 14th February.

Seydewitz makes the extraordinary claim that the Nobles, father and son, had "given the order for the destruction of Dresden." Mutschmann had been forewarned—although, to his chagrin, given only twenty-four hours' notice instead of the three days he had allegedly been promised. Even Goebbels knew, Seydewitz claims, pointing to transcripts of telephone conversations that were "still available after the end of the war" but, mysteriously, no longer at the time of writing (1955).

This hysterical Stalinist rumor-mongering had a profound effect on the Dresden population, isolated as it was and unable to access other points of view or cross-check the "facts" stated by their Communist rulers. Apart from the story of Mutschmann's being warned, there is also the rumor that the Allied bomber fleets were guided to Dresden by lights in the tower window of the Villa San Remo. These myths are as common in the bomber war as were the folk stories of lanterns tied to the tails of donkeys to lure ships onto the rocks when smugglers ruled, or in earlier wars of traitors inside besieged cities who placed lights to let the enemy know that the stronghold could be entered. The notion that the RAF and the USAAF could only find a city of almost three-quarters of a million inhabitants with the aid of a lamp in the window of the tower, or that they had no idea where the densely populated parts of the city were, or where its great buildings and monuments were located (information available in any prewar tourist guide), bespeaks a staggering, almost bizarre naivete.

Seydewitz's account eventually reached a truly, dementedly, fantastic climax. He had Noble gleefully watching the bombing of Dresden from his palatial villa overlooking the city, like the villain in a James Bond film gloating over his evil handiwork. All that was missing was the white Persian cat purring on his lap:

> We can be sure that Mister Noble waited, on that night of Shrove Tuesday, for the "Christmas trees" to appear over the night-dark silhouette of Dresden and show the bombers their target. We can be sure that Mister Noble then watched from window of the veranda of the Villa San Remo, and enjoyed the cruel spectacle of the leaping flames and the collapse of the priceless monuments . . .

What seems to have actually happened was that the Nobles, who in many ways had been seriously disadvantaged during the war, losing control of their business in all but name and being subjected to various restrictions, remained at liberty for less than two months after VE-Day. They flew the American flag, thinking it some kind of guarantee, and extended hospitality to American officers who visited Dresden during the immediate postwar honeymoon. In July 1945 Charles and John were arrested by the Russian secret police. The Soviet authorities seized their factory and used it to produce Praktica cameras for the Red Army. As for the Nobles, they spent more than a year at the Münchner Platz jail and some months in Buchenwald concentration camp near Weimar, which the Soviets rapidly reopened for the imprisonment of their own enemies. Finally, after being held without charge for five years, they were tried separately and convicted on charges of spying for the Americans.

Sentenced to fifteen years, young John was sent to a Siberian gulag. Charles A. Noble was released from an East German prison in 1952. He spent the following years back in America, campaigning for his son's release. John was released in 1955 after a personal intervention by President Eisenhower. Seydewitz's story was a propaganda broadside, without material evidence, intended to justify the imprisonment of the two men and inspire yet more anti-Western feeling in the captive GDR.

Curiously, the story was specifically denied by Walter Weidauer, the former Communist high burgomaster of Dresden, in his book *Inferno Dresden* (1965). Exotic spy-novel accounts of meetings between Noble and the Nazi leadership, not to mention the camera manufacturer's clandestine radio contact with Wall Street, are also conspicuous by their absence (though the Nobles' arrest by the Russians was justified because of "war crimes"). Nevertheless, the entire Noble conspiracy saga appeared again in the 1982 edition of Seydewitz's book, published (as all books in the GDR had to be) only after government approval—even though the previously equally definite story of the American strafing of civilians had been entirely cut.

Some Dresdeners still believe the tale of the traitor in the Villa San Remo. A recent book of personal narratives about Dresden contains an account by a man who experienced the firestorm as a child. He describes his family seeking refuge afterward at an uncle's flat out in Weisser Hirsch—an area he casually describes as having been spared because an American spy lived there, at the Villa San Remo. "Allegedly," he adds, "but I'm sure that's how it was [*es wird schon so gewesen sein*]."

For the son of a heartless American spy, John Noble certainly evinced a surprising commitment to the city he had supposedly "betrayed." In 1992 he returned to Dresden, with his younger brother George, and claimed back the factory and the house that had been confiscated by the Communists after 1945. For five years they tried to rebuild the factory, producing a 180-degree panoramic camera of revolutionary design, but the business had to be sold in 1997, along with the villa. The camera, however, is still in production. Mr. Noble, a deeply religious man who found God in the gulag, maintains that, far from being "spared" by the Allied bombers, the Villa San Remo suffered partial damage to its roof, and the family struggled to put out fires that could have destroyed the entire building. Charles and John Noble were recently celebrated in a local Dresden press survey as belonging among the "100 Dresdeners of the Century."

If Weidauer's account of this legend, for all its politically conditioned limitations, restores an element of reality, the former high burgomaster was also capable of spreading rumors of his own. Among these is the story of how the Soviet army saved Dresden from the atomic bomb.

In October 1963 the editor of the German scientific magazine *Physikalische Blätter* was granted an interview with Werner Heisenberg, Nobel Prize–winning nuclear physicist and head of the Reich's abortive attempt to build an atomic bomb. In the course of the talk, Heisenberg allegedly told the journalist that in July 1944 he was visited by an adjutant to Reich Marshal Göring, commander of the Luftwaffe. This emissary informed him that the Americans had issued a threat, via German diplomatic channels in neutral Lisbon: If the Germans did not agree to make peace within the next few weeks, an atom bomb would be dropped on Dresden. The adjutant wanted to know, on the Reich marshal's behalf, if Heisenberg thought that this was a plausible scenario—to which the scientist, well aware that whatever answer he gave could have grave consequences, replied that he considered it improbable but not completely impossible. He immediately qualified this already ambivalent advice further by pointing out that the development of such a weapon to the point of use would demand an enormous industrial effort, which he did not think the Americans were yet in a position to make.

Nothing came of the alleged conversation. Göring, it seems, didn't really believe that the Americans had the atom bomb either. Certainly Germany did not surrender, then or for almost a year to come. So, if this bizarre threat was actually made, what was the significance? Was it a real threat? Or were the Americans testing the Germans' nerve?

The Heisenberg story would be irrelevant if Weidauer didn't then press on, massing circumstantial evidence, to assert that, in fact, the apocalyptic Thunderclap discussions of that summer were conducted by the same committee that was responsible for the atomic bomb, ergo the two were plausibly connected. Not only that, but the atomic bomb had, by agreement, to be tested on an undamaged city. And another thing: After the war, General Groves, who was in charge of the military side of the development of the atomic bomb, stated that at the end of December 1944 President Roosevelt had told him that "he wished for us to be prepared to drop the bomb on Germany, if we had the first bombs before the end of the war in Europe." Ergo, Dresden would have been the perfect target. Ergo, Dresden must have been the planned target. Lurching quickly toward the conclusion of his alarmingly unstable tightrope of reasoning, Weidauer then asserts that the city was saved only by the gallantry of the Red Army, which in January 1945 advanced so quickly that Dresden was suddenly too close to the front to be atom-bombed without harming Soviet forces. Thus the city was preserved from the fate of Hiroshima and Nagasaki.

The follow-through punch comes, of course, with the assertion that, when it became impossible to use the atomic bomb on the deliberately preserved "experimental" city of Dresden, in order to spite and frighten the advancing Russians, a massive conventional attack was substituted at the last moment. Ergo, the Dresden firestorm.

There are a number of contradictions in all this. First, the Americans did not have a usable bomb in the summer of 1944. In that regard, both the scientist and the Luftwaffe commander were quite right. But the enemy was closer to its goal than Göring may have thought and Heisenberg may have feared. Second, as early as September 1944, there is evidence that the main choice facing President Roosevelt was whether the bomb should be dropped on Japan or kept back as an instrument of threat. Germany was not really part of the discussion. Third, by December 1944 Roosevelt knew that August 1945 was the most likely time when a bomb would become available for use. Fourth—and as a consequence—if the sick, exhausted Roosevelt did say such a thing to Groves, it must have been because he was tempted at that time (perhaps under the shock of the unexpectedly fierce German counteroffensive in the Ardennes) into believing that the nightmare in Europe might indeed drag on into the autumn of 1945, when the bomb would be usable.

Weidauer's evidence is thin, his logic tenuous, but of course he has a cold war propaganda point to make. If the Anglo-Americans were prepared to destroy Dresden with an atomic bomb, and were deterred only by the rapid Soviet advance (which, puzzlingly, the western Allies were at the time doing all they could to assist), what horrors might the "imperialists" not inflict on the peaceful Communist world twenty years later?

The story of Dresden as the planned first target for the atomic bomb appears to have been around since shortly after the war, but Walter Weidauer is the source that Dresdeners refer to. Even among educated people it is a confidently asserted fact: "You know, if the war hadn't ended when it did, then the Allies planned to drop the atomic bomb on Dresden . . ."

As Weidauer, setting the scene for the party's other propagandists, reminded his East German readers:

> If, despite all this, Dresden—which according to the above facts was marked out as the target for the first atomic bomb—was spared the fate of Hiroshima and Nagasaki, then we have to thank in foremost place the soldiers, officers and generals of the Soviet Army. Their swift advance, their glorious deeds, which led to the unconditional surrender of Hitler-Germany on the 8th May 1945 . . . excluded the possibility of the first atom bomb's being dropped on Dresden. This we should never forget!

ENDNOTES

Chapter 1: Bood and Treasure

11 "the English were treasured": Pastor Dr. Karl-Ludwig Hoch, interview by author Dresden-Löschwitz, February 2002. Also for the following.

Chapter 2: The Twin Kingdom

20 "a small city made of wood": Johann Christian Hasche, *Eine Umständliche Beschreibung Dresdens*, quoted in *Dresden: Die Geschichte der Stadt*, ed. Dresdner Geschichtsverein (Dresden, 2002), p. 90.

25 "the most terrible day": Letter reprinted in *Dresden in der Goethezeit*, ed. Günter Jäckel (Berlin, 1990), p. 48.

27 Frederick's characteristic use of surprise: Jacek Stazhewski "Der Sachsenhof in Warschau (1756–63)" in *Dresdner Hefte 50: Polen und Sachsen: Zwischen Nähe und Distanz* (Dresden, 1997), p. 66ff.

29 "Dresden no longer exists": Quoted in *Dresden: Die Geschichte der Stadt*, p. 102.

Chapter 3: Florence on the Elbe

30 "a German Florence": Quoted in Olaf B. Rader, "Dresden," in *Deutsche Erinnerungsorte III*, eds. Etienne François and Hagen Schulze (Munich, 2001), p. 459.

34 the best of old and new: Wolfgang Zimmer, "Bevölkerungsentwicklung und Sozialstruktur in der Stadt nach 1871," in *Dresdner Hefte 61: Industriestadt Dresden? Wirtschaftswachstum im Kaiserreich* (Dresden, 1997), p. 21.

Chapter 4: The Last King of Saxony

36 "crowded living conditions": *Dresden: Die Geschichte der Stadt*, p. 175.

37 the Pan-German League: *Dresden: Die Geschichte der Stadt*, p. 188.

37 a huge new group: The DNHV mushroomed from 160 members in 1894 to

more than 100,000 in 1907 and continued its rapid growth until 1914. See Dirk Stegmann, *Die Erben Bismarcks: Parteien und Verbände in der Spätphase des Wilhelminischen Deutschlands* (Cologne/Berlin, 1970), p. 41.

38 shortages and hunger: Robin Neillands, *The Bomber War:* Arthur Harris and the Allied Bomber Offensive 1939–1945 (London, 2001), p. 13.

38 fourteen thousand were killed: Claus-Christian W. Szejnmann, *Vom Traum zum Alptraum: Sachsen in der Weimarer Republik* (Leipzig, 2000), p. 27. The MDR (Central German Radio) book and website on Saxon history suggest as many as 220,000, plus 19,000 "missing." If the proportions for Saxony as a whole and Dresden in particular are similar, the lower figure seems more likely for actual, directly attributable war casualties. If the higher figure is true, that would mean that almost a third of those Saxons who went to war never returned. Figures for Dresden are from *Dresden: Die Geschichte der Stadt*, p. 198.

38 a compromise peace: Prince Ernst Heinrich von Sachsen, *Mein Lebensweg: vom Königsschloss zum Bauernhof* (Dresden/Basel, 1995), p. 102ff.

42 "a bringer of hope": Gertraud Freundel, interview by author, Dresden, October 2001.

43 the Nazi vote in Dresden: Figures from Gunda Ulbricht, "Die Wahlen in Dresden 1932/1933," in *Dresden unterm Hakenkreuz*, ed. Reiner Pommerin (Cologne/Weimar/Vienna, 1998), p. 39ff. In Breslau the Nazi vote was 43.5 percent. See Norman Davies and Roger Moorhouse's work on Breslau/Wroclaw, *Microcosm: Portrait of a Central European City* (London, 2002), p. 337.

43 *Der Freiheitskampf:* Ralf Krüger, "Presse unter Druck. Differenzierte Berichterstattung trotz nationalsozialistischer Presselenkungsmassnahmen: die liberalen 'Dresdner Neueste Nachrichten' und das NSDAP-Organ, 'Der Freiheitskampf' im Vergleich," in *Dresden unterm Hakenkreuz*, ed. Reiner Pommerin (Cologne/Weimar/Vienna, 1998), p. 44.

Chapter 5: The Saxon Mussolini

46 Mutschmann was born: Information about Mutschmann's career from Karl Höffkes, *Hitlers Politische Generale: Die* Gauleiter *des Dritten Reiches* (Tübingen, 1986), p. 242ff, and Peter Hüttenberger, *Die Gauleiter: Studie zum Wandel des Machtgefüges in der NSDAP* (Stuttgart, 1969) p. 217 (for biographical details) and passim.

47 published a searing critique: For this and other Nazi defectors' unflattering views of Mutschmann see Clemens Vollnhalls, ed., *Sachsen in der NS-Zeit* (Leipzig, 2002), p. 39.

47 "no gods other than himself": Josef Goebbels, *Tagebücher* 1924–1945 (Munich, 1999), June 25, 1937, entry.

48 "everyone is afraid": Victor Klemperer, *The Klemperer Diaries 1933–45: I Shall Bear Witness Until the Bitter End*, abridged and trans. Martin Chalmers (London, 2000), p. 5.

48 Von Killinger escaped the fate: Von Killinger, who was credited with organizing the coup that brought the Romanian fascist Antonescu to power, died at the hands of pro-Russian militia when the German embassy in Bucharest was stormed in September 1944. See the profile and selection of German and British press cuttings in WO 208/4480, PRO, London.

51 "he showed us his bare back": Otto Griebel, *Ich War Ein Mann der Strasse: Lebenserinnerungen Eines Dresdner Malers* (Altenburg, 1995), p. 209.

52 Liebmann's death: See Norbert Haase, Stefi Jersch-Wenzel, and Hermann Simon, eds. (Bearbeitetet von Marcus Gryglewski), *Die Erinnerung hat ein Gesicht: Fotografien und Dokumente zur nationalsozialistischen Judenverfolgung in Dresden 1933–45* (Leipzig, 1998), p. 100.

52 Sachs's death: Vollnhalls, ed., *Sachsen in der NS-Zeit*, p. 189.

000 "to encourage the others":Vollnhalls, ed., *Sachsen in der NS-Zeit*, p. 191.

000 the party leadership . . . was corrupt: Carsten Schreiber, "Täter und Opfer: Der Verfolgungsapparat im NS-Staat," in *Sachsen in der NS-Zeit*, ed. Clemens Vollnhalls (Leipzip, 2002), p. 179.

Chapter 6: A Pearl with a New Setting

55 "Dresden is a pearl": Quotation from Matthias Gretzschel, *Dresden im Dritten Reich*, reprinted in ed. *Landeszentrale für politische Bildung Hamburg und Dresden im Dritten Reich: Bombenkrieg und Kriegsende* (Hamburg, 1993), p. 96.

55 "and bless the deed": This and succeeding descriptions of Hitler's visit to Dresden in 1934 from Günter Jäckel's autobiographical essay, "Für den herrlichen Führer danken wir Dir," taken from proofs of *Lebensbilder III (Die finstere Zeit)* provided to the author by Günter Jäckel, March 2002.

58 the director . . . wrote with a hint of triumph: Copy of original document supplied to the author by Holger Starke of the Dresden City Museum. See also his article, "Vom Werkstättereal zum Industriegelände: Die Entwicklung des Industriegebietes an der Königsbrücker Strasse in Dresden vor der Entstehung der Albertstadt bis zur Auflösung der Industrieanlagen Nord (1873–1952)," in *Dresdner Geschichtsbuch 5* (Altenburg, 1999), p. 150ff.

59 the largest garrison: Manfred Beyer, "Dresden als Keimzelle des militärischen Widerstandes: Die Garnison in der NS-Zeit," in *Dresdner Hefte 53: Dresden als Garnisonstadt* (1998), p. 52ff.

59 slowly remilitarized: Starke, "Vom Werkstättereal zum Industriegelände," p. 171ff.

59 new commercial buildings: See Matthias Lerm, *Abschied vom Alten Dresden: Verluste historischer Bausubstanz nach 1945* (Rostock, 2000), p. 14ff.

60 Hitler's improvements: For details of planning after 1933 see Matthias Lerm, "Konzepte für den Umbau der Stadt Dresden in den 30er und frühen 40er Jahren des 20. Jahrhunderts" in *Vorträge und Forschungsberichte: 4 Kolloquium zur dreibändigen Stadtgeschichte 2006 vom 18. März 2,000*, p. 31ff.

Chapter 7: First the Synagogue Burns, Then the City

62　Jewish population in 1933: Breslau figure from Davies and Moorhouse, *Micro-cosm*, p. 367; Berlin figure from Leonard Gross, *The Last Jews in Berlin* (London, 1983), p. 11; demographic breakdown for Saxon cities from Vollnhalls, ed., *Sachsen in der NS-Zeit*, p. 202 ff.

62　"plying their trade": Uwe Ullrich, *Zur Geschichte der Juden in Dresden* (Dresden, 2001), p. 10.

62　recompense for Lehmann: Quotations from Ullrich, *Zur Geschichte der Juden in Dresden*, p. 10ff.

63　a sign forbade Jews and dogs: See HATiKVA, ed. *Spurensuche: Juden in Dresden, Ein Begleiter durch die Stadt* (Dresden, 1996), p. 35.

63　Bernard Hirschel: Ingrid Kirsch, "Das Ringen um die Gleichstellung der Dresdner Juden und ihre Religionsgemeinde von 1830 bis 1871," in *Dresdner Hefte Nr 45* (rev. and reprinted 2000), p. 19.

65　Jewish population: Figures from Gerald Kolditz, "Zur Entwicklung des Antisemitismus in Dresden während des Kaiserreichs," in *Dresdner Hefte 45* (rev. and reprinted 2000), p. 44.

66　the growth in prejudice varied: Observations on marriage patterns from W. E. Mosse, *The German-Jewish Economic Elite 1820–1935: A Socio-Cultural Profile* (Oxford/New York, 1989), p. 182. Remarks on prejudice against Jews from p. 225.

67　a one-day boycott: For details of the boycott see Haase, Jersch-Wenzel, and Simon, eds., *Die Erinnerung hat ein Gesicht*, p. 101ff.

67　the Nazi thugs brandished signs: See Günther Kirsch, "Die gesetzliche und aussergesetzliche Judenverfolgung in Dresden und Sachsen in den ersten Monaten der nationasozialistischen Herrschaft," in *Historische Blätter H. 4* (1994), p. 10.

68　spit at for two hours: See Henny (Wolf) Brenner, *"Das Lied ist aus": Ein jüdisches Schicksal in Dresden* (Zürich/Munich, 2001), p. 69.

69　the local Nazis had put signs: "Die Johannstadt wieder als Vorbild," in *Der Freiheitskampf 2* (August 1935).

70　Goebbels embarked on a hate-filled speech: See Michael Burleigh, *The Third Reich: A New History* (London, 2000), p. 325–26.

71　The synagogue collapsed: Not quite all was lost. A young Aryan fireman, Alfred Neugebauer, rescued the golden Star of David after it toppled from the tower of the synagogue and secured it in his home. He returned it to the (by then tiny) Jewish community after the war. It was placed back on the new synagogue (consecrated 2002). Herr Neugebauer, now eighty-five years old, was an honored guest at this ceremony.

72　"uniformed SA people": Griebel, *Ich war ein Mann der Strasse*, p. 239f.

72　Hirsch's fur store: Günter Jäckel, interview by author, Dresden-Kleinschachwitz, February 2002.

73 "the synagogues that caught fire": Quoted in HATiKVA, ed., *Spurensuche:* p. 41.

75 "a pair of fine Nazis": Jäckel, interview. The publicity surrounding Jäckel's election as chair of the Goethe Society reached the ears of his school friend, now a retired academic living in the Pacific Northwest of America but still in touch with events in his old hometown. The "two Günthers" now talk regularly on the transatlantic phone connection—as befits retired citizens— "when it's cheap."

75 "this fire will return": Griebel, *Ich war ein Mann der Strasse,* p. 240.

Chapter 8: Laws of the Air

77 "a sea of flame": Cajus Becker, *The Luftwaffe War Diaries* (London, 1967), p. 75.

77 postponed because of bad weather: John Buckley, *Air Power in the Age of Total War* (Bloomington, Ind., 1999), p. 127.

77 refuge in a roadside ditch: Nicholas Bethell, *The War That Hitler Won: September 1939* (London, 1972), p. 104.

78 "putting out no flags": Klemperer, *Klemperer Diaries*, p. 296f.

78 "when I asked": From Günter Jäckel's essay, "Dieser 1. September 1939," presented by Jäckel to the author.

79 a balloon laden with explosives: See Stephen A. Garrett, *Ethics and Airpower in World War II: The British Bombing of German Cities* (New York, 1996), p. 3f. Also for the following accounts of early aerial bombing.

81 "frightfulness": Quoted in Norman Longmate, *The Bombers* (London, 1983), p. 29.

82 "the combatant spirit of the people": Quoted in Longmate, *Bombers,* p. 32. Also for Trenchard's remarks.

85 General Guilio Douhet's views: See Buckley, *Air Power in the Age of Total War,* p. 75ff, for this and discussion of Douhet's apocalyptic theories on air war.

87 "a more total war": Buckley, *Air Power in the Age of Total War,* p. 3.

89 70 percent of the town was destroyed: See Jörg Friedrich, *Das Gesetz des Krieges* (Munich, 2002), p. 720f (and notes) for discussion of the moral aspects of the raid and references to specialist historians' accounts. Friedrich states the number of victims at three hundred, Neillands in *Bomber War* cites one thousand victims, from Hugh Thomas's book *The Spanish Civil War*.

90 *his* Guernica: Story of visit to Guernica recounted to the author by the historian concerned, Matthias Neutzner, February 2002.

Chapter 9: Call Me Meier

92 Baldwin's 1932 dictum: See Denis Richards, *The Hardest Victory: Bomber Command in the Second World War* (London, 1994), p. 8.

92 feared . . . 150,000 casualties: See Longmate, *Bombers,* p. 54.

93 Nickel leaflets: Methods and problems of "Nickel" leaflet dropping described in Richards, *Hardest Victory*, p. 23ff. Text of the "These . . . Are Your Leaders!" leaflet reproduced in German in Bethell, *War That Hitler Won*, p. 204, and in English (with the other leaflets quoted) in Longmate, *Bombers*, p. 75f.

94 Noël Coward remarked: Cited in Bethell, *War That Hitler Won*, p. 208.

95 *Feuertaufe:* The film actually had precisely the opposite effect in many cases. In theaters where the movie was shown, German women, especially, were heard expressing sympathy for Polish women and children.

95 second great example: For the German invasion of Holland and Belgium, see Len Deighton, *Blood, Tears and Folly: An Objective Look at World War II* (London, 1993), p. 185f. Deighton also presents a cool, commonsense analysis of the bombing of Rotterdam (p. 365).

96 attempts by British bombers: See Richards, *Hardest Victory*, p. 45ff (also for the May 15/16 raid against German industrial targets); Deighton, *Blood, Tears and Folly*, p. 363f.

100 heroes' welcomes: Quoted from the essay by Hermann Rahne, "Zur Geschichte der Dresdner Garnison im Zweiten Weltkrieg 1939 bis 1945" in *Verbrannt bis zur Unkenntlichkeit: Die Zerstörung Dresdens 1945*, ed. Friedrich Reichert (Dresden, 1994), p. 127.

100 "colossal anger": Goebbels's reaction quoted in Jörg Friedrich, *Der Brand: Deutschland im Bombenkrieg* (Munich, 2002), p. 69.

101 "a splendid propaganda device": Goebbels, *Tagebücher*, September 11, 1940, entry.

Chapter 10: Blitz

105 "fire gained ground": Norman Longmate, *Air Raid: The Bombing of Coventry* (London, 1976), p. 93.

105 the Warwickshire Hospital: These and other details of the raid from Longmate, *Air Raid, passim* but especially p. 85f for the destruction of the cathedral.

106 568 civilians died: Figures from Longmate, *Air Raid*, p. 181f. (details of bombs dropped) and p. 190 (casualty figures).

106 "a concentrated attack": Quote from the postwar autobiography of Sir Arthur Harris, *Bomber Offensive* (London, 1947), p. 83.

106 "though the flames licked": From *Britain under Fire* (London, 1941), p. 12.

108 according to Albert Speer: Albert Speer, *Inside the Third Reich* (London, 1970), p. 388f.

109 Arthur Harris wrote: Harris, *Bomber Offensive*, p. 51f.

110 average daily death toll: German casualty figures for 1944 and 1945 in Friedrich, *Der Brand*, p. 68.

Chapter 11: Fire and the Sword

114 Mr. Southam explained: Quotes from the author's interview with Vaughan Southam B.Sc. (Cantab.), interview by author, Bristol (England), May 2001.

118 two hundred heavy bombers: Quoted in Max Hastings, *Bomber Command* (London, 1979, rev. ed. 1999), p. 111.

119 the morale of the enemy: Quoted in Martin Middlebrook, *The Battle of Hamburg: The Firestorm Raid* (London, 1980), p. 24.

120 Harris's schooling: See Dudley Saward, *"Bomber" Harris* (London, 1984), p. 4ff, and Henry Probert, *Bomber Harris, His Life and Times* (London/Mechanicsburg, 2001), p. 23ff.

120 "he had learned": Probert, *Bomber Harris*, p. 31, and p. 36ff for Harris's career with the RFC.

122 Harris made a reputation: Harris, *Bomber Offensive*, p. 19ff for his account of the Northwest Frontier fiasco and Iraq, including quotations.

124 Harris had optimistically believed: For his high hopes and subsequent disappointment, see Harris, *Bomber Offensive*, pp. 92–95.

126 "the first German city": Harris's comments in Harris, *Bomber Offensive*, p. 105. Description of the raid in Richards, *Hardest Victory*, p. 119f.

127 "the Führer declares": Goebbels, *Tagebücher*, April 27, 1942, entry.

128 "Baedeker Raids": Quoted in Niall Rothnie, *The Baedeker Blitz: Hitler's Attack on Britain's Historic Cities* (Shepperton, 1992), p. 11.

128 "The Thousand Plan tonight": Longmate, *Bombers*, p. 220.

129 damage to Cologne: Quoted in Longmate, *Bombers*, p. 224. See also the slightly less impressive account of the thousand-bomber raid's effect on Cologne in Friedrich, *Der Brand*, p. 87ff.

129 "we have stationed": Goebbels, *Tagebücher*, May 31, 1942, entry.

131 window: Account of the approach and sequence of attack from Middlebrook, *Battle of Hamburg*, p. 117ff. unless otherwise indicated. For further accounts of Hamburg ref. Martin Middlebrook *passim* and Longmate, *Bombers*, p. 256ff.

132 "a glass marble in a barrel of peas": Quotation from Middlebrook, *Battle of Hamburg*, p. 128.

132 damage to Hamburg: Longmate, *Bombers*, p. 269.

133 "to a total halt": Speer, *Inside the Third Reich*, p. 389.

134 "the destruction of a metropolis": Goebbels, *Tagebücher* July 29, 1943, entry.

134 panic that resulted: See Olaf Groehler, *Der Bombenkrieg gegen Deutschland* (Berlin, 1990), p. 199.

Chapter 12: The Reich's Air Raid Shelter

135 the sirens began to wail: Information on early alarms and stray bombs from Götz Bergander, *Dresden im Luftkrieg: Vorgeschichte, Zerstörung, Folgen* (Würzberg, 1998, rev. ed.), p. 9ff.

136 Dresden had been excluded: Bergander, *Dresden im Luftkrieg,* p. 95.

137 air raid security gone crazy: Bergander, *Dresden im Luftkrieg,* p. 96.

138 "I do not dispute": Bergander, *Dresden im Luftkrieg,* p. 99ff, for the correspondence between Obergruppenführer von Woyrsch (SS chief for the Upper Elbe, based in Dresden) and Reichsführer Himmler, August 1943 onward.

140 the December attack: Groehler, *Der Bombenkrieg gegen Deutschland,* p. 205.

140 the firestorm in Kassel: Details and figures of the Kassel raid from Friedrich, *Der Brand,* p. 117ff.

142 Rumpf's report: Quotes from Rumpf and the economic warfare officer from Groehler, *Der Bombenkrieg gegen Deutschland,* p. 208. For other details see this and a lecture by Olaf Groehler reprinted in "Bombenkrieg gegen Leipzig 1940–1945," in *Texte des Leipziger Geschichtsvereins e.V. Heft 4* (1994).

143 civilian casualties . . . were much lower: Groehler, "Bombenkrieg gegen Leipzig," p. 18.

144 "danger exists": Quote and the story of the failed evacuation of Dresden's children in Matthias Neutzner, ed., *Martha Heinrich Acht: Dresden 1944/45* (Dresden, 2000), p. 23ff.

145 "I want to be where you are": Frau Anita (Kurz) John and Frau Nora Lang, interview by author, Dresden-Johannstadt, February 2002.

146 Rhenania Ossag: For details of the Freital-Gittersee raid of August 24, 1944, Bergander, *Dresden im Luftkrieg* p. 25ff.

146 Robert Lee commented: Robert Lee's recollections of Dresden and his time as a POW in Freital from papers donated to Imperial War Museum, London.

147 "trust in the leadership": Gittersee *Lagebericht* of August 30, 1945, quoted in Neutzner, ed., *Martha Heinrich Acht,* p. 45.

Chapter 13: A City of No Military or Industrial Importance?

148 three-letter manufacturing codes: *Oberkommando des Heeres: Liste der Fertigunskennzeichen für Waffen, Munition und Gerät* (Berlin 1944, reprinted 1977 and 1999).

148 "very incomplete": Starke, "Vom Werkstättenareal zum Industriegelände," p. 181.

149 "from 1923 I worked": Interview with Rolf W. for IG "13. Februar" e. V. reproduced in Neutzner, ed., *Martha Heinrich Acht,* p. 53.

149 including torpedo parts: Notes regarding the company's prewar business in Hauptstaatsarchiv Dresden (references, HStaD) *Findbuch,* p. IIf. Report of

March 27, 1944, regarding changeover to armament production in HStaD Fa. Maschinenfabrik Richard Gäbel, Dresden, 1888–1947 Nr 46. , correspondence with torpedo-testing station in Gotenhafen (Gdynia) Nr. 26.

150 wartime instructions included: Memorandum in company file in HStaD Nr 46.

150 J. C. Müller profits: See Beiträge zur Betriebsgeschichte VEB Tabak und Industriemaschinen Dresden (1956) in HStaD 11683 Nr 27.

150 fabulously rich Quandt family: Goebbels was married to the former wife of a member of this dynasty, to which his stepson, Harald Quandt, was one of the heirs. The Quandt family survived the war, unlike many of their employees. Its members remain at time of writing the largest private shareholders in the BMW motor company.

150 "not to a good place": Ilana Turner, telephone interview by author, Tel Aviv, September 2002.

152 craftsmen at Hellerau: Information regarding the war production at Hellerau from Dr. Olaf Przybilski, Technical University, Dresden, with reference to essay by Virginie Przybilski, "Fremd- und Zwangsarbeit in Dresden zur Zeit des Zweiten Weltkrieges," on which he collaborated.

153 Radio-Mende's workforce: For the story of Radio-Mende see Starke, "Vom Werkstättenareal zum Industriegelände," p. 178ff.

153 "City Disinfecting Institution": Haase, Jersch-Wenzel, and Simon, eds., *Die Erinnerung hat ein Gesicht,* p. 10.

155 Klemperer's views of the armaments Jews: Klemperer, *Klemperer Diaries,* p. 610.

155 fate of the Hellerberg Jews: Haase, Jersch-Wenzel, and Simon, eds., *Die Erinnerung hat ein Gesicht,* p. 184f and p. 186ff for reproduction of the list of the so-called Ostabwanderung (east emigration) of Dresden Jews on March 3, 1943, kept in the records of the Dresden Jewish Community.

156 "everyone would still be alive": Heinz Meyer to Rudolf Apt 9.9.45 quoted in Haase, Jersch-Wenzel, and Simon, eds., *Die Erinnerung hat ein Gesicht,* p. 135.

156 "twenty-minute journey": Haase, Jersch-Wenzel, and Simon, eds., *Die Erinnerung hat ein Gesicht,* p. 136.

158 "fine human being": Quote from Brenner, "Das Lied ist aus," p. 59.

158 Henny Wolf's eyesight: Information about close work at Zeiss-Ikon from Frau Henny Wolf Brenner, interview by author, Weiden October 2001.

158 closing of the Jewish Department: For the documents regarding this discussion at the RSHA in Berlin, see Haase, Jersch-Wenzel, and Simon, eds., *Die Erinnerung hat ein Gesicht,* p. 140f.

158 "the fear was worse": Brenner, "Das Lied ist aus," p. 85.

160 size of the Dresden directorate: See Matthias Neutzner, "'Der Wehrmacht so nahe verwandt'—Eisenbahn in Dresden 1939 bis 1945" in *Dresdner Geschichtsbuch 5* (Altenburg, 1999), p. 199ff.

161 "work rhythm of Dresden": Neutzner, "Der Wehrmacht so nahe verwandt," p. 205.

162 this was an important junction: Generalbetriebsleitung (Ost). The Generalbe-triebsleitungen (East, West, South, and North) represented the tier of railway management above the directorates, and were responsible for coordinating long-distance rail movements. The Generalbetriebsleitung (Ost) in Berlin was the structure mainly responsible for the movements to and from the concentration and extermination camps.

163 "would shuttle five times": Neutzner, "Der Wehrmacht so nahe verwandt," p. 211.

163 "no sympathetic gaze": Prisoner quoted in Heiner Lichtenstein, *Mit der Reichsbahn in den Tod* (Cologne, 1985), p. 81.

163 twenty-eight military trains: Neutzner, "Der Wehrmacht so nahe verwandt," p. 203f.

163 "Dresden was an armed camp": Letter from Colonel Harold E. Cook to the *Vancouver Sun*, n.d., reprinted in Alan Cooper, *Target Dresden* (Bromley, 1995), p. 245.

164 "a good target": From a document of the railway directorate quoted in Neutzner, "Der Wehrmacht so nahe verwandt," p. 213.

164 inspection of the air raid shelter arrangements: See Peter Reichler, *Dresden Hauptbahnhof* (Egglham, 1998), p. 54.

164 "air raid shelter of the Hauptbahnhof": Neutzner, "Der Wehrmacht so nahe verwandt," p. 214.

164 "by the last winter": Bergander, *Dresden im Luftkrieg*, p. 108. Also for the suc-ceeding quotation.

Chapter 14: Ardennes and After

170 lost forty-seven thousand men: Friedrich, *Der Brand,* p. 145f.

173 a lengthy analysis: Probert, *Bomber Harris,* p. 289, for this and subsequent arguments regarding Bomber Command's support role in Overlord.

173 "effects of strategic bombing": Harris's memorandum on "The Employment of the Night Bomber Force in Connection with the Invasion of the Continent from the UK" of January 13, 1944, quoted at length in Saward, *"Bomber" Harris,* p. 247. The paper was circulated to Portal, Leigh-Mallory, and General Montgomery.

174 Martin Middlebrook's words: See Martin Middlebrook, *The Berlin Raids: RAF Bomber Command Winter 1943–44* (London, 1988), p. 306ff for the fig-ures on casualties, p. 325 for his verdict.

174 ULTRA: ULTRA signified—to those "in the know"—material gained from German (and Japanese) coded radio communications. The British had early in the war cracked the German "ENIGMA" code, and were therefore privy to much secret information of enormous, even decisive, benefit in their wartime decision making.

174 "your very good letter": Correspondence between Harris and Churchill quoted in Probert, *Bomber Harris,* p. 305.

176 cities that remained unbombed: Probert, *Bomber Harris,* p. 308.

176 "the lack of success": Harris's letter to Portal, January 18, 1945, quoted in Probert, *Bomber Harris,* p. 311.

177 crisis when it came to fuel supplies: See Ronald C. Cook and Roy Conyers Nesbit, *Target: Hitler's Oil—Allied Attacks on German Oil Supplies 1939–45* (London, 1985), p. 165ff.

177 "attack on Germany's oil industry": Ralph Bennett, author of *Behind the Battle: Intelligence in the War with Germany 1939–45,* quoted in Probert, *Bomber Harris,* p. 313.

178 "education of a bomber crew": Harris, *Bomber Offensive,* p. 98.

Chapter 15: Thunderclap and Yalta

179 Bill Cavendish-Bentinck: For the background of Cavendish-Bentinck, see Patrick Howarth, *Intelligence Chief Extraordinaire: The Life of the Ninth Duke of Portland* (London, 1986), p. 9ff.

180 "heavy air attacks on Berlin": PRO, London, CAB 81/93.

180 code-named Thunderclap: *Bombing of Berlin,* note by the Secretary, January 22, 1945, PRO, London, CAB 81/127.

181 "a shattering effect": Quoted in Longmate, *Bombers,* p. 331.

181 "immense devastation": Memorandum from Portal for chiefs of staff quoted in Sir Charles Webster and Noble Frankland, *The Strategic Air Offensive against Germany 1939–1945,* vol. 3 (London, 1961). p. 55. See also for the authors' remarks.

181 general report on bombing: *Strategic Bombing in Relation to the Present Russian Offensive,* Report by the Joint Intelligence Sub-Committee, January 25, 1945. JIC (45) 31 (0), PRO, London, CAB 81/127.

183 added to the list: Probert, *Bomber Harris,* p. 318, citing the official history.

183 use of "baste": Definitions from *Collins English Dictionary* (British ed., 1998). David Irving is the chief accuser, and clearly believes Churchill was using a culinary metaphor. He remarks that, since Breslau was a fortress, "the retreat was entirely an evacuation of non-combatants. Historians to come will accordingly be pardoned for finding the Prime Minister's choice of words in questionable taste . . ." See David Irving, *Apocalypse 1945: The Destruction of Dresden* (London, n.d., Focal Point ed.), p. 114.

184 universal alarm: See Davies and Moorhouse, *Microcosm,* p. 13ff, for the announcement of Russian advance and the consequent panic-stricken evacuation.

184 his message was clear: Irving, *Apocalypse 1945,* p. 114.

184 retreat to the west: Anthony Beevor, *Berlin, the Downfall* (London, 2002), p. 60f.

185 main communications centres: Quoted at length in Saward, *"Bomber" Harris,* p. 283.

185 letter to Harris: Text of Bottomley's letter to Harris reproduced in Sir Charles Webster and Noble Frankland, *The Strategic Air Offensive against Germany 1939–45,* vol. 4; *Annexes and Appendices,* (London, 1961), p. 301.

187 "Evacuation Areas": PRO, London, CAB 121/003.

189 the fate of Hamburg and Kassel: Groehler, *Der Bombenkrieg gegen Deutschland,* p. 398f. Also for the discussion of casualty figures.

190 "hinder the enemy": NARA, Washington, D.C., RG 43, World War II Conferences, Yalta (Crimea) Conference, Box 4, appendix including translation of General Antonov's Statement.

190 "very much involved": Hugh Lunghi's statement in interview for CNN series *Comrades,* July 1996 (text online at the George Washington University National-al Security Archive website), confirmed by Major Lunghi in telephone conversation with the author, April 2002.

191 Portal's signal: Text of signal in Saward, *"Bomber" Harris,* p. 287f.

192 Harris was so displeased: See Irving, *Apocalypse 1945,* p. 127. The latter is a reported statement, having been retailed to Irving by a third-party witness. Saundby, Harris's deputy, also later told a Canadian television interviewer that the bombing of Dresden had been queried.

Chapter 16: Intimations of Mortality

193 "huge losses": Neutzner, ed., *Martha Heinrich Acht,* pp. 27–29.

195 this was the 111th alarm: Information from comprehensive list of air raid alarms in Dresden printed in Bergander, *Dresden im Luftkrieg,* p. 403ff.

195 numerous detonations: See Neutzner, ed., *Martha Heinrich Acht,* p. 37 for details of the damage.

196 "rumbling and rushing in the air": Klemperer, *Klemperer Diaries,* p. 797.

197 officially, no bombs: See Neutzner, ed., *Martha Heinrich Acht,* p. 42. See also Klemperer, *Klemperer Diaries,* p. 798.

197 "no islands of peace": Quoted in Neutzner, ed., *Martha Heinrich Acht,* p. 42.

197 "we could not help ourselves": Hoch, interview.

198 "I prefer the bombers": Klemperer, *Klemperer Diaries,* p. 799.

198 "worries me greatly": Klemperer, *Klemperer Diaries,* p. 801.

199 "back on the offensive": See *Dresdner Zeitung,* December 23, 1944.

199 "free and open gaze": *Gendarmerieposten* Dresden-Gittersee, police reports quoted in Neutzner, ed., *Martha Heinrich Acht,* p. 63.

199 special coupons: Neutzner, ed., *Martha Heinrich Acht,* p. 63.

200 "the only Christmas trees": Klemperer, *Klemperer Diaries,* p. 815.

200 that final Christmas: From correspondence of Dora Baumgärtl, April 7, 1945, in Stadtarchiv Dresden, Collection of 13.Februar e.V.

201 October and January raids: See Neutzner, ed., *Martha Heinrich Acht,* p. 67ff. and Bergander, *Dresden im Luftkrieg,* p. 62ff, for this and the details of the Second Bomber Division's attack on central Germany.

201 a mix of high-explosive and incendiary bombs: Bombing details from Bergander, *Dresden im Luftkrieg,* p. 64f.

202 no mention: Cited in Bergander, *Dresden im Luftkrieg,* p. 65.

202 the Americans' own photographic evaluation: See Bergander, *Dresden im Luftkrieg,* p. 67.

202 "life is granted": *Dresdner Zeitung,* January 29, 1945.

202 fantastic article: "Unvorstellbar, was London auszuhalten hat," *Dresdner Zeitung,* January 19, 1945.

203 "like a doll's house": John and Lang, interview.

203 Dresden's flak: For the rise and fall of flak in Dresden, 1940–45, see the chapter in Bergander, *Dresden im Luftkrieg,* p. 48ff.

204 Steffen was hospitalized: Steffen Cüppers, interview by author, Dresden, February 2001. Cüppers's gun was converted to ground use and acted as conventional artillery in the "Ruhr pocket" during April 1945. They fought to the last. On April 17 the emplacement—manned entirely by eighteen- and nineteen-year-olds—was overrun by American infantry. Two boys were killed, both still registered as pupils at Dresden high schools.

205 yet Chemnitz too: Bergander, *Dresden im Luftkrieg,* p. 56.

206 high fatality rate: Groehler, *Der Bombenkrieg gegen Deutschland,* p. 396ff. See also Martin Middlebrook and Chris Everitt, *The Bomber Command War Diaries: An Operational Reference Book* (Leicester, 2000), p. 653.

206 "a latter-day Carthage": Burleigh, *Third Reich,* p. 755.

Chapter 17: Time and Chance

208 February bombings: Details of RAF operations from Middlebrook and Everitt, *Bomber Command War Diaries,* p. 662.

208 official history reported: Wesley Frank Craven and James Lea Cate, eds. *The Army Air Forces in World War II,* vol. 3 (Chicago, 1951), p. 729 and for the following details.

209 Churchill's travels: Details of Churchill's living arrangements at Yalta from Lord Moran, *Winston Churchill: The Struggle for Survival 1940–1945* (London, 1966), p. 244, and life aboard the *Franconia,* p. 255ff.

209 "all the work": Lesley "Uncle Will" Hay, interview by author, Weybridge, Surrey, April 2002.

210 unwelcome news: Miles Tripp, *The Eighth Passenger: A Flight of Recollection*

and Discovery (London, 1969, rev. ed. 1993), p. 76. Sadly, Mr Tripp died in 2000, as the author was beginning work on this book.

211   a shortage of high-explosive bombs: See Harris, *Bomber Offensive,* p. 238f. See also his letter to Sir Norman Bottomley of March 29, 1945: "All H.E. is seriously limited in supply. Incendiaries are not. All these factors must therefore also be considered . . ." This and the other letters reproduced in full in Saward, *"Bomber" Harris,* p. 292ff.

212   the most powerful forces: Figures from Middlebrook and Everitt, *Bomber Command War Diaries*, summaries of raids for the above period.

212   "less than two hundred heavies": Memorandum of February 17, 1945, PRO, London, AIR 24/309.

213   cloud cover was expected: Forecast from Bomber Command Report on Night Operations 13th/14th February 1945, March 22, 1945, PRO, London, AIR 24/309.

213   approval for the night raid: Irving, *Apocalypse 1945,* p. 129.

213   "a military necessity": Harris, *Bomber Offensive,* p. 242.

214   the Dresden briefing: Hay, interview.

214   aircraft going in the first wave: General details on Five Group from the official RAF historical website, www.raf.mod.uk.

215   lethal effectiveness: Middlebrook and Everitt, *Bomber Command War Diaries,* p. 580. See also the chapter on the bombing of Darmstadt in Hastings, *Bomber Command*, p. 303ff.

215   "reason for the raid": John Aldridge, interview by author, Norfolk, September 2002.

216   Dresden target information files: Zone map and information sheets, PRO, London, AIR 40/1680.

217   separate target information sheet: Target Information Map for Dresden (Südvorstadt), PRO, London, AIR 34/595.

217   its own target file: Target Operation GS 257, PRO, London, AIR 34/602.

219   "met officer comes on": Hay, interview.

219   "a better target": Aldridge, interview.

220   "we knew our job": Hay, interview. Descriptions from Hay interview and others.

220   the procedure: Hay, interview.

220   the middle of the squadron: 49th Squadron Reports and Operations Record Book, PRO, London, AIR 27/483.

222   feints and decoy screens: Hay, interview.

222   Freya: An early-warning radar, which provided the German defenders the presence and course of the incoming bombers, but not their altitude.

222   Mandrel screen: A radar-jamming facility, ground- or air-borne. Like window, it had been around for some time, but by 1945 likewise served as just another trick in the bombers' bag.

223 according to the operational reports: Information and quotations from No. 6 Group (RCAF) Operations Record Book, p. 22, in PRO, London, AIR 25/131, and No. 8 Group (Pathfinders) Operations Record Book, p. 987 (*Attack on Braunkohle Benzin AG, Böhlen*), in PRO, London, AIR 25/154.

Chapter 18: Shrove Tuesday

225 a question of pracicality: Information about the *Verteidigungsbereich Dresden* in Hermann Rahne, "Die 'Festung Dresden' von 1945," in *Dresdner Hefte* 41, p. 19ff.

225 Elbe line: Quoted in Rahne, "Die 'Festung Dresden' von 1945," p. 20.

226 article that took the form of a dialogue: "Taktik für Hausfrauen," *Der Freiheitskampf*, February 14, 1945.

228 safe assembly points: See Irving, *Apocalypse 1945*, p. 103.

229 no permit to reside in Dresden: See Neutzner, ed., *Martha Heinrich Acht*, p. 72f.

230 "short-term accommodation": Neutzner, ed., *Martha Heinrich Acht*, p. 73.

230 "on the station": Bergander, *Dresden im Luftkrieg*, p. 212ff, for this and his following observations.

231 "city was not overflowing": Frau Annerose Hennig, letter to Dresden City Museum, April 17, 1993, as reproduced in Friedrich Reichert, ed., *Verbrannt bis zur Unkenntlichkeit: Die Zerstorung Dresdens 1945* (Dresden, 1994), p. 51.

231 experiences with refugees: Hoch, interview. Christoph Adam tells of his experience as a welfare worker, interview by the author. Dresden, November 2001.

231 "refugees, yes": Freundel, interview.

232 deportation order: Facsimile of deportation order in Haase, Jersch-Wensel, and Simon, eds., *Die Erinnerung hat ein Gesicht*, p. 180.

233 an interview with the Gestapo: Brenner, "*Das Lied ist aus*," p. 69f.

235 "last of the fighter-bombers": Account of Günter Jäckel's experiences in France from his memoir, "Die Dunkle Zeit: France-Comté 1944," in *Ostragehege 3*, p. 64ff. Details of his wound and his return to Dresden, based on Jäckel, interview.

236 "lay a Rhinelander": Jäckel, interview.

236 Hannelore Kuhn was born the middle child: Author interview with Fritz and Hannelore Kuhn. Dresden, October 2001.

237 there was no school: Author interview with Nora Lang.

238 almost exactly the same situation: Author interview with Anita Kurz.

239 "Dresden had been preserved": Adam, interview.

240 "my friend Fritz": Details from an untitled memoir by Günther Kannegiesser originating from the early 1990s, supplemented by an interview by the author, Dresden-Wachwitz, February 2002.

240 "large numbers of enemy aircraft": Like many other boys of the time, Günther and Fritz used to tune in to the Flak Radio Station Vom Horizont and had

cracked its grid code (which used a combination of letters and numbers—hence Martha Heinrich 8 for Dresden). If they were home, they would use this to plot the enemy aircraft movements in their school atlas.

Chapter 19: "Tally-Ho!"

242   the master bomber's duties: Irving, *Apocalypse 1945,* p. 141.

243   most likely target: Irving, *Apocalypse 1945,* p. 156.

243   "enemy combat units": Sequence of Local Air Raid Leadership's communications and warnings taken from Bergander, *Dresden im Luftkrieg,* p. 125ff. Also the following references of this kind unless otherwise indicated.

243   downed by unknown fire: The first represents Irving's view in *Apocalypse 1945,* p. 156, the second Bergander, *Dresden im Luftkrieg,* p. 128. In fact, Bergander thinks it possible that all the flak had actually been wthdrawn from Klotzsche.

244   "Achtung! Achtung!": Quoted in Bergander, *Dresden im Luftkrieg,* p. 126.

244   sector bombing: Described in Irving, *Apocalypse 1945,* p. 84.

245   at less than eight hundred feet: See Irving, *Apocalypse 1945,* p. 159f, and Cooper, *Target Dresden,* p. 150.

245   "the first bombs dropped": Bergander, *Dresden im Luftkrieg,* p. 126.

247   "an arc of thirty-two degrees": Hay, interview.

249   all of 49 squadron: Details of 49 Squadron's attack in squadron and individual aircraft reports in PRO, London, AIR 27/483.

Chapter 20: "Air Raid Shelter the Best Protection"

252   Nora and Anita's recollection: John and Lang, interview.

252   Günther and Siegfried's: experiences Kannegiesser, interview. At seventy-two, he still has headaches from the tiny pieces that embedded themselves in his temples almost sixty years ago.

253   "very good air raid shelter": Götz Bergander, interview by author, Belin, October 2001.

254   "series of whistling sounds": Griebel, *Ich war ein Mann der Strasse,* p. 254f.

255   some passengers were trampled: See Reichler, *Dresden Hauptbahnhof,* p. 54.

256   undershooting: Sir Arthur Harris, *Despatch on War Operations 23 February, 1942, to 8 May, 1945,* ed. Sebastian Cox (London, 1995), p. 81.

257   bomb loads: Figures regarding bomb loads and mix of bombs from *Bomber Command Intelligence Narrative of Operations No. 1007 (Night 13th/14th February),* PRO, London, AIR 14/3422.

258   "even in those places": "Kann der Selbstschutz auch Grossbrände bekämpfen?" in *Der Freiheitskampf,* October 23, 1943.

258 "the best protection": "Tagesspiegel: Luftschutzkeller bester Schutz" in *Der Freiheitskampf,* December 21, 1944.

259 Birke's eerie journey: Cited in Bergander, *Dresden im Luftkrieg,* p. 168. Bergander edited Herr Birke's recollections.

260 shut down the night shift: Brenner, *"Das Lied ist aus,"* p. 86.

261 Birke's interrogation: Cited in Bergander, *Dresden im Luftkrieg,* p. 166. Also for Georg Feydt's observations.

261 the deteriorating situation: Quoted in Bergander, *Dresden im Luftkrieg,* p. 166.

262 the raid was over: The Wolf family's experiences of the night Dresden was bombed are recounted in Brenner, *"Das Lied ist aus,"* p. 88ff.

263 ordered to another location: Quoted in Bergander, *Dresden im Luftkrieg,* p. 168, from Walter Weidauer, *Inferno Dresden* (Berlin, 1965), p. 106. Weidauer, an East German source, simply cites "im Staatsarchiv Dresden" and shows a photograph of the document. Neither Bergander nor this author has seen the document.

263 Mutschmann's whereabouts: See Weidauer, *Inferno Dresden,* p. 102. Though a highly tendentious and propagandizing writer—Weidauer was Communist high burgomaster of Dresden for many years and a leading Communist Party activist—he did have access to many documents and eyewitness accounts. He asserts that many of these locate the elusive gauleiter in his bunker at home rather than "at the head of his troops" in the air raid command shelter. Clearly, Mutschmann would not have been keen to advertise the fact that he had spent the raid safely ensconced with his family in their state-of-the-art private bunker while tens of thousands of their fellow Dresdeners died terrible deaths.

264 Johannstadt hospital complex: Annette and Jenni Dubbers, *Johannstadt: Aus der Geschichte eines Stadtteils,* (Dresden, 2000), p. 39f.

265 nurses did their best: See the article "Erkannt am rosa Kärtchen" in *Sächsische Zeitung,* February 13, 2002, which launched an appeal by two women who were orphaned that night. The publication of the article resulted in other children who had been saved—but who had lost their mothers—being brought together. A website, www.ueberlebendekinderdresden.de, is at the center of the ongoing project and presents several personal stories.

265 "curtains were burning": John, interview.

265 "my parents decided": Lang, interview.

266 "this first attack": Adam, interview.

268 "on the Striesener Strausse": Kannegiesser, interview and memoir. Bad Schandau is a small town on the Elbe about twenty miles southeast of Dresden. According to the fire officer Alfred Birke, quoted by Götz Bergander, all of this group perished in the second attack. Their known movements put the time young Günther saw them at after 1 A.M.

269 "Dresden was an inferno": This description is from Herr Kühnemund's letter to Alexander McKee for his book *The Devil's Tinderbox: Dresden 1945* (London,

1982, 2000). The letter is in Mr. McKee's papers in the Imperial War Museum, London. McKee's partial quotation, cited here, is in *Devil's Tinderbox,* p. 140f.

270 "there's a fire": Hannelore Kuhn, interview.

270 "fire-fighting forces": Rumpf quoted in Bergander, *Dresden im Luftkrieg,* p. 166.

272 "young woman had her baby": Nora Lang, quoted in Matthias Neutzner, ed., *Lebenszeichen: Dresden im Luftkrieg 1944/45* (Dresden, 1994), p. 9f.

Chapter 21: The Perfect Firestorm

274 cheered Miles Tripp: Tripp, *Eighth Passenger*, p. 80f.

275 Derek Jackson's recollections: letter and telephone interview with Derek Jackson, Manchester, July 2002.

275 watch and wait: Hay, interview.

275 further confusion: See Bomber Command Report on Night Operations 13th/14th February 1945 (Report No. 837), PRO, London, AIR 24/309. Other details of the second wave's approach from Bergander, *Dresden im Luftkrieg,* p. 132f, Irving, *Apocalypse 1945,* p. 174f, and Cooper, *Target Dresden* p. 158.

276 "new bomber force": Quoted in Bergander, *Dresden im Luftkrieg,* p. 133.

276 some disappointment felt: Irving, *Apocalypse 1945,* p. 175.

278 number of bombs: Figures collated (slightly differently) in Bergander, *Dresden im Luftkrieg,* p. 368f. Originals in Bomber Command Intelligence Narrative of Operations No. 1007 (Night 13th/14th February), PRO, London, AIR 14/3422.

280 "forty miles from Dresden": Tripp, *Eighth Passenger*, p. 83.

280 sixty or seventy miles: Cooper, *Target Dresden,* p. 158f.

280 waiting vainly for orders: Bergander, *Dresden im Luftkrieg,* p. 136.

281 signaling to the British: Irving, *Apocalypse 1945,* p. 179f.

281 Tripp made a decision: Tripp, *Eighth Passenger,* p. 83.

282 so he didn't: McKee in *Devil's Tinderbox* also cites (p. 202) Tripp's account, but assumes that like Admiral Nelson at the Battle of Copenhagen, Tripp set a "blind eye" to the master bomber's instructions. Tripp denied this vehemently in a later edition of his own book. De Wesselow's silence, he felt, enabled him to follow his conscience and not wreak further havoc in a city already beyond salvation (though there is actually no indication that he did "miss" the city). Tripp is, however, adamant that he would have followed specific instructions from the master bomber, no matter how personally unpalatable. In fact, Tripp's opinions about the bombing of Dresden were complex. He by no means belonged to the "war crime" camp on the issue.

282 Hicks's trial under fire: Doug Hicks, interview by author, Oshkosh, July 2002 and subsequent correspondence.

282 the second wave's work: Bomber Command Intelligence Narrative of Operations No. 1007 (Night 13th/14th February), PRO, London, AIR 14/3422.

Chapter 22: Catastrophe

286 Interview with and written memoir from Rudolf Eichner, Dresden, February 2002.

287 died of simple asphyxiation: See Neutzner, "Der Wehrmacht so nah verwandt," p. 216.

287 "through a long passage": Cited in Irving, *Apocalypse 1945,* p. 226.

287 Gertraud Freundel's family: Freundel, interview and correspondence.

290 "sea of flame": Berthold Meyer, "Flucht durch die brennende Blochmannstrasse," in *Dresdner Hefte 41, Dresden—Das Jahr 1945,* p. 49. Meyer, born in 1921, was bombed out of his homes in Bremen and Hannover and came to Dresden as a student at the Technical University. His report was written in the early part of 1945.

290 Anita escapes from the basement: John and Lang, interview.

291 Margret's escape: Margret Freyer, cited in Alexander McKee, *Devil's Tinderbox,* p. 171.

293 Hans Schröter's letter: Letter from Hans Schröter to Frau Ganze, August 5, 1945, reproduced in Reichert, ed., *Verbrannt bis zur Unkenntlichkeit,* p. 50f.

294 the vast tank: Description of the Altmarkt reservoir and the events of the night from Max Seydewitz, *Die unbesiegbare Stadt: Zerstörung und Neuaufbau von Dresden* (Berlin, 1955, rev. ed. 1982), p. 76ff.

295 out of a water container: Irving, *Apocalypse 1945,* p. 251.

295 Margret's search: Margret Freyer, cited in McKee, *Devil's Tinderbox,* p. 175.

296 destroyed by enemy aircraft fire: One of several references to "strafing" during the British raids. None of the RAF aircraft at Dresden was designed or equipped to do this, and by this stage in the night such an undertaking would have been not just impossible but utterly suicidal. Several Lancasters did crash in the city, however, and perhaps one such doomed aircraft had screamed overhead at low altitude just as a bomb struck the fire truck, creating the impression of an attack.

296 the commander's report: HStaD *Feuerwehrpolizei Bereitschaft 9* reports.

297 Berthold's journey to the river: Quoted from Meyer, *"Flucht durch die brennende Blochmannstrasse,"* p. 49f.

298 "explosions went on": Kannegiesser, interview.

299 ruins of the "B" wing: Seydewitz, *Die unbesiegbare Stadt,* p. 93f.

299 Vogelwiese: The Vogelwiese (bird meadow) was an area of the riverside meadows where fairs, amusements, and sporting activities were staged.

299 Günther's journey to the Elbe: Kannegiesser, memoir.

300 "much more shocking": Nora Lang, quoted in Neutzner, ed., *Lebenszeichen,* p. 10.

301 Nora reaches the Elbe: Lang, interview.

301 "We experienced the second attack": Adam, interview.

302 "soon we realized": Brenner, *"Das Lied ist aus,"* p. 92.

303 they hurried downstairs: Klemperer, *Klemperer Diaries*, p. 836f.

306 "no one could get them out": Brenner, *"Das Lied ist aus,"* p. 88.

306 "after nine o'clock": Turner, interview.

307 notorious "T4" program: See Thomas Schilter, *Unmenschliches Ermessen: Die nationalsozialistische "Euthanasie"-Tötungsnastalt Pirna-Sonnenstein 1940/41* (Leipzip, 1999).

308 orchestra and the choir: See Seydewitz, *Die unbesiegbare Stadt*, p. 41.

308 at the Universelle factory: See Traute Richter, "Die Schliessung des Theaters (Briefe an die Eltern September 1944)" in *Dresdener Hefte Nr 41*, p. 41ff.

309 "Sarrasani management announced": Details of Sarrasani and quotations from Ernst Günther, *Sarrasani, Wie er wirklich war* (Berlin, 1991), p. 290ff.

311 the Chinese acrobat: Interview with the acrobat's widow included in the German documentary series *Der Jahrhundertkrieg*, episode on the bomber war against Germany.

312 "middle of the square": Irving, *Apocalypse 1945*, p. 215, for the quotation and his own remarks.

314 "scarcely got into the hiding place": Prince Ernst Heinrich von Sachsen, *Mein Lebensweg*, p. 283.

Chapter 23: Ash Wednesday

316 "breakfast of bacon and eggs": Quoted in McKee, *Devil's Tinderbox*, p. 209.

317 an awful lot of American aircraft: Figures in Bergander, *Dresden im Luftkrieg*, p. 140.

318 the bomb loads to be carried: Bomb loads from Bergander, *Der Luftkrieg in Dresden*, p. 139.

318 "Dresden Marshaling Yard": Report in NARA, Washington, D.C., First Air Division, Intops Summary No. 290.

318 "marshaling yard in Chemnitz": Narrative Reports of Bomber Groups in NARA, Washington, D.C., First Air Division, Intops Summary No. 290.

318 definitive report: First Division report of Operations, Dresden, February 14, 1945, NARA, Washington, D.C.

319 "reason I remember": Quoted in McKee, *Devil's Tinderbox*, p. 213.

319 "like an embryo": Quoted in McKee, *Devil's Tinderbox*, p. 214f.

320 the dogfights that ensued: Returning from this mostly successful operation, more than three hundred German aircraft were shot down by their own artillery. Someone had forgotten to tell the German antiaircraft defenses about the Bodenplatte project.

320 "we were not briefed": Major Alden P. Rigby, telephone interview by the author, October 2002.

321 a supposedly temporary detour: For the saga of First Bombardment Wing's "detour," see Bergander, *Dresden im Luftkrieg,* p. 141f.

323 "I was awoken": Günther Kannegiesser, personal memorandum Dresden, February 2002.

324 "a bust of Hitler": Fritz and Hannelore Kuhn, interview by the author.

325 where there was less smoke: Intops Summary No. 290, NARA, and Report 41st CBW "A" Group, in NARA, Washington, D.C.

325 "thirty-six A/C": Narrative Report of Mission (plus individual reports of aircrew) for 303rd Bombardment Group, NARA, Washington, D.C.

326 some confusion ensued: Re: Field Order No.629, 457th—Dresden—February 14, 1945, NARA, Washington, D.C.

326 "due to dense contrails": This account on the trials of the 305th from Report of Operation, Dresden, of February 25 from HQ First Division, plus report 40 CBW February 16, 1945, both in NARA, Washington, D.C.

327 "standing in the factory yard": Bergander, interview.

328 "biblical pillars of fire": Jäckel, interview.

328 dropping heavy crates: Günter Jäckel is referring to the five-hundred-pound canisters of M17 incendiaries, which the Flying Fortresses carried in such unusually high numbers for the daylight raid on Dresden.

329 "we had this phosphor": Fritz and Hannelore Kuhn, interview. Also for the following.

330 "due to cloud cover": Intops Report No. 290, NARA, Washington, D.C.

330 the most serious damage: Cited in Bergander, *Dresden im Luftkrieg,* p. 148f.

332 for special persecution: Bergander, *Dresden im Luftkrieg,* p. 151.

Chapter 24: Aftermath

333 "my father came to me": John, interview.

334 "they had gone": Frau A. Kleinstück, interview by the author, Ottendorf-Akrilla, October 2001.

336 he would be back at school: Adam, interview.

336 they found refuge: Brenner, *"Das Lied ist aus,"* p. 93ff.

337 textile industry had declined: See Michael C. Schneider, "Die Wirtschaftsentwicklung von der Wirtschaftskrise bis zum Kriegsende" in *Sachsen in der NS-Zeit,* ed. Clemens Vollnhalls (Leipzig, 2002), p. 72f.

337 another crowded night: Bomber Command Intelligence Narrative of Operations No. 1008 14th/15th February 1945, PRO, London, AIR 14/3422.

338 abstract application of ethics: Tripp, *Eighth Passenger,* p. 90.

338 through thick cloud: Intops Summary No. 290, NARA, Washington, D.C. See also Bergander, *Dresden im Luftkrieg,* p. 154.

338 "bombing . . . appeared scattered": Bomber Command Intelligence Narrative no. 1008, PRO, London, AIR 14/3422.

338 glow of fires: Operations Record Book, RAF Bomber Command 1 Group, 14/15 February 1945, PRO, London, AIR 25/3.

338 "over the continent": Tripp, *Eighth Passenger,* p. 91.

339 without a scratch: Chemnitz was, however, to be bombed by the RAF again on the night of March 5–6, 1945, with 760 aircraft. This time, a third of the city's built-up area was destroyed and thirty-seven hundred of the city's inhabitants died. Among the important factories destroyed was the Siegmar plant, which made engines for German tanks. Sir Arthur Harris was nothing if not persistent.

340 "Frauenkirche is still standing": Fritz and Hannelore Kuhn, interview.

340 bricked up from the outside: For details see Wolfram Jäger and Dieter Rosenkrantz, "Der letzte Trümmerberg Dresdens sagt aus" in Reichert, ed. *Verbrannt bis zur Unkenntlichkeit,* p. 136ff.

341 the cloud stayed thick: See Bergander, *Dresden im Luftkrieg,* p. 158f.

341 display problems: Bergander, *Dresden in Luftkrieg,* p. 159.

342 they were picked up by police: See Seydewitz, *Die unbesiegbare Stadt,* p. 103ff. and "Die ganze Komödie dauerte zwei Tage." Gerichtliche Vergeltungsmaßnahmen gegenüber oppositionellen Tschechen aus dem "Protektorat Böhmen und Mähren" in Haase, Norbert, and Birgit Sack, eds. *Die Strafjustiz der Diktaturen und der historische Ort. Leipzig,* 2001. p. 84.

343 "this was the last straw": Bergander, interview.

343 "city gone": Lebenszeichen von Himmstädt, Ali dated 15.2.45 from Stadtarchiv Dresden Collection 13. Februar 1945 e.V. Handschriften.

343 "we just stared": Lang, interview.

Chapter 25: City of the Dead

347 "attacks were obviously very severe": Bergander, *Dresden im Luftkrieg,* p. 175f.

347 rescue and clearance: *Schlussmeldung über die vier Luftangriffe auf den LS-Ort Dresden am 13. 14. und 15. Februar 1945 vom 15. März 1945,* copy in HstA D and also reproduced in Weidauer, *Inferno Dresden,* p. 206ff (hereafter S-S-Final Report).

348 special powers: See Groehler, *Der Bombenkrieg gegen Deutschland,* p. 197.

348 armed with full powers: Bergander, *Dresden im Luftkrieg,* p. 165ff.

348 "rumor-mongering": *Der Freiheitskampf,* February 17, 1945, p. 2 in the cut-out-and-keep section "Was jeder wissen muss!" (What everyone must know): "Warm food will be . . . available to all. No one need worry about nourishment."

348 immediately executed: SS Final Report.

349 "Many holes were dug at once": Kurt Vonnegut, *Slaughterhouse-Five,* New York, 1969, p. 272f.

349 corpse-filled cellars: Papers of R. C. Dunford, former prisoner of war, in Imperial War Museum, London, 86/251.

349 basements had been emptied: Diary of Alec White, 1944–45 in Stadtarchiv Dresden 2001.131.

350 It became imperative": Ellgering quoted in Götz Bergander, *Dresden im Luftkrieg,* p. 179.

351 "huge grill racks": "Riesige Roste." See Bergander, *Dresden im Luftkrieg,* p. 180.

351 former staff from . . . Treblinka: See Groehler, *Der Bombenkrieg gegen Deutschland,* p. 412.

351 lengthy, meticulous report on air attack: SS Final Report.

351 situation report: Der Chef der Ordnungspolzei, Berlin den 22. März 1945 Betr: Luftangriffe auf das Reichsgebiet Lagemeldung 1404 (hereafter Situation Report 1404); copy in HStaD file *Feuerschutzpoilizei Dresden,* original at Bundesarchiv Koblenz.

351 "the total number of dead": SS Final Report.

352 saddest category of death: Lists of dead, street by street and house by house, in HStaD file *Feuerschutzpolizei Dresden.*

352 result of a pact: Prince Ernst Heinrich von Sachsen, *Mein Lebensweg,* p. 287.

353 "heat was like in an oven": Letter "Lieber Hans!" dated March 29, 1945, and letter of December 1997 to Günther Kannegiesser, copies supplied to the author by Herr Kannegiesser.

354 a death toll: SD reports cited in Bergander, *Dresden im Luftkrieg,* p. 314f.

354 government had "hushed up": See Longmate, *Air Raid,* p. 220.

355 list of streets: Matthias Neutzner, "Wozu leben wir nun noch? Um zu warten, bis die Russen kommen?" in Dresdner Hefte 41, p. 11.

355 "special case": For this and the following, see Neutzner, "Der Wehrmacht so nah verwandt," p. 217ff. See also Bergander, *Dresden im Luftkrieg,* p. 177.

356 for trains traveling southeast: See SS Final Report.

356 "back to normal": Neutzner, "Der Wehrmacht so nahe verwandt," p. 220. On March 2 the USAAF would launch another damaging attack on the Dresden railway system. This time, the railway bridge was worse hit. For a few days the line was once more impassable. The technical brigade planned to import nine thousand workers to put the Dresden system back in full use yet again, but by that stage this was a truly unrealistic ambition. Official accounts from the American side overemphasized the damage done to Dresden as a railway center in February 1945. German announcements erred in the opposite direction. The truth lies somewhere in between. Vital shipments could be brought through, albeit more slowly, but restoration of the pre–February 13 service was never more than partial.

356 remained simply "missing": Railway report of February 18, 1945, cited in Neutzner, "Der Wehrmacht so nah verwandt," p. 220.

356 twelve thousand dwellings: Situation Report 1404.

357 low military death toll: Situation Report 1404.

357 "all the barrack buildings": Hoch, interview.

357 two hundred factories: Figures for factories and damage from SS Final Report.

358 industry worst affected: Stadtarchiv Dresden Kriegsschädenamt (OB 1432) 11.4.45, cited in Reichert, ed., *Verbrannt bis zur Unkenntlichkeit,* p. 60.

358 Zeiss-Ikon's plants: Situation Report 1404.

359 effect on armaments: The exact estimate was arrived at by Dresden city archivist Friedrich Reichert, Reichert, ed., *Verbrannt bis zur Unkenntlichkeit,* p. 60.

359 "suffered seriously": Figures on Balda-Werke and estimates of damage to the Balda factory from Herbert Blumtritt, *Geschichte der Dresdner Fotoindustrie* (Stuttgart, 2001) p. 90f.

359 now said to be suitable: Stadtarchiv Dresden Kriegsschädenamt (OB 1431) 11.4.45, cited in Reichert, ed., *Verbrannt bis zur Unkenntlichkeit,* p. 60.

Chapter 26: Propaganda

360 "a big double attack": NARA, Washington, D.C., RG331, Box 15, Public Relations Division Briefings, February 14, 1945.

361 "first of all": Transcript of press briefing in NARA, Washington, D.C., RG331. Also for the following references to journalists' questions and Grierson's answers.

362 "story was passed": Memo, Lieutenant Colonel Richard H Merrick, Hotel Scribe, to Colonel Dupuy, American Express Building, Story by Cowan, AP, Re Bombing, February 18 1945, in NARA, Washington, D.C., RG331.

362 "this is entirely horrifying": Cecil King, *With Malice Towards None: A War Diary.* Entry for February 17, 1945, p. 290.

363 "the form of a bonus": Copy of cable from Reuters correspondent Steen in NARA Washington, D.C., RG331.

363 reorganize the entire press department: See memorandum from Deputy Chief of Air Staff (General D. M. Schlatter) to Bedell-Smith (and the latter's reply), in NARA, Washington, D.C., RG331.

363 "what happened on that evening": Stokes' speech in *Hansard* Parliamentary Debates. Commons. vol. 408, Col. 1901. See also Irving's detailed account in *Apocalypse 1945,* p. 312ff.

364 a finishing school: Cook and Nesbit, Target: Hitler's Oil, p. 182.

365 "Dresden also is being smashed": Stuart Ball, ed., *Parliament and Politics in the Age of Churchill and Attlee: The Headlam Diaries 1935–1951* (Cambridge, 1999), p. 447.

365 a personal reassurance: Webster and Frankland, *Strategic Air Offensive,* vol. 3, p. 113.

366 "he immediately said": See Len Giovannitti and Fred Freed, *The Decision to Drop the Bomb* (London, 1967), p. 41.

366 "Suddenly I was out there": Günter Jäckel, interview.

367 reportedly shaking with rage: Ian Kershaw, *Hitler 1936–1945: Nemesis* (London, 2000), p. 779.

367 with the same bitter determination: *Trial of the Major War Criminals before the International Military Tribunal,* vol. 16. Proceedings, June 21, 1946.

367 "Without Baggage": Cited in Erich Kästner's diary entry for March 8, 1945, in Erich Kästner, *Notabene 1945: Ein Tagebuch* (Zürich, 1961), p. 52.

368 fell down or became irreparable: See Speer, *Inside the Third Reich,* p. 428f.

368 terrible "phosphor" burns: Copies of these from the records of the German Embassy in Berne, Switzerland from archive of the German Foreign Office (*Archiv des Auswärtigen Amtes*), in the author's possession, courtesy of Dr. H. Schnatz of Koblenz.

368 the "terror" aspects: German Foreign Office document *Betr.: Bombenterror* dated January 16, 1945 (date-stamped by Berne Mission January 27) courtesy of Dr. H. Schnatz of Koblenz.

368 "his benighted name": German Foreign Office document *Manifest gegen Luftmarschal Harris,* dated February 7, 1945 (date-stamped by Berne Mission February 12) courtesy of Dr. H. Schnatz of Koblenz.

369 "wholly possible for the residential areas": Copy of DNB press release of February 16, 1945, from Germany Embassy, Berne, courtesy of Dr. H. Schnatz of Koblenz.

369 "Dresden, Victim of the Air War": Telegram from Foreign Office, Berlin, February 19, 1945 to Embassy Berne (registered there February 20, 1945), courtesy of Dr. H. Schnatz of Koblenz.

370 columns of fleeing vehicles: Bergander, *Dresden im Luftkrieg,* p. 214.

370 an extra zero: See Weidauer, *Inferno Dresden,* pp. 111–112.

371 "150,000 dead": Fritz and Hannelore Kuhn, interview.

371 his own private estimate: *Trial of the Major War Criminals before the International Military Tribunal,* vol. 17. Proceedings, June 29, 1946 (afternoon). The evidence was given by senior Propaganda Ministry official Hans Fritzsche, who was present at that conference.

Chapter 27: The Final Fury

373 the greatest tonnage: Figures from Middlebrook and Everitt, *Bomber Command War Diaries,* p. 645.

373 he devoted more effort: See especially Sebastian Cox's remarks in his introduction to Harris, *Despatch on War Operations,* p. xxiii.

373 not . . . entirely risk-free: Some fell victim to the usual foul-ups, collisions and accidents, but most were shot down by flak or enemy fighters. Eighteen Lan-

casters of 5 Group (7 percent) were lost over the Lützkendorf oil plant on March 14–15, 1945 (as John Aldridge, who took part in that raid, reminded the author), while on March 16–17 German night fighters got in among No.1 Group's bomber stream on its way to Nuremburg and shot down twenty-four aircraft, 10.4 percent of the group's strength.

374 six days of fierce fighting: Friedrich, *Der Brand,* p. 313ff.

374 "stopped production so effectively": Cited in Webster and Frankland, *Strategic Air Offensive,* vol. 3, p. 264.

375 these were never used again: Ian V. Hogg, *German Secret Weapons of the Second World War,* (London/Mechanicsburg, 1999), p. 43.

375 "moment has come": Text of the Churchill memorandum in PRO, London, CAB 121/3, Bombing Policy in North-West Europe.

378 "The feeling, such as there is, over Dresden could be easily explained by a psychiatrist.": Henry Probert, *Bomber Harris,* p. 322.

378 a fairly rare event: Field Marshal Lord Alanbrooke, *War Diaries 1939–45* (London, 2001), p. 679 (entry for March 29, 1945).

379 notoriously prone to self-contradiction: R. G. Casey, *Personal Experience 1939–46* (London, 1962), p. 166 (diary entry for June 27, 1943). Casey, a conservative Australian politician, was British representative in the Middle East with ministerial rank in 1942–43. He responded flatly to Churchill that "we hadn't started it, and . . . it was them or us."

379 "more harm to ourselves": Text of revised memorandum and the chiefs of staff's response in Webster and Frankland, *Strategic Air Offensive,* vol. 3, p. 117f.

380 most straightforwardly justifiable: For the estimated April 17 casualty figures, see Neutzner, ed., *Martha Heinrich Acht,* p. 103.

381 its long-held aim: See Bergander, *Dresden im Luftkrieg,* p. 249ff. The chapter is entitled "The Forgotten Attack." The oddity is, many Dresdeners have forgotten the April 17 raid ever took place. Some have since denied its existence pointblank, so powerfully does the horror of the firestorm overshadow the rest of the city's suffering and distort the shape of its victims' memories. Bergander remarked to the author: "The 17 April raid is denied by many people in Dresden. I had a discussion with a woman who said it didn't happen. What can you do?"

Chapter 28: The War Is Over. Long Live the War

383 just west of the city: Rahne, "Die 'Festung Dresden' von 1945," p. 29.

383 "Dresden was the only city": Hoch, interview.

384 "Russians went into the houses": From *Eine Familienchronik 1946,* quoted in Cornelia Adam, "Vergewaltigungen in Dresden nach 1945," in *Dresdner Hefte 53, Dresden als Garnisonstadt* p. 61ff.

384 a letter from a Herr G.: Letter from Herr G. to the High Burgomaster of Dresden quoted in Adam, "Vergwaltigungen in Dresden nach 1945."

385 a rationing system: *Dresden: Die Geschichte der Stadt,* p. 238.

385 for the average German: See Vollnhalls, ed., *Sachsen in der NS-Zeit,* p. 236.

386 "stunned and silent population": Seydewitz, *Die unbesiegbare Stadt,* p. 208f.

387 "famously have you fought": Excerpts from the Special Order of the Day reproduced in Probert, *Bomber Harris,* p. 344, and for Harris's reactions to apparent slights in this period.

387 reluctant to say anything: Probert, *Bomber Harris,* p. 345.

387 service medals: Those serving sixty hours or more with an operational squadron between September 3, 1939, and June 5, 1994, were awarded the Air Crew Europe Star. After the D-Day landings they were entitled to the France and Germany Star, as were all forces on ground, sea or air directly involved in the conquest of the continent.

388 "the Defence Medal and no other": Quoted in Probert, *Bomber Harris,* p. 348.

389 weapons of the future: John Colville, *On the Fringes of Power: 10 Downing Street Diaries 1939–1955* (London, 1985), p. 564 (entry for February 23, 1945).

390 "anything that makes": Quoted in Friedrich Reichert, "Zur Rezeptions-geschichte des 13. Februar 1945"in *Verbrannt bis zur Unkenntlichkeit: Die Zerstorung Dresden 1945,* (Dresden, 1994), p. 151. And for the following words of Major Broder.

391 "one of Germany's armories": W. A. Ruben, "Abschaum der Menschheit" in *Tageszeitung für die deutsche Bevölkerung,* June 2, 1945, quoted in Bergander, *Dresden im Luftkrieg,* p. 294f.

391 "our sole vow": *Sächsische Zeitung,* February 13, 1947 quoted in Reichert, "Zur Rezeptionsgeschichte," p. 151.

392 "not with glory but with dishonor": Sächsische Zeitung, February 12, 1949, cited by Reichert, "Zur Rezeptionsgeschichte," p. 152.

392 in the street fighting: See Beevor, *Berlin: the Downfall,* p. 424.

393 Truman and Eisenhower: Cited by Reichert "Zur Rezeptionsgeschichte," p. 153.

393 in 1954 the death toll: See Bergander, *Dresden im Luftkrieg,* p. 295.

Chapter 29: The Socialist City

395 population in August 1945: *Dresden: Die Geschichte der Stadt,* p. 237.

396 "at least seventy years": Figures and Conert's prognosis in Lerm, *Abschied vom Alten Dresden,* p. 35.

Chapter 30: The Sleep of Reason

402 the extent of German conquests: Bergander, *Dresden im Luftkrieg,* p. 347.

402 their promise of "precision bombing": Richard G. Davis, *Carl A. Spaatz and the Air War in Europe* (Washington, D.C., 1993), p. 569.

404 "once the conflict was over": Richard Overy, *Why the Allies Won* (London, 1995), p. 22.

404 "dimensions of the bombing offensive": W. G. Sebald, *On the Natural History of Destruction: With Essays on Alfred Andersch, Jean Améry and Peter Weiss* (German: *Luftkrieg und Literatur*) (London, 2003), p. 18.

404 "damage to private homes": Goebbels, *Tagebücher*, April 6, 1943, entry.

405 "with respect to the air war": Groehler, *Der Bombenkrieg gegen Deutschland*, p. 392.

406 less than half the size: See Middlebrook and Everitt, *Bomber Command War Diaries*, p. 669.

406 also carefully planned: Groehler, *Der Bombenkrieg gegen Deutschland*, p. 404.

406 "Dresden, the seventh largest city": Hastings, *Bomber Command*, p. 254.

407 "a significant effect on the Russians": Quotes (in German) in Groehler, *Der Bombenkrieg gegen Deustschland*, p. 404.

407 "we respected the Russian army": McKee, *Devil's Tinderbox*, p. 119.

407 "still more impress": De Wesselow, quoted by McKee, *Devil's Tinderbox*, p. 165.

409 into five figures: Friedrich, *Der Brand*, p. 175f.

409 "an exciting variation": Craven and Cate, eds., *Army Air Forces in World War II*, vol. 3, p. 741f, and for an account of the raid from an American point of view.

409 a special case: See Friedrich, *Der Brand*, p. 515 ff, and Groehler, *Der Bombenkrieg gegen Deutschland*, p. 306ff.

410 interrogation of Mutschmann: Text of Mutschmann's interrogation in Weidauer, *Inferno Dresden*, p. 107f.

411 "how should one treat them": Quoted in Beevor, *Berlin; the Downfall*, p. 34.

411 the same number of Soviet citizens: Figure of Russian air raid dead cited by Richard Overy in "There Was a War On" (review of Robin Neilland's *The Bomber War*) in *The Literary Review*, February 2001.

411 systematic bombing of Stalingrad: See Anthony Beevor, *Stalingrad* (London, 1998), p. 106.

412 "if the war is lost": Speer, *Inside the Third Reich*, p. 588.

413 "no longer believed in victory": Bergander, *Dresden im Luftkrieg*, p. 349.

414 frankly admitted defeat: Speer, *Inside the Third Reich*, p. 567.

414 "the indirect effects": See Overy, *Why the Allies Won*, p. 130ff.

415 most popular destination: See map in Groehler, *Der Bombenkrieg gegen Deutschland*, p. 288.

415 surprising secret location: Copy of Radio-Mende document courtesy of Holger Starke of the Dresden City Museum.

417 "what began as routine": Bergander, *Dresden im Luftkrieg*, p. 350.

Afterword: Commemoration

420 "how many died": *Wieviele starben? Wer kennt die Zahl?/An deinen Wunden sieht man die Qual/der Namenlosen, die hier verbrannt/im Höllenfeuer aus Menschenhand.*

422 such a provocative act: It is not customary in Britain to erect statues of living people. Harris died in 1984. In 1988 a statue had been erected outside St. Clement Danes—a church that had been almost entirely destroyed by the Luftwaffe in 1941and had strong RAF connections—to honor Lord Dowding, head of the RAF's Fighter Command during the Battle of Britain. Bomber Command veterans thought it fair that Harris (and thereby Bomber Command) should also be honored on the same site. Over the next three years, funds were raised by public subscription (mainly through the Bomber Command Association) and a commission awarded to the same sculptor who had created the Dowding memorial. By 1992 the statue was ready.

425 "disturbing the Dresden legends": Bergander, *Dresden im Luftkrieg,* p. 187.

Appendix A: The "Massacre on the Elbe Meadows"

429 "almost entirely ineffective": Intops Report No. 290, NARA, Washington, D.C.

429 "didn't really leave altitude": Rigby, interview.

430 "the city itself": Quotation from Axel Rodenberger, *Der Tod von Dresden: Bericht vom Sterben einer Stadt* (Dortmund, 1953), p. 140f.

431 further stoked the fires: Max Seydewitz, *Zerstörung und Wiederaufbau von Dresden (Destruction and Reconstruction of Dresden)* (East Berlin, 1955).

431 "fired at the people on the ground": Seydewitz, *Zerstörung und Wiederaufbau von Dresden,* p. 80.

431 "Mustang fighter": Irving, *Apocalypse 1945,* p. 236f.

432 account by the cantor: Seydewitz, *Die unbesiegbare Stadt,* p. 87.

432 most recently revised edition: The online version of *Apocalypse 1945,* available from Mr. Irving's website, www.fpp.co.uk.

433 "low across the city": Irving, *Apocalypse 1945,* p. 191f, for this and following.

433 an attack on a vehicle: Bergander, *Dresden im Luftkrieg,* p. 200.

433 his account of the alleged massacre: Irving, *Apocalypse 1945,* p. 191.

434 "he saw no low-flying fighters": Letter from Günther Kannegiesser to Götz Bergander, December 22, 1994; confirmed in conversation with this author, February 2002.

434 "during the midday raid": Werner Ehrlich, letter of March 2, 1985, in archive of Götz Bergander, Berlin.

435 "'B' group fighters": Irving, *Apocalypse 1945,* p. 190f.

437 the specific orders: Intops Report No. 290, NARA, Washington, D.C. Also see the discussion in Bergander, *Dresden im Luftkrieg,* p. 198f. Also for the following quotation.

437 "this then is the order": Bergander, *Dresden im Luftkrieg,* p. 199.

437 "went to deck for strafing": Oakland Summary—F.O. 1622A, February 14, 1945, in NARA, Washington, D.C.

438 a mistake was admitted: Headquarters First Division, Report on Operations, NARA, Washington, D.C.

438 "uneventful escort to target": Fighter Flash Reports, February 14, 1945, (352nd A and B Groups), in NARA, Washington, D.C.

439 "they were immediately bounced": 306th Group Narrative—Mission to Dresden, February 14, 1945, in NARA, Washington, D.C. This is also mentioned in Intops Summary No. 290 (C. Intelligence. 1. Enemy Air Opposition), NARA, Washington, D.C.

439 numerous authoritative articles: Dr. Helmut Schnatz, *Der Luftkrieg im Raum Koblenz 1944/45. Eine Darstellung seines Verlaufs, seiner Auswirkungen and Hintergründe.* Boppard a.Rh., 1981.

440 "ablaze with phosphor": Quote from "Feuersturm," special edition of *Die* Wochenpost, *February 9, 1996, quoted in Helmut Schnatz,* Tiefflieger über Dresden? Legenden und Wirklichkeit *(Cologne, 2000) p. 129.*

442 research among the records: E-mail from Herr Krause, who has researched the archives of the Brockhaus-Verlag, Leipzig, publishers of the 1982 edition, in possession of the author.

Appendix B: Counting the Dead

443 "word for word": Rodenberger, Der Tod von Dresden, p. 168.

445 "open use can be made": Quoted in Bergander, *Dresden im Luftkrieg,* p. 222.

446 and then some: Irving, *Apocalypse 1945,* p. 199.

446 a diary entry claiming: It is possible that the diarist, a doctor, had been shown the faked estimate. McKee does not discuss this possibility.

446 "might easily be doubled": McKee, *Devil's Tinderbox,* p. 322.

447 "greater than the 35,000": Seydewitz, *Der unbesiegbare Stadt,* p. 155.

447 ad hoc counting methods: See Reichert, ed., *Verbrannt bis zur Unkenntlichkeit,* p. 55ff.

448 "thousands of victims": Irving, *Apocalypse 1945,* p. 322.

448 no bodies have been recovered: Friedrich Reichert, *Verbrannt bis zur Unkenntlichkeit,* p. 61.

Appendix C: Legends of the Fall

449 two enterprising German brothers: See John Noble's autobiography, *I Found God in Soviet Russia* (London, 1959), p. 16ff.

449 an exchange: Details unless otherwise indicated from Blumtritt, *Geschichte der Dresdner Fotoindustrie,* p. 73ff.

450 the Noble factory produced: Blumtritt, *Geschichte der Dresdner Fotoindustrie,* and Seydewitz, *Der unbesiegbare Stadt,* p. 188.

450 Blumtritt . . . expresses suspision: Blumtritt, *Geschichte der Dresdner Fotoindustrie.*

450 "Noble's tasks": For the entire Noble conspiracy story see Seydewitz, *Die unbesiegbare Stadt,* p. 189ff.

453 an American spy lived there: Petra Jacoby, ed., *Leben in Dresden 1920–1990: Erzählte Geschichte* (Erfurt, 2000) p. 30, account of Ralph Hoxhold (born 1935).

454 "100 Dresdeners of the Century": *Mit "Prakitflex" und "Praktica" zu Weltruhm* in *Dresdner Neueste Nachrichten* (100 Dresdner des Jahrhunderts)in www.dnn-online.de. See also Noble, *I Found God in Soviet Russia,* and for the postwar fate of the Noble Camera Workshops and its owners, Blumtritt, *Geschichte der Dresdner Fotoindustrie,* p. 129ff.

454 the Americans had issued a threat: Weidauer, *Inferno Dresden,* p. 63.

455 the main choice facing: See Giovannitti and Freed, *Decision to Drop the Bomb,* p. 20f, for this and the following point.

456 "if, despite all this": Weidauer, *Inferno Dresden,* p. 69.

SOURCES AND BIBLIOGRAPHY

SOURCES

Public Record Office, London

Target Information Sheets and Maps: Dresden.

Dresden: Südvorstadt air armaments: target information.

Dresden: Cartonnagenindustrie paper factory: target information.

Bomber Command HQ, February 15, 1945. Immediate Interpretation Report No. K3742.

Bomber Command Intelligence Narrative of Operations, No. 1007, February 14, 1945.

Night Raid Report No. 837. Bomber Command Report on Night Operations, February 13-14, 1945.

Bomber Command Intelligence Narrative of Operations, No. 1008, February 14-15, 1945.

Bomber Command Intelligence Reports and Intelligence Narrative of Operations, February 1945.

Cabinet Papers Chiefs of Staff, Strategic Bombings in Reaction to the Present Russian Offensive.

Operations Record Book, No. 1 Group, February 13-14, 1945.

Bomber Command No. 1 Group, Appendices for February 1945.

Bomber Command No. 3 Group, Appendices for February 1945.

Bomber Command No. 5 Group, Appendices for February 1945.

Bomber Command No. 6 Group, Appendices for February 1945.

Bomber Command No. 8 Group, Appendices for February 1945.

Interpretation Report No. K4171, April 19, 1945.

Operations Record Book, Forty-ninth Squadron, January-December 1945.

National Archives and Record Service, Washington D.C.

Headquarters, Eighth Air Force, AAF Station 101. Intops Summaries, 1944–45.

First Air Division Bomber Groups as cited.

Fighter Groups 78th, 339th, 353rd, 357th. Combat Mission Reports, Field Orders, Mission Reports, etc. as cited.

RG 331, SHAEF Public Relations Briefings.

RG 43, World War II Conferences, Yalta (Crimea) Conference.

Microfilm 5018, 5019, Field Orders (with Annexes); Mission Reports; Divisional, Wing, and Squadron Reports.

Sächsisches Hauptstaatsarchiv, Dresden

COMPANY RECORDS

Universelle-Werke J. C. Müller & Co.
Clemens Müller
Richard Gäbel
Oskar Hantzsche
Siemens-Wärme
Radio-Mende

FEUERSCHUTZPOLIZEI DRESDEN

Einsatzberichte
Totenlisten

Stadtarchiv Dresden

Sammlung Interessengemeinschaft, "Februar 13 1945" e.V.
Diary of Alec White

Imperial War Museum, London

Papers, R. C. Dunford
Papers, Robert Lee

Interviews and Correspondence

BRITAIN AND COMMONWEALTH

John Aldridge
Lesley Hay
Doug Hicks
John Hurst
Derek Jackson
Fred Jones
Hugh Lunghi
Vaughan Southam
John Whitely

GERMANY

Dr. Christoph Adam
Götz Bergander
Henny Brenner (née Wolf)
Steffen Cüppers
Rudolf Eichner
Gertraud Freundel
Dr. Karl-Ludwig Hoch
Professor Dr. Günter Jäckel
Anita John (née Kurz)
Günther Kannegiesser
Frau A. Kleinstück
Hannelore and Fritz Kuhn
Nora Lang

UNITED STATES

Major Alden P. Rigby

ISRAEL

Ilana Turner

BIBLIOGRAPHY

Works referred to in the text

Dresden

CITY HISTORY

Beyer, Manfred. "Dresden als Keimzelle des militärischen Widerstandes: Die Garnison in der NS-Zeit." In *Dresdner Hefte 53: Dresden als Garnisonstadt*. 1998.

Blumtritt, Herbert. *Geschichte der Dresdener Fotoindustrie*. Stuttgart, 2001.

Dresdner Geschichtsverein, eds. *Dresden: Die Geschichte der Stadt*. Dresden, 200.

Dubbers, Annette and Jenni. *Johannstadt Aus der Geschichte eines Stadtteils*. Dresden, 2000.

François, Etienne, and Hagen Schulze, eds. *Deutsche Erinnerungsorte III*. Munich, 2001

Günther, Ernst *Sarrasani Wie er wirklich war*. Berlin, 1991

Haase, Norbert, and Birgit Sack, eds. *Münchner Platz Dresden Die Strafjustiz der Diktaturen und der historische Ort*. Leipzig, 2001

Jäckel, Günter. *Dresden in der Goethezeit*. Berlin, 1990.

Jacoby, Petra, ed. *Leben in Dresden*. Erfurt, 2000.

Lagler, Kurt-Joachim, and Heinz Weise, eds. *Dresdner Kuriosa*. Dresden, 2001.

Lerm, Matthias. *Abschied vom Alten Dresden: Verluste historischer Bausubstanz nach 1945*. Rostock, 2000.

———. *Konzepte für den Umbau der Stadt Dresden in den 30er und frühen 40er Jahren des 20. Jahrhunderts in Vorträge und Forschungsberichte: 4 Kolloquium zur dreibändigen Stadtgeschichte 2006 vom 18. März 2000*.

Neutzner, Matthias. "'Der Wehrmacht so nahe verwandt'—Eisenbahn in Dresden 1939 bis 1945." In *Dresdner Geschichtsbuch 5*. Altenburg, 1999.

Pommerin, Reiner, ed. *Dresden unterm Hakenkreuz*. Cologne/Weimar/Vienna, 1998.

Reichier, Peter. *Dresden Hauptbahnhof*. Eggiham, 1998.

Schilter, Thomas. *Unmenschliches Ermessen: Die nationalsozialistische "Euthanasie"-Tötungsanstalt Pirna-Sonnenstein 1940/41*. Leipzig, 1999.

Starke, Holger. "Vom Werkstättereal zum Industriegelände: Die Entwicklung des Industriegebietes an der Königsbrücker Strasse in Dresden vor der Entstehung der Albertstadt bis zur Auflösung der Industrieanlagen Nord (1873–1952)." In *Dresdner Geschichtsbuch 5*. Altenburg, 1999.

Szejnmann, Claus-Christian W. *Vom Traum zum Alptraum: Sachsen in der Weimarer Republik*. Leipzig, 2000.

Vollnhals, Clemens, ed. *Sachsen in der NS-Zeit*. Leipzig, 2002.

JEWISH LIFE IN DRESDEN

Haase, Norbert; Stefi Jersch-Wenzel; and Hermann Simon, eds. Bearbeitet von Marcus Gryglewski. *Die Erinnerung hat ein Gesicht: Fotografien und Dokumente zur nationalsozialistischen Judenverfolgung in Dresden 1933-45*. Leipzig, 1998.

Gilbert, Martin. *The Boys: Triumph over Adversity*. London, 1996.

HATiKVA, ed. *Spurensuche Juden in Dresden Ein Begleiter durch die Stadt*. Dresden 1996.

Heinrich, Jürgen. *Die Synagoge zu Dresden*. Taucha, 2001.

Kirsch, Günther. "Die gesetzliche und aussergesetzliche Judenverfolgung in Dresden und Sachsen in den ersten Monaten der nationalsozialistischen Herrschaft," in *Historische Blätter. Aus Politik und Geschichte* (1994), vol. 4.

Kirsch, Ingrid. "Das Ringen um die Gleichstellung der Dresdner: Juden und ihre Religionsgemeinde von 1830 bis 1871." In *Dresdner Hefte 45: Zwischen Integration und Vernichtung*. rev. ed., 2000.

Kolditz, Gerald. "Zur Entwicklung des Antisemitismus in Dresden während des Kaisserreichs." In *Dresdner Hefte 45: Zwischen Integration und Vernichtung*. Rev. ed., 2000

Lichtenstein, Heiner. *Mit der Reichsbahn in den Tod*. Cologne, 1985.

Mosse, W. E. *The German-Jewish Economic Elite 1820–1935: A Socio-Cultural Profile*. Oxford/New York, 1989.

Ullrich, Uwe. *Zur Geschichte der Juden in Dresden*. Dresden, 2001.

DRESDEN 1945

German

Adam, Cornelia. "Vergewaltigungen in Dresden nach 1945." In *Dresdner Hefte 53: Dresden als Garnisonstadt* 1998.

Bergander, Götz. *Dresden im Luftkrieg: Vorgeschichte, Zerstörung, Folgen*. Rev. ed. Würzburg, 1998,.

Hoffmann, Daniel. *Der Knabe im Feuer: Ein Erlebnisbericht von Dresdens Untergang*. Berlin, 1957.

Kempowski, Walter. *Der Rote Hahn: Dresden im Februar 1945*. Munich, 2001.

Landeszentrale für politische Bildung. *Hamburg und Dresden im dritten Reich: Bombenkrieg und Kriegsende*. Hamburg 1993.

Meyer, Berthold. "Flucht durch die brennende Blochmannstrasse." In *Dresdner Hefte 41, Dresden—Das Jahr 1945*.

Neutzner, Matthias, ed. *Lebenszeichen: Dresden im Luftkrieg 1944/45*. Dresden, 1994.

———. ed. *Martha Heinrich Acht: Dresden 1944/45*. Dresden, 2000.

Rahne, Hermann. "Die 'Festung Dresden' von 1945," in *Dresdner Hefte 41, Dresden—Das Jahr 1945*.

Reichert, Friedrich, ed. *Verbrannt bis zur Unkenntlichkeit: Die Zerstörung Dresdens 1945*. Dresden, 1994.

Rodenberger, Axel. *Der Tod von Dresden: Bericht vom Sterben einer Stadt*. Dortmund, 1953.

Schnatz, Helmut. *Tiefflieger über Dresden: Legende oder Wirklichkeit?* Cologne, 2000.

Seydewitz, Max. *Die unbesiegbare Stadt: Zerstörung und Neuaufbau von Dresden*. Berlin, 1955, rev. ed. 1982.

Weidauer, Walter. *Inferno Dresden*. Berlin, 1965.

English

Cooper, Alan. *Target Dresden*. Bromley, 1995.

Irving, David. *Apocalypse 1945: The Destruction of Dresden*. Focal Point, London, n.d.

———*The Destruction of Dresden*. London, 1963.

McKee, Alexander. *The Devil's Tinderbox: Dresden 1945*. London, 1982, 2000

Russell, Alan, ed. *Why Dresden?* Arundel, 1998.

MEMOIRS/FICTION

Brenner, Henny (Wolf). *"Das Lied ist aus": Ein jüdisches Schicksal in Dresden*. Zürich/Munich, 2001.

Griebel, Otto. *Ich War ein Mann der Strasse: Lebenserinnerungen Eines Dresdner Malers*. Altenburg, 1995.

Kästner, Erich. *Als Ich ein kleiner Junge war*. Zürich, 1957.

——. *Notabene 1945: Ein Tagebuch*. Zürich, 1961.

Klemperer, Victor. *The Klemperer Diaries 1933–45: I Shall Bear Witness Until the Bitter End*. Abridged and trans. from the German ed. by Martin Chahners. London, 2000.

Kokoschka, Oskar. *My Life*. London, 1974.

Noble, John. *I Found God in Soviet Russia*. London, 1959.

Peter, Richard. *Dresden—Eine Kamera Klagt An*. Dresden, 1949.

Sachsen, Prince Ernst Heinrich von. *Mein Lebensweg: vom Königsschloss zum Bauernhof*. Dresden/Basel, 1995.

Vonnegut, Kurt. *Slaughterhouse-Five*. New York, 1969.

Germany at War

Becker, Cajus. *The Luftwaffe War Diaries*. London, 1967.

Beevor, Anthony. *Berlin, the Downfall*. London, 2002.

——. *Stalingrad* London, 1998.

Bethell, Nicholas. *The War That Hitler Won: September 1939*. London, 1972.

Burleigh, Michael. *The Third Reich: A New History*. London, 2000.

Davies, Norman, and Roger Moorhouse. *Microcosm: Portrait of a Central European City*. London, 2002.

Fest, Joachim. *Speer: The Final Verdict*. London, 2001.

Friedrich, Jörg. *Der Brand: Deutschland im Bombenkrieg*. Munich, 2002.

——. *Das Gesetz des Krieges*. Munich, 1993.

Gellately, Robert. *Backing Hitler: Consent and Coercion in Nazi Germany*. Oxford, 2001.

Goebbels, Josef. *The Goebbels Diaries, 1939–41*. Ed. and trans. by Fred Taylor. London, 1982.

——. *Tagebücher* 1924–1945 (5 vols.). Munich, 1999.

Gross, Leonard. *The Last Jews in Berlin*. London, 1983.

Groehier, Olaf. *Der Bombenkrieg gegen Deutschland*. Berlin, 1990.

——. "Bombenkrieg gegen Leipzig 1940–1945," In *Texte des Leipziger Geschichtsvereins e.V. Heft 4*. (1994).

Höffkes, Karl. *Hitlers Politische Generale: Die Gauleiter des Dritten Reiches*. Tübingen, 1986.

Huttenberger, Peter. *Die Gauleiter: Studie zum Wandel des Machtgefüges in der NSDAP*. Stuttgart, 1969.

Johnson, Eric. *The Nazi Terror: Gestapo, Jews and Ordinary Germans*. London, 2000.

Kershaw, Ian. *Hitler 1936–1945: Nemesis*. London, 2000.

Sebald, W. G. *Luftkrieg und Literatur/On the Natural History of Destruction: With essays on Alfred Andersch, Jean Améry and Peter Weiss*. London, 2003.

Speer, Albert. *Alles, was ich weiss*. Munich, 1999.

———. *Inside the Third Reich*. London, 1970.

Stegmann, Dirk. Die *Erben Bismarcks: Parteien und Verbände in der Spätphase des Wilhelminischen Deutschlands*. Cologne/Berlin, 1970.

The Allies at War

Alanbrooke, Field Marshal Lord. *War Diaries 1939–45*. London, 2001.

Ball, Stuart, ed. *Parliament and Politics in the Age of Churchill and Attlee: The Headlam Diaries 1935-1951*. Cambridge, 1999.

Britain Under Fire. With a foreword by J. B. Priestley. London, 1941.

Buckley, John. *Air Power in the Age of Total War*. Bloomington, Ind., 1999.

Casey, R. G. *Personal Experience 1939–46*. London, 1962.

Colville, John. *On the Fringes of Power: 10 Downing Street Diaries 1939–1955*. London, 1985.

Cook, Ronald C., and Roy Conyers Nesbit. *Target: Hitler's Oil-Allied Attacks on German Oil Supplies 1939–45*. London, 1985.

Craven, Wesley Frank, and James Lea Cate, eds. *The Army Air Forces in World War II*, vol. 3. Chicago, 1951.

Davis, Richard G. *Carl A. Spaatz and the Air War in Europe*. Washington, D.C., 1993.

Deighton, Len. *Blood, Tears and Folly: An Objective Look at World War II*. London, 1993.

Frankland, Noble. *History at War*. London, 1998.

Freeman, Roger A. *The Mighty Eighth*. London, 1970, 2000.

Garrett, Stephen A. *Ethics and Airpower in World War II. The British Bombing of German Cities*. New York, 1996.

Giovannitti, Len, and Fred Freed. *The Decision to Drop the Bomb*. London, 1967.

Harris, Sir Arthur. *Bomber Offensive*. London, 1947.

———. *Despatch on War Operations 23 February, 1942, to 8 May, 1945*. Ed. and with an introduction by Sebastian Cox. London, 1995.

Hastings, Max. *Bomber Command*. London, 1979, rev. ed. 1999.

Hogg, Ian V. *German Secret Weapons of the Second World War*. London/Mechanicsburg, 1999.

Howarth, Patrick. *Intelligence Chief Extraordinaire: The Life of the Ninth Duke of Portland*. London, 1986.

King, Cecil. *With Malice Towards None: A War Diary*. London, 1970.

Longmate, Norman. *Air Raid: The Bombing of Coventry*. London, 1976.

———. *The Bombers*. London, 1983.

Middlebrook, Martin. *The Battle of Hamburg: The Firestorm Raid*. London, 1980.

———. *The Berlin Raids: RAF Bomber Command Winter 1943–44*. London, 1988.

Middlebrook, Martin, and Chris Everitt. *The Bomber Command War Diaries: An Operational Reference Book*. Leicester, 2000.

Moran, Lord. *Winston Churchill: The Struggle for Survival 1940–1945*. London, 1966.

Neillands, Robin. *The Bomber War: Arthur Harris and the Allied Bomber Offensive 1939–1945*. London, 2001.

Overy, Richard. *Why the Allies Won*. London, 1995.

Probert, Henry. *Bomber Harris, His Life and Times*. London/Mechanicsburg, 2001.

Richards, Denis. *The Hardest Victory: Bomber Command in the Second World War*. London, 1994.

Rothnie, Niall. *The Baedeker Blitz: Hitler's Attack on Britain's Historic Cities*. Shepperton, 1992.

Saward, Dudley. *"Bomber" Harris*. London, 1984.

Tripp, Miles. *The Eighth Passenger: A Flight of Recollection and Discovery*. London, 1969, rev. ed. 1993.

Ward, John. *Beware of the Dog of War: An Operational Diary of 49th Squadron*. Belper, 1997.

Webster, Sir Charles, and Noble Frankland. *The Strategic Air Offensive against Germany 1939–1945*, vol. 3. London, 1961.

———. *The Strategic Air Offensive against Germany 1939–45*, vol. 4: Annexes and Appendices. London, 1961.

Whiting, Charles. *The Three-Star Blitz: The Baedeker Raids and the Start of Total War 1942–43*. London, 1987.

INDEX

Grateful acknowledgment is made for permission to reprint material from previously published works:

IN ENGLISH

Frank Cass/Crown Copyright for *Despatches on War Operations* by Sir Arthur Harris (ed. with an Introduction by Sebastian Cox). Introduction copyright © 1995 by Sebastian Cox.

Extracts from *The Eighth Passenger* by Miles Tripp are reproduced by permission of Peters Fraser Dunlop on behalf of the Estate of Miles Tripp. Copyright © 1969 by Miles Tripp.

Greenhill Books for *Bomber Harris His Life and Times* by Henry Probert. Copyright © 2001 by Henry Probert.

David Higham Associates Ltd for *Bomber Offensive* by Sir Arthur Harris. Copyright © by Estate of Sir Arthur Harris.

Quotations from David Irving's "Apocalypse 1945: The Destruction of Dresden" by permission of Focal Point Publications (free download at fpp.co.uk/books/Dresden). Copyright © 1995 by Focal Point, London. Internet Edition copyright © 1999 by Focal Point, London.

William Burr/National Security Archive for the CNN/*Cold War* series interview with Hugh Lunghi, National Security Archive website (http://www.gwu.edu/~nsarchiv/coldwar/interviews/episode-1/lunghi5.html). Copyright © 1996 by National Security Archive.

Cassell Publishing for *The Battle of Hamburg* by Martin Middlebrook. Copyright © 1980 by Martin Middlebrook.

Random House (United States and UK) for *Slaughterhouse-Five* by Kurt Vonnegut, published by Jonathan Cape. Reprinted by permission of the Random House Group Ltd. Copyright © 1969 by Kurt Vonnegut Jr.

Random House (UK) for *The Bombers* by Norman Longmate, published by Hutchinson. Reprinted by permission of the Random House Group. Copyright © 1983 by Norman Longmate.

Random House (UK) for *Air Raid: The Bombing of Coventry* by Norman Longmate, published by Hutchinson. Reprinted by permission of the Random House Group. Copyright © 1976 by Norman Longmate.

Why the Allies Won by Richard Overy, first published by W.W. Norton & Co. Reproduced by permission of the author c/o Rogers, Coleridge & White Ltd, 20 Powis Mews, London W11 1JN. Copyright © 1995 by Richard Overy.

Inside the Third Reich by Albert Speer, reprinted with the permission of Scribner, an imprint of Simon & Schuster Adult Publishing Group, from *Inside the Third Reich* by Albert Speer, translated by Richard and Clara Wilson. Copyright © 1969 by Verlag Ullstein GmbH. English translation copyright © 1970 by Macmillan Publishing Company.